Biological Membranes: Structure, Biogenesis and Dynamics

NATO ASI Series

Advanced Science Institutes Series

A series presenting the results of activities sponsored by the NATO Science Committee, which aims at the dissemination of advanced scientific and technological knowledge, with a view to strengthening links between scientific communities.

The Series is published by an international board of publishers in conjunction with the NATO Scientific Affairs Division

A Life Sciences B Physics	Plenum Publishing Corporation London and New York
C Mathematical and Physical Sciences D Behavioural and Social Sciences E Applied Sciences	Kluwer Academic Publishers Dordrecht, Boston and London
F Computer and Systems Sciences G Ecological Sciences H Cell Biology I Global Environmental Change	Springer-Verlag Berlin Heidelberg New York London Paris Tokyo Hong Kong Barcelona Budapest

NATO-PCO DATABASE

The electronic index to the NATO ASI Series provides full bibliographical references (with keywords and/or abstracts) to more than 30000 contributions from international scientists published in all sections of the NATO ASI Series. Access to the NATO-PCO DATABASE compiled by the NATO Publication Coordination Office is possible in two ways:

- via online FILE 128 (NATO-PCO DATABASE) hosted by ESRIN, Via Galileo Galilei, I-00044 Frascati, Italy.
- via CD-ROM "NATO Science & Technology Disk" with user-friendly retrieval software in English, French and German (© WTV GmbH and DATAWARE Technologies Inc. 1992).

The CD-ROM can be ordered through any member of the Board of Publishers or through NATO-PCO, Overijse, Belgium.

Series H: Cell Biology, Vol. 82

Biological Membranes: Structure, Biogenesis and Dynamics

Edited by

Jos A. F. Op den Kamp

Centre for Biomembranes and Lipid Enzymology
Padualaan 8
3584 CH Utrecht
The Netherlands

Springer-Verlag
Berlin Heidelberg New York London Paris Tokyo
Hong Kong Barcelona Budapest
Published in cooperation with NATO Scientific Affairs Division

Proceedings of the NATO Advanced Study Institute on Structure, Biogenesis and Dynamics of Biological Membranes, held at Cargèse, France, June 14–26, 1993

Additional material to this bood can be downloaded from http://extra.springer.com.

ISBN-13:978-3-642-78848-2 e-ISBN-13:978-3-642-78846-8
DOI: 10.1007/978-3-642-78846-8

Library of Congress Cataloging-in-Publication Data. Biological membranes: structure, biogenesis, and dynamics / edited by Jos A.F. Op den Kamp. p. cm. – (NATO ASI series. Series H, Cell biology; vol. 82) "Proceedings of the NATO Advanced Study Institute on Structure, Biogenesis, and Dynamics of Biological Membranes, held at Cargèse, France, June 14-26, 1993" – T.p. verso. "Published in cooperation with NATO Scientific Affairs Division." Includes bibliographical references and index. ISBN-13:978-3-642-78848-2
1.Membranes (Biology)–Congresses.2.Membrane lipids–Congresses. 3. Membrane proteins–Congresses. 4. Biological transport–Congresses. I. Kamp, Jos A. F. op den (Jos Arnoldus Franciscus), 1939- . II. North Atlantic Treaty Organization. Scientific Affairs Division. III. NATO Advanced Study Institute on Structure, Biogenesis and Dynamics of Biological Membranes (1993: Cargèse, France) IV. Series. QH601.B493 1994 574.87'5–dc20 94-2915

Softcover reprint of the hardcover 1st edition 1994

Typesetting: Camera ready by authors
31/3145 - 5 4 3 2 1 0 - Printed on acid-free paper

PREFACE

The Advanced Study Institute on "Structure, Biogenesis and Dynamics of Biological Membranes, held in Cargèse from June 14-26, 1993, has been dealing with four major topics in membrane biochemistry today: lipid dynamics and lipid-protein interactions, protein translocation and insertion, intracellular traffic and protein structure and folding. The lecturers discussed these topics starting from several disciplines, including biochemistry, cell biology, genetics, and biophysics. This way an interdisciplinary and very interesting view on biological membrane systems was obtained.

At first an extensive overview of - mainly biophysical - techniques which can be used to study dynamic processes in membranes was presented. Sophisticated approaches such as ESR and NMR have been applied succesfully to unravel details of specific lipid-protein interactions. X-ray analysis provides detailed structural information of several proteins and the possible implications for protein functions. Information obtained this way is complemented by studies on mechanisms and kinetics of protein folding. The latter information is indispensable when discussing protein translocation and insertion : processes in which folding and unfolding play essential roles.

Extensive insight was offered in the complicated machinery of phospholipid biosynthesis. In particular, the application of sophisticated genetic techniques has allowed a better understanding of the mechanisms regulating the synthetic machinery and detailed studies on a variety of mutants, lacking one or more of the essential enzymes, have resulted in the beginning of a better understanding of specific functions of individual phospholipid species.

Mutants are furthermore indispensible in the search for mechanisms of protein insertion and translocation. In particular, easily accessible systems like yeast and bacteria are manipulated these days with a variety of approaches, yielding an enormous variety of abnormalities in structure, general behaviour, function, regulation etc. The study of these mutants has led already to surprisingly detailed information about the mechanisms involved and more in particular about the many components which are active and essential to lead newly formed proteins to their final destinations. Several examples were presented including the formation and insertion of toxins in bacterial systems.

Finally, intracellular protein and lipid traffic were discussed in detail. This complex phenomenon involves a large number of cell compartments and different tracks along which compounds can traffic into or out of the cell interior. Data on mechanisms and kinetics of various trafficking pathways were backed up by structural details of subcellular compartments and particles involved.

November 1993

Jos A.F. Op den Kamp

CONTENTS

ROLE OF PHOSPHOLIPIDS IN CELL FUNCTION

William Dowhan
Department of Biochemistry and Molecular Biology
University of Texas Medical School
Houston, Texas 77225
U. S. A.

A primary and essential role for phospholipids is defining the permeability barrier of the cell membrane and all internal organelles. However, due to the molecular diversity of phospholipids, individual phospholipid species must play a more dynamic role in cell function than simply defining the physical properties of the membrane. In addition to forming the membrane permeability barrier, phospholipids are intermediates in the synthesis of or direct precursors to other cellular components as well as regulatory molecules which affect cell physiology. Because of this pleiotropic requirement for phospholipids in maintaining normal cell function, it has often been difficult to assign a specific role to a particular phospholipid or group of phospholipids through *in vivo* studies. There are no direct assays for phospholipid function, as there are for enzymes, so the functions of individual phospholipids have come to light in many cases incidental to the study of a particular cellular process *in vitro* rather than by direct study of a particular phospholipid. In most cell types it is difficult to effect systematic and extensive alteration of the normal phospholipid composition which has made it difficult to verify *in vivo* the physiological significance of functions deduced from *in vitro* biochemical approaches alone. Utilization of classical genetic approaches has established the essential role of specific phospholipids for cell viability but has not provided a precise understanding at the molecular level for their requirement (Raetz and Dowhan, 1990). Application of more direct approaches of modern molecular genetics has made possible the design of mutants in which phospholipid metabolism and therefore phospholipid composition can be more precisely controlled (Dowhan,

NATO ASI Series, Vol. H 82
Biological Membranes:
Structure, Biogenesis and Dynamics
Edited by Jos A. F. Op den Kamp

1992). With such mutants it has been possible to establish new *in vivo* roles for phospholipids which has lead to biochemical studies to define the molecular basis for these functions.

Escherichia coli has been the primary organism of choice for designing mutants with defined phospholipid composition because alteration of phospholipid metabolism in eukaryotic cells, with their greater complexity, potentially could lead to numerous pleiotropic effects. Background on the biochemistry and genetics of phospholipid metabolism in *E. coli* (Shibuya, 1992; Raetz and Dowhan, 1990; Vanden Boom and Cronan, 1989; Raetz, 1986) can be found in several reviews.

Overview of Phospholipid Metabolism in *E. coli*

Figure 1 illustrates the pathway in *E. coli* for the synthesis of glycerol-based phospholipids which make up the bilayer structure of the inner cytoplasmic membrane and the inner leaflet of the outer membrane of Gram negative bacteria; the outer leaflet of the outer membrane is composed of glucosamine-based phospholipids which are reviewed elsewhere (Raetz, 1990). The genes responsible for this metabolic scheme are distributed over the *E. coli* genome and are not found in organized operons or under any apparent coordinate regulation. Genes for every step of this pathway except for Step 7 have been defined by mutation and cloning. Of the cloned genes, a gene sequence has been published for all except Steps 2 and 8. Purification schemes for isolation of homogeneous enzyme have been developed for every step except Steps 2, 7 and 8. The genes *pgpA* and *pgpB* listed for Step 7 encode the enzymatic activity shown (Icho and Raetz, 1983), but do not encode the primary phosphatase responsible for this step since construction of a double null mutant in these two listed genes does not result in the expected accumulation of phosphatidylglycero-*P* and cessation in phosphatidylglycerol and cardiolipin biosynthesis (Funk *et al.*, 1992). Based on genetic evidence the remaining steps are catalyzed by a single enzymatic activity necessary to provide sufficient product for cell viability.

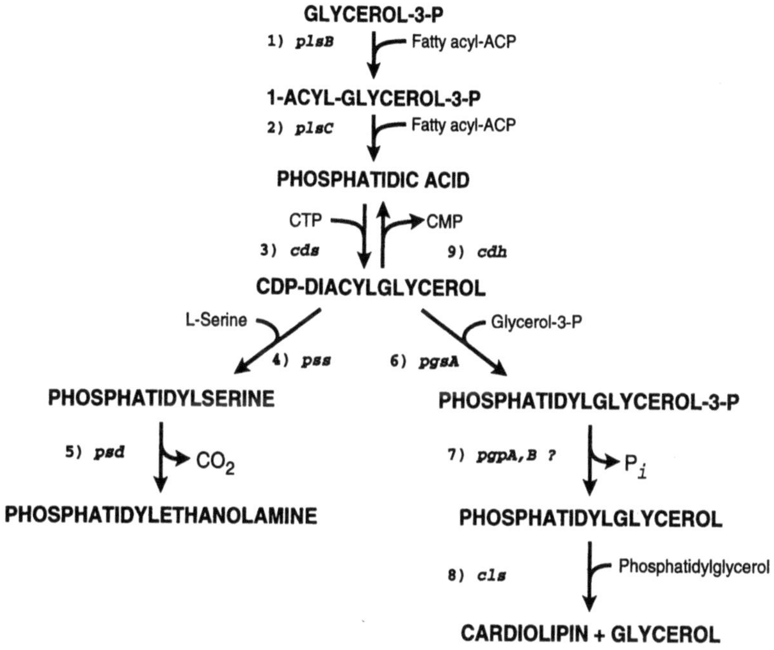

Figure 1. Pathway of phospholipid metabolism in *E. coli* and the associated genes (Dowhan, 1992). The name of each gene is listed with the respective step catalyzed by the following enzymatic activities: 1) Glycero-*P* acyltransferase; 2) Acylglycero-*P* acyltransferase; 3) CDP-diacylglycerol synthase; 4) Phosphatidylserine synthase; 5) Phosphatidylserine decarboxylase; 6) Phosphatidylglycero-*P* synthase; 7) Phosphatidylglycero-*P* phosphatases; 8) Cardiolipin synthase; 9) CDP-diacylglycerol hydrolase.

The committed step to phospholipid biosynthesis (Step 1) is regulated in a complex manner which depends primarily on the availability of fatty acyl CoA precursors (Jackowski *et al.*, 1991). The factors which determine the final phospholipid head group composition, and therefore the charge properties of the membrane matrix, are not known but must be associated with the regulation of Steps 4 (leading to a zwitterionic product) and 6 (leading to acidic products) and the turnover of the final products, phosphatidylethanolamine (PE, 70-75%), phosphatidylglycerol (PG, 20-25%, and cardiolipin (CL, 5-10%). The substrate for this branch point is synthesized by the product of the *cds* gene and is a mixture of both the ribose and deoxyribose derivative (Sparrow and Raetz, 1985); both derivatives are equally good substrates for the next reactions (Raetz, 1986). The only mutant thus far isolated in this step is one which is sensitive to growth in medium above pH 8. Under all growth conditions this mutant accumulates abnormal levels of phosphatidic acid which increases with grown above pH 8 (Ganong *et al.*, 1980); it is not known whether the lethal phenotype is due to accumulation of phosphatidic acid *per se* or accumulation of phosphatidic acid at pH 8.

Phosphatidylserine (PS) synthase catalyzes the committed step to the synthesis of PE and is the only enzyme of this pathway which is not tightly associated with the inner membrane of *E. coli* (Louie and Dowhan, 1980) or other Gram negative organisms (Dutt and Dowhan, 1977). Conditional lethal temperature sensitive mutants in the *pss* gene (Step 4) stop growing when the level of PE reaches 30-40% of the total phospholipid except in the presence of 20-50 mM Mg^{2+} which suppresses the growth phenotype without suppressing the abnormal phospholipid metabolism (Raetz *et al.*, 1979; Ohta and Shibuya, 1977); the remaining phospholipid pool in growing cells at the restrictive temperature is primarily made up of PG and CL, but the level of CL now approaches 25-30%. Conditional lethal temperature sensitive mutants in the *psd* gene (Step 5) have similar properties to *pss* mutants, and also stop growing when PE reaches 30-40% but in this case at the expense of accumulated PS and not the major acidic phospholipids (Hawrot and Kennedy, 1978). Mg^{2+} in the growth medium also suppresses the growth phenotype of the *psd2* mutant and results in about 20% of the phospholipid pool being PG + CL and 80% of the pool being PS + PE which is similar to the distribution of the final products of Step 4 and 6 in wild type cells, but now PE only represents 5% of the total pool. The isolation and characterization of these *pss* and *psd* mutants established the requirement for PE under normal growth conditions and showed that the branching of metabolism after Step 3 appears not to be regulated by the level of the final or initial products of the branch point enzymes. The molecular basis for suppression by divalent metal ion, the molecular basis for the PE requirement under normal growth conditions and the question of whether PE is necessary under all growth conditions remain unanswered using these mutants.

Early studies of the requirement for acidic phospholipids based on the isolation and phenotypic characterization of mutants in the *pgsA* and *cls* loci were not definitive. Strains carrying point mutants *pgsA10* (Nishijima and Raetz, 1979) and *pgsA3* (Miyazaki *et al.*, 1985) had nearly undetectable *in vitro* enzymatic activity, but grew normally; *pgsA10* mutants only showed a small reduction in the acidic phospholipid pool and although this pool in *pgsA3* mutants was reduced by nearly 10-fold (PE increase accounted for the decline), it required a second suppressor mutation in order to be viable (Asai *et al.*, 1989; Heacock and Dowhan, 1987). The suppressor mutation turned out to be in the *lpp* gene which encodes the major outer membrane lipoprotein; this protein requires PG as a precursor in a posttranslational modification step necessary for proper assembly of the protein and integrity of the outer membrane, but not for cell viability (Wu *et al.*, 1983). Construction of a null allele (*pgsA30*) of this locus by insertion of a drug marker did result in a lethal phenotype which could not be suppressed by mutations in the *lpp* locus or to date by any other mutation or by growth conditions (Heacock and Dowhan, 1987). On the other hand strains carrying either point mutants or null mutants (constructed by insertion of a drug marker) of the *cls* gene (Step 8) are viable with few distinguishing phenotypes (Nishijima *et al.*, 1988). Therefore, the PG is essential for cell viability and CL may be important for some cell functions but is not essential in otherwise wild type backgrounds. The molecular basis for the PG requirement and the partial suppression of this requirement by a mutation in the *lpp* gene could not be addressed with these mutants.

Role of Acidic Phospholipids in Cell Function

In order to address the role of acidic phospholipids in cell function a series of mutants were constructed in which the level of PG + CL could by systematically varied by addition of a specific non-metabolized inducer to the growth medium. The prototype of this strain (HDL1001) carried a null allele (*pgsA30::kan*) of the chromosomal copy of the *pgsA* gene and a single copy of a gene fusion of the structural gene with the regulated promoter of the *lac* operon of *E. coli* (ϕ[*lacOP-pgsA*]1); the gene fusion is also in single copy integrated into the chromosomal *lac* locus (Heacock and Dowhan, 1989). In this strain the steady state level of the PGP synthase (Step 6) can be regulated by the level of the gratuitous inducer of the *lac* operon, IPTG; in order to have a linear relationship between enzyme level and IPTG the lactose permease must be inactivated (*lacY*::Tn*9*). The growth rate (Figures 2 and 4), the level of acidic phospholipids

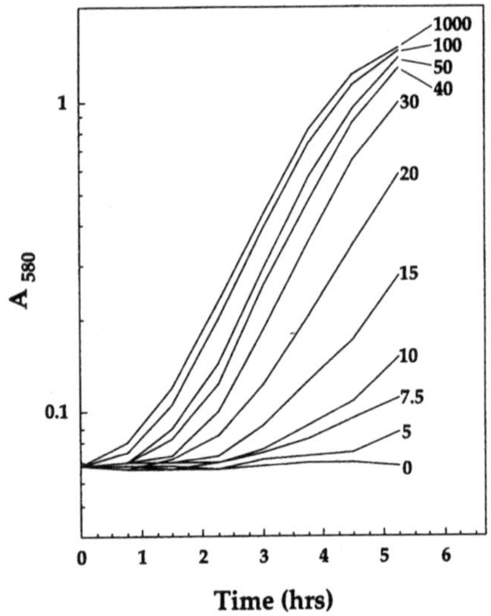

Figure 2. Growth dependence of strain HDL1001 on IPTG (μM) as measured by the increase of absorbance (A) of the cultures (Heacock and Dowhan, 1989).

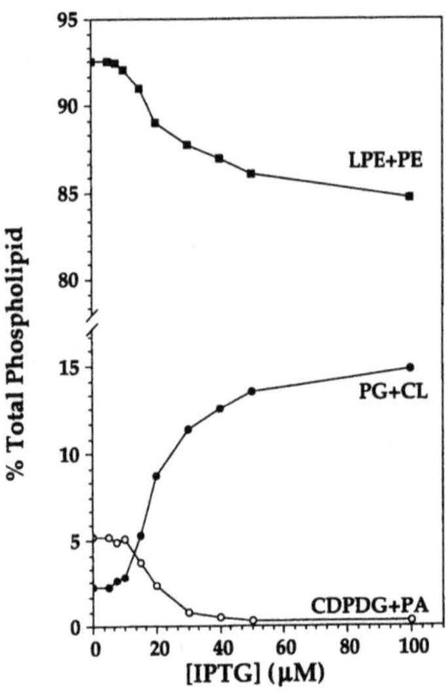

Figure 3. Dependence of phospholipid composition of strain HDL1001 on IPTP. (L, lyso; CDP-DG, CDP-diglyceride) (Heacock and Dowhan, 1989).

(Figure 3), and the level of PGP synthase activity (Figure 5) for strain HDL1001 are all proportional to the concentration of IPTG in the growth medium. A derivative (strain HDL11) of strain HDL1001 carrying the *lpp2* mutation which suppresses the growth phenotype of strains carrying the *pgsA3* allele is also independent of IPTG, but the phospholipid composition and enzyme activity of this strain is still dependent on IPTG (Kusters *et al.*, 1991). Although the basis for the suppression by the *lpp* mutation is still unknown, the expression of the gene fusion is not completely shut off in the absence of IPTG which makes it similar phenotypically to the *pgsA3* allele which still expresses a low level of activity.

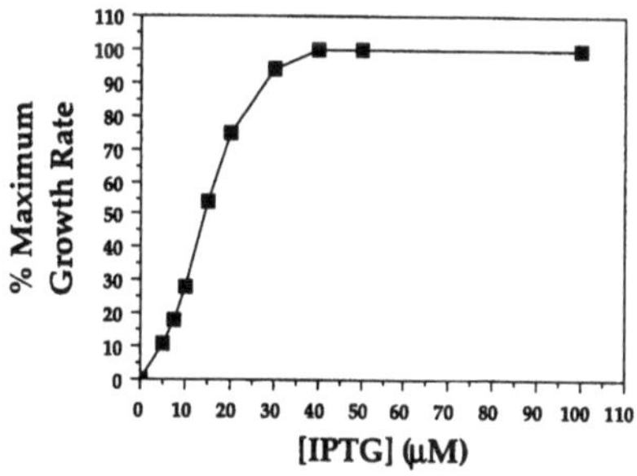

Figure 4. Growth rate of strain HDL1000 as a function of IPTG (Heacock and Dowhan, 1989).

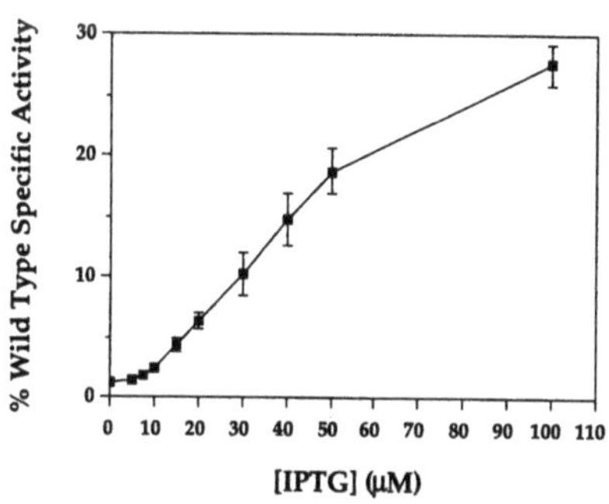

Figure 5. Dependence of the specific activity of the synthase of strain HDL1001 on IPTG (Heacock and Dowhan, 1989).

With the ability to systematically vary the acidic phospholipid composition in growing cells, it is possible to correlate *in vivo* the properties of any cellular process with a particular phospholipid composition. Important but nonessential processes can be drastically altered without adversely affecting other cellular processes such as cell energetics, growth or viability. Even essential processes may be progressive affected with decreasing acidic phospholipid content and be identified before resulting in numerous pleiotropic affects. Finally, there may be several layers of PG involvement for which there are different possible growth condition or genetic suppressors which could support growth at different PG levels. One example is the *lpp* suppressor. If other suppressor mutations can be identified and characterized, they would provide a means for uncovering other functions of acidic phospholipids. Once strong in vivo evidence for a phospholipid requirement is established, in vitro biochemical studies can be initiated (which are on a firm physiological base) to determine the molecular basis for phospholipid involvement. Thus far these mutants and this approach has uncover a requirement for acidic phospholipids in the organization and function of the membrane associated complex responsible for the export of many proteins out of the cytoplasm for assembly into the outer membrane. This complex is composed of a cytoplasmic component (*secA* gene product), a membrane component (*secY* gene product) and the presursor protein to be translocated across the membrane. This work is well covered and documented elsewhere (Lill *et al.*, 1990; Kusters *et al.*, 1991; Hendrick and Wickner, 1991). A recent study (Van der Goot *et al.*, 1993) has provided in vivo support for the in vitro evidence that acidic phospholipids play a role in the assembly of colicin A into inner membrane; in vivo this assembly occurs from the outside of the cell and results in a lethal K^+ pore the properties of which were also dependent on acidic phospholipids.

Intiation of DNA replication and acidic phospholipids--In vitro evidence supports the involvement of the acidic phospholipid component of membranes in forming the complex which initiates DNA replication in *E. coli*. In wild type *E. coli* RNA primers are made at several sites (*oriK*) on the chromosome including the site of normal initiation of DNA replication, *oriC* (se Figure 6). At *oriC*, the soluble DnaA protein initiates the formation of a complex of proteins which then proceed first to initiate synthesis and then elongation. The DnaA protein can bind to DNA in its ATP or ADP form (both with high affinity), but is only competent when associate with ATP; during initiation ATP is converted to ADP inactivating the DnaA protein. In studies describing the reconstitution of DNA initiation *in vitro* using *oriC* plasmids (Sekimizu and Kornberg, 1988; Yung and Kornberg, 1988;

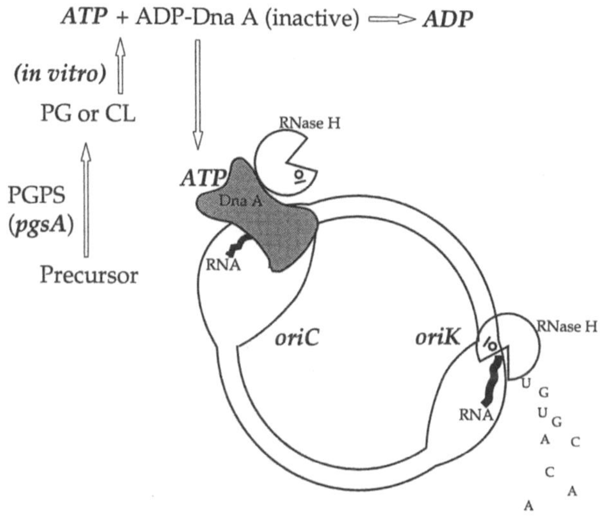

Figure 6. DnaA protein-dependent initiation at *oriC* in the presence of RNase H with the postulated requirement for acidic phospholipids and the *pgsA* gene product.

Crooke *et al.*, 1992), addition of PG and/or CL allowed continuous rounds of initiation to occur by facilitating the exchange of tightly bound ADP (inactive form) for ATP (active form).

Normally, *in vivo* the RNA primers at the *oriK* sites are rapidly degraded (Figure 6) by RNase H (*rnh* gene product). Since *rnh* mutants can initiate DNA replication at the *oriK* sites (stable DNA replication) independent of *oriC* and DnaA protein (Kogoma and von Meyenburg, 1983; Horiuchi *et al.*, 1984), *rnh* mutants act as suppressors of mutations in functions dependent on *oriC* and/or DnaA protein (Figure 7). Stable DNA replication is also dependent on certain functions of the RecA protein (*recA* gene). Therefore, if one of the processes contributing to the growth arrest in strain HDL1001 in the absence of IPTG (low acidic phospholipid content) is dysfunctional DnaA protein-dependent initiation of DNA replication at *oriC*, then a *rnh* mutation should be a suppressor, in a RecA protein-dependent manner, of the IPTG requirement of strain HDL1001 for growth (Figure 7). Preliminary evidence (Xia and Dowhan, unpublished observation) suggests the predicted relationship between acidic phospholipids, *oriC*-DnaA protein, and *rnh-recA* mutants exists (Figure 8).

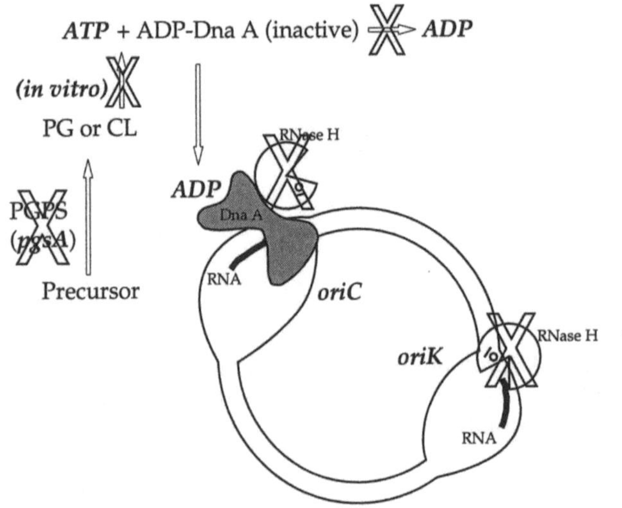

Figure 7. DnaA protein-independent initiation of DNA replication at an *oriK* site in the absence of RNase H and the *pgsA* gene product.

A measure of suppression of the requirement for acidic phospholipids is the ability of strain HDL1001 and its derivatives to form colonies on rich media agar plates in the absence of IPTG; some colonies always form under these conditions due to selection for other mutations such as constitutive expression from the *lacOP* promoter. In results of the experiment summarized in Figure 8, a *recA*+ strain similar to HDL1001 in its other genetic and phenotypic properties was used. In the first panel colony formation by this strain is highly dependent on IPTG. Introduction of the *rnh* mutation (no RNaseH) into this strain resulted in IPTG independence; the surviving colonies were shown to be dependent on IPTG for acidic phospholipid synthesis. Introduction of a *recA* mutation into the *rnh* strain or transformation of the *rnh* strain with a plasmid-borne copy of the *rnh+* gene resulted in IPTG dependence for colony formation. These preliminary results are consistent with a requirement for acidic phospholipids in forming a membrane associated

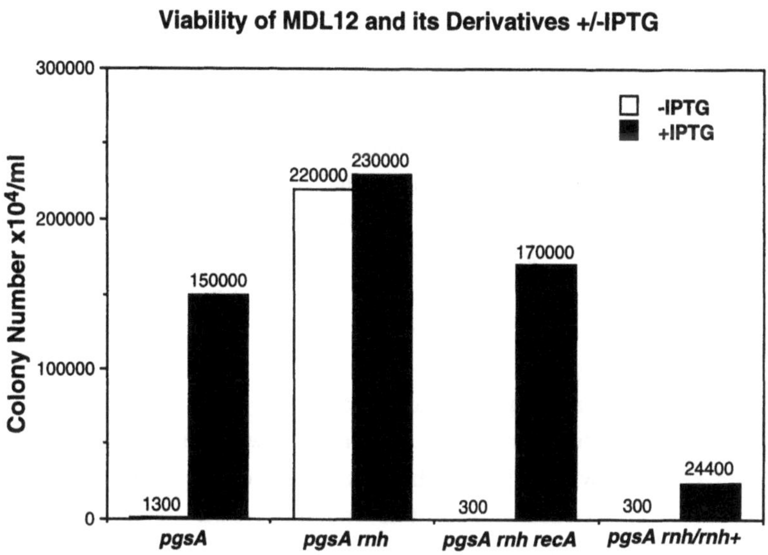

Figure 8. Suppression of IPTG-dependent growth (requirement for acidic phospholipids) by a mutation at the *rnh* locus (RNase H) in a recA dependent manner.

complex from normally cytoplasmic complexes as suggested by the in vitro data of Crooke *et al.* (1992). These result illustrate another example of acidic phospholipids functioning as a nucleation site for organizing cytoplasmic components on the membrane surface; in both cases a cytoplasmic ATPase was involved.

Role of phosphatidylethanolamine in *E. coli*

To address questions left unanswered by point mutants in the *pss* gene whose lethal phenotype could be suppressed by Mg^{2+} in the growth medium, a null allele of the gene (*pss93::kan*) was constructed (DeChavigny *et al.*, 1991). As expected, strains (AD90 and AD93) carrying this allele lost viability when the PE content reached 30-40% as was the case with the temperature sensitive mutants. Unexpectedly, addition of any of several divalent metal ions (Ca^{2+} > Mg^{2+} >Sr^{2+}) at 10-50 mM to the media not only suppressed the growth phenotype, but supported growth of the strains without PE in the order of effectiveness indicated. Mg^{2+} is normally high in the cytoplasm (50-100 mM) while Ca^{2+} and Sr^{2+} are maintained in the μmolar range by an active proton-Ca^{2+} antiport system which excludes these toxic ions from the cell (Gangola and Rosen, 1987). The membrane protein to phospholipid ratio of these mutants grown in the presence of divalent metal ions is normal with no new phospholipids appearing; therefore, the two acidic phospholipids now make up the membrane bilayer. CL now makes up 43% of the phospholipid in Mg^{2+} and Sr^{2+} grown cells and a lower level of 27% in Ca^{2+} grown cells. A detailed analysis of the phase properties of the phospholipids extracted from this mutant grown in the presence of different divalent metal ions suggests that one of the functions of the divalent metal ions is too provide an environment (presumably on the outside of the inner membrane) for CL to potentially assume a nonbilayer structure (Rietveld *et al.*, 1993). PE can assume nonbilayer structures in the absence of divalent metal ions, but CL only assumes this structure in the presence of the 3 divalent metal ions which support growth of this mutant (Vasilenko *et al.*, 1982). Previous studies of strains with mutations in both the *pss* and *cls* genes suggested that cells could not survive without both PE and CL (Shibuya *et al.*, 1985) while null mutations in the *cls* gene are not lethal (DeChavigny *et al.*, 1991). Therefore, one role of PE, for which CL and divalent metal ion can substitute, may be purely structural in providing nonbilayer potential to the normal membrane bilayer of cells. Removal of divalent metal ion from strain AD93 results in rapid loss of viability (even upon readdition of ions) with accompanying cell lysis also supporting a structural role for PE.

Being able to maintain cell viability in the absence of PE allows the investigation of nonessential, but important, processes which might require PE. An obvious location of such processes would be the cell membrane which provides the membrane barrier necessary for solute transport and energy transduction. Replacing PE with PG and CL must alter the surface negative charge density of the membrane which might be expected to influence the assembly and function of membrane associated complex.

Electron transport and energy transduction--The group of mutant strains of *E. coli* (AD93, AD90, AH930) lacking PE, but able to grow in the presence of Mg^{2+} was used for an extended biochemical study of energy transducing reactions in these membranes in comparison with wild type strains with normal phospholipid composition. Using this

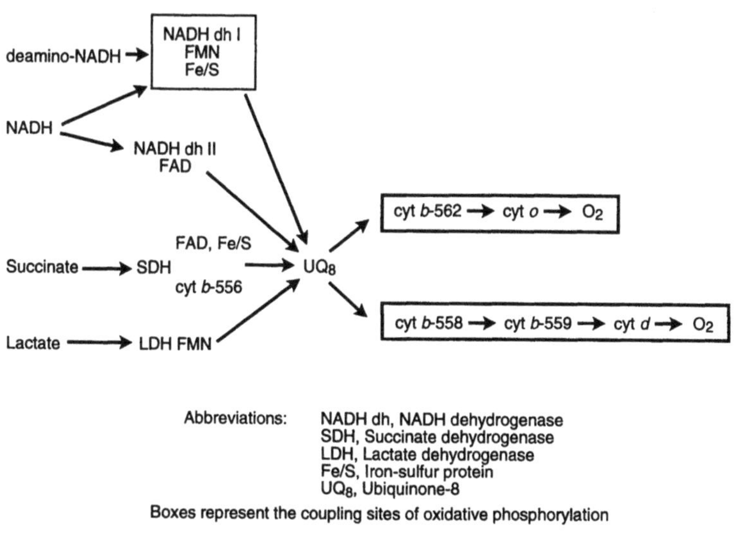

Figure 9. Constitutive components of the aerobic electron transport chain of *E. coli*.

approach an *in vivo* requirement for PE in the transfer of electrons from NADH dehydrogenase II (NADH specific) to UQ_8 (natural CoQ acceptor) during the process of whole chain electron transport-dependent oxidation of NADH (Mileykovskaya and Dowhan, 1993) was demonstrated (see Figure 9). The fact that cells lacking PE, which normally accounts for 75% of the *E. coli* phospholipid, are viable predicts that there are no major alterations in the general structural features of the membrane. In fact, inverted membrane vesicles from mutant and wild type strains were very similar in there ability to generate and maintain a pH gradient and membrane potential (positive inward) when given either succinate or lactate as the source of electrons in the presence of oxygen as the electron acceptor (whole chain electron transport) or ATP as an energy source (Figure 10). In addition the rate of oxidation of lactate and succinate with molecular oxygen as acceptor was comparable for all vesicles (Table I). However when NADH was used as the source of electrons, the magnitude of the pH gradient was reduced by 40-60% of wild type and whole chain coupled activity in mutant membranes was only 10-20% of wild type levels. This initial result suggested a defect in the utilization of NADH as an electron donor. As shown in Figure 9, *E. coli* contains a Type I NADH dehydrogenase, in which redox reactions are directly coupled to proton gradient formation, and a Type II NADH dehydrogenase, in which the redox reaction is not directly coupled but transfers electrons which are utilized by the downstream cytochromes in energy coupling reactions; because of the liability of the Type I enzyme, the enzyme operative in the isolated inverted membrane vesicles is the more stable Type II enzyme. Use of artificial electron acceptors such as DCIP showed that the level of total NADH dehydrogenase II was normal in the mutant. Supplementation of membrane vesicles with the less hydrophobic artificial UQ_1 analog restored, in a dose response manner, the magnitude of the pH gradient, the membrane potential, and the whole chain coupled activity when NADH was used as the donor. In addition the determination of the endogenous UQ_8 pool using HPLC purification and mass spectral analysis showed that the content of this electron carrier in both wild type and mutant membranes was the same. Spectral analyses also showed the normal content of cytochromes in the mutant membranes except for cytochrome *d* which was not induced to its normal level in late log cells.

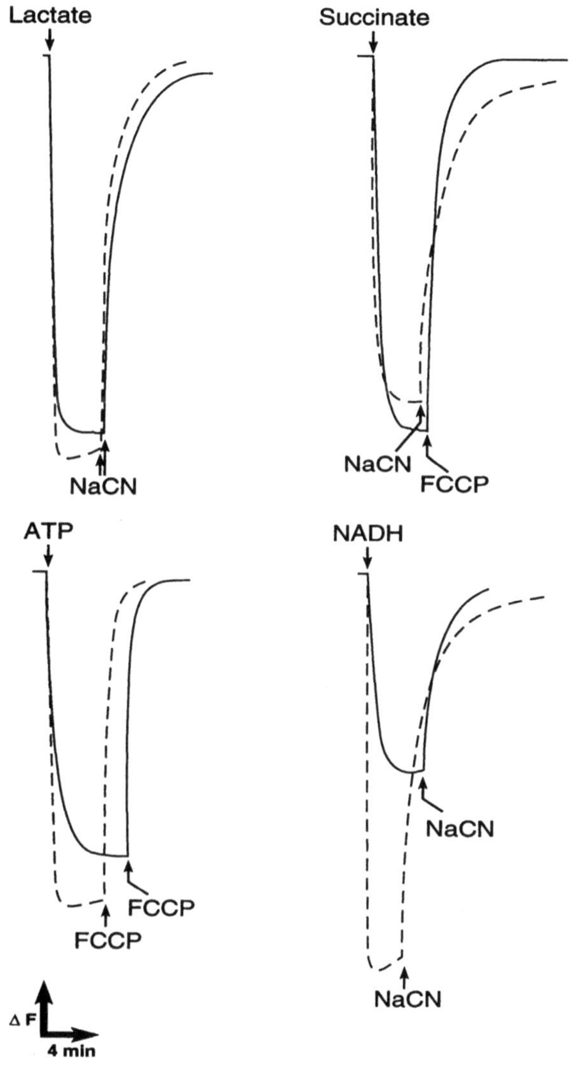

Figure 10. Formation of a ΔpH gradient (positive inward), as measured by fluorescence changes (ΔF), across inverted membrane vesicles made from wild type (dotted line) and mutants lacking PE (solid line) by either whole chain electron transfer to oxygen (succinate, lactate, NADH) or by ATP hydrolysis.

Table I. Enzyme Activities of the Oxidase and Reductase Systems in Membranes from strains W3899 (wild type) and AD93

Activity	W3899	AD93	% of wild type activity
Lactate oxidase[a]	530	513	97
Succinate oxidase[a]	637	642	100
NADH oxidase[a]	890	230	26
NADH:DCIP reductase[b]	248	242	98
NADH:Q_1 reductase[c]	890	908	102

[a] ngat of oxygen per min per mg of protein; [b] nmol DCIP per min per mg of protein; [c] nmol of NADH per min per mg of protein.

Study by steady-state kinetic of NADH oxidation in mutant membranes showed no major change in accessibilityto NADH as the result of the alteration of the surface charge at the membrane-water interphase. Therefore, PE is necessary for the specific optimum interaction of the NADH dehydrogenase Type II with its natural acceptor UQ_8. The molecular basis for this requirement is now under investigation using in vitro biochemical approaches.

Lactose permease--Chen and Wilson (1984) showed that reconstitution of transport function of the isolated lactose carrier (*lacY* gene product) *in vitro* in phospholipid vesicles required PE which could not be replaced by other zwitterionic phospholipids such as phosphatidylcholine. This requirement for PE has been verified

(Bogdanov and Dowhan, 1993) with *in vivo* studies of the *lacY* gene product in strains lacking PE (strain AD93). The *lac* operon (either in single copy or from a multicopy number plasmid) is fully inducible in strain AD93 as evidenced by β-galactosidase activity. The lactose permease is also synthesized and stably integrated into the membrane as quantified by titration with specific antibody. However, the energy coupled accumulation of substrate mediated by the permease is

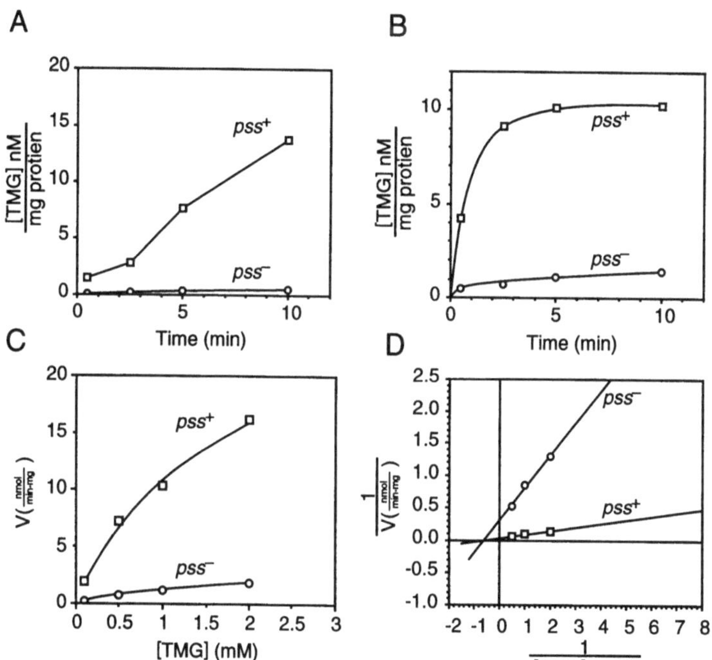

Strain	V_{max}(nmol/min-mg)	Km (mM)
pss+	20	0.8
pss−	2	0.8

Figure 11. TMG uptake by whole cells of wild type (*pss*+) and mutants lacking PE (*pss*-) induced for lactose permease either from a single copy (A) or from a multicopy number plasmid (B). Panels C and D are the initial velocity of uptake versus TMG concentration of the cells used in panel B.

defective in whole cells and isolated inverted membrane vesicles from strain AD93 lacking PE (Figure 11). The Km for residual TMG (analogue of lactose) uptake in strain AD93 is unchanged from wild type cells, but the rate of active transport is at least 10-fold reduced and never reaches the same level of accumulation as seen in wild type cells. Energy independent facilitated diffusion of substrate across the membrane still occurs but efflux of TMG from loaded inverted membrane vesicles is not stimulated by ATP hydrolysis. The carrier is membrane associated, does function as a facilitator of lactose transport, but appears unable to accumulate substrate against a concentration gradient. Therefore, lack of PE appears to uncouple transport from metabolic energy and utilization of the proton gradient in what is normally a proton-substrate cotransport process. Current experiments are aimed at analyzing the organization of the multiple transmembrane domains of this carrier in the mutant cells.

Acknowledgement:

This work was supported in part by United States Public Health Service Grants GM 20487 and GM 35143 from the National Institutes of General Medical Sciences.

REFERENCES:

Asai Y, Katayose Y, Hikita C, Ohta A, Shibuya I (1989) Suppression of the lethal effect of acidic-phospholipid deficiency by defective formation of the major outer membrane lipoprotein in *Escherichia coli*. J Bacteriol 171:6867-6869

Bogdanov M, Dowhan W (1993) Phosphatidylethanolamine (PE) is required fro in vivo function of the lactose permease of *Escherichia coli*. FASEB J 7:A518

Chen C-C Wilson TH (1984) The phospholipid requirement for activity of the lactose carrier of *E. coli*. J Biol Chem 259:10150-10158

Crooke E, Castuma CE, Kornberg A. (1992) The chromosome origin of *E. coli* stabilizes DnaA protein during rejuvenation by phospholipids. J Biol Chem 267:16779-16782

DeChavigny A, Heacock PN, Dowhan W (1991) Sequence and Inactivation of the *pss* gene of *Escherichia coli.* J Biol Chem 266:5323-5332

Dowhan, W (1992) Strategies of generating and utilizing phospholipid synthesis mutants in *Escherichia coli.* Method in Enzymol (Dennis EA, Vance DE, eds.) 209:7-20

Dowhan W, Heacock PN (1987) Construction of a lethal mutation in the synthesis of the major acidic phospholipids of *Escherichia coli.* J Biol Chem 262:13044-13049

Dutt A, Dowhan W, (1977) Intracellular distribution of enzymes of phospholipid metabolism in several Gram-negative bacteria. J Bacteriol 132:159-165

Funk CR, Zimniak L, Dowhan W (1992) The *pgpA* and *pgpB* genes of *Escherichia coli* are not essential: Evidence for a third phosphatidylglycerophosphate phosphatase. J Bacteriol 174:205-213

Ganong BR, Leonard JM, Raetz CRH (1980) Phosphatidic acid accumulation in the membranes of *Escherichia coli* mutants defective in CDP-diglyceride synthetase. J Biol Chem 255:1623-1629

Gangola P, Rosen BP (1987) Maintenance of intracellular calcium in *Escherichia coli.* J Biol Chem 262:12570-12574

Hawrot E, Kennedy EP (1978) Phospholipid composition and membrane function in phosphatidylserine decarboxylase mutants of *Escherichia coli.* J Biol Chem 253:8213-8220

Heacock PN, Dowhan W (1989) Alteration of the phospholipid composition of *Escherichia coli* through genetic manipulation. J Biol Chem 264:14972-14977

Hendrick JP, Wickner W (1991) SecA protein needs both acidic phospholipids and SecY/E protein for functional high-affinity binding to the *Escherichia coli* plasma membrane. J Biol Chem 266:24596-24600

Horiuchi T, Maki H, Sekiguchi M. (1984) RNase H-defective mutants of *E. coli*: A possible discriminatory role of RNase H in initiation of DNA replication. Mol Gen Genet 195:17-22

Icho T, Raetz CRH (1983) Multiple genes for membrane-bound phosphatase in *Escherichia coli* and their action on phospholipid precursors. J Bacteriol 153:722-730

Jackowski S, Cronan, Jr. JE, Rock CO (1991) Lipid metabolism in procaryotes. *In* Biochemistry of lipids, lipoproteins and membranes. (Vance DE, Vance J, eds) Elsevier Sci. Publ. B. V., Amsterdam, 43-85

Kogoma T, von Meyenburg K (1983) The origin of replication, *oriC*, and the *dnaA* protein are dispensable in stable DNA replication (*sdrA*) mutants of *E. coli* K-12. EMBO J 2:463-468

Kuge O, Nishijima M, Akamatsu Y (1991) A cloned gene encoding phosphatidylserine decarboxylase complements the phosphatidylserine biosynthetic defect of a Chinese hamster ovary mutant. J Biol Chem 266:6370-6376

Kusters R, Dowhan W, de Kruijff B (1991) Negatively charged phospholipids restore prePhoE translocation across phosphatidylglycerol-depleted *Escherichia coli* inner membranes. J Biol Chem 266:8659-8662

Lill R, Dowhan W, Wickner W (1990) The ATPase activity of SecA is regulated by acidic phospholipids, SecY, and the leader and mature domains of precursor proteins. Cell 60:271-280

Louie C, Dowhan, W (1980) Investigations on the association of phosphatidylyserine synthase with the ribosomal component of *Escherichia coli*. J Biol Chem 255:1124-1127

Mileykovskaya EI, Dowhan W (1993) Alterations in the electron transfer chain in strains of *Escherichia coli* lacking phosphatidylethanolamine. J Biol Chem, in press

Miyazaki C, Kuroda M, Ohta A, Shibuya I (1985) Genetic manipulation of membrane phospholipid composition in *Escherichia coli*: *pgsA* mutants defective in phosphatidylglycerol synthesis. Proc Natl Acad Sci USA 82:7530-7534

Nishijima S, Asami Y, Uetake N, Yamogoe S, Ohta A, Shibuya I (1988) Disruption of the *Escherichia coli cls* gene responsible for cardiolipin synthesis. J Bacteriol 170:775-780

Nishijima M, Raetz CRH (1979) Membrane lipid biogenesis in *Escherichia coli*: Identification of genetic loci for phosphatidylglycerophosphate synthetase and construction of mutants lacking phosphatidylglycerol. J Biol Chem 254:7837-7844

Ohta A, Shibuya I (1977) Membrane phospholipid synthesis and phenotypic correlation of an *Escherichia coli pss* mutant. J Bacteriol 132:434-443

Raetz CRH (1986) Molecular genetics of membrane phospholipid synthesis. Annu Rev Genet 20:253-295

Raetz CRH (1990) Biochemistry of endotoxins. Annu Rev Biochem 59:129-70.

Raetz CRH, Dowhan W (1990) Biosynthesis and function of phospholipids in *Escherichia coli*. J Biol Chem 265:1235-1238

Raetz CRH, Kantor GD, Nishijima M, Newman KF (1979) Cardiolipin accumulation in the inner and outer membranes of *Escherichia coli* mutants defective in phosphatidylserine synthetase. J Bacteriol 139:544-551

Reitveld, AG, Killian A, Dowhan W, de Kruijff B (1993) Polymorphic regulation of membrane phospholipid composition in Escherichia coli. J Biol Chem, in press

Shibuya I (1992) Metabolic regulations and biological functions of phospholipids in *Escherichia coli.* Prog Lipid Res 31:245-299

Shibuya I, Miyazaki C, Ohta A (1985) Alteration of phospholipid composition by combined defects in phosphatidylserine and cardiolipin synthases and physiological consequences in *Escherichia coli.* J Bacteriol 161:1086-1092

Sekimizu K, Kornberg A (1988) Cardiolipin activation of dnaA protein, the initiation protein of replication in *E. coli.* J Biol Chem 263:7131-7153

Sparrow CP, Raetz CRH (1985) Purification and properties of the membrane-bound CDP-diglyceride synthetase of *Escherichia coli.* J Biol Chem 260:12084-12091

Vanden Boom T, Cronan, Jr. JE (1989) Genetics and regulation of bacterial lipid metabolism. Annu Rev Microbiol 43:317-343

Van der Goot FG, Didat N, Pattus F, Dowhan W, Letellier L (1993) Role of acidic lipids in the translocation and channel activity of colicins A and N in *Escherichia coli.* Eur J Biochem 213:217-221

Vasilenko I, de Kruijff B, Verkleij AJ (1982) Polymorphic phase behavior of cardiolipin from bovine heart and from *Bacillus subtilis* as detected by ^{31}P-NMR and freeze-fracture. Biochim Biophys Acta 684:282-286

Wu HC, Tokunaga M, Tokunaga H, Hayashi S, Giam C-Z (1983) Posttranslational modification and processing of membrane lipoproteins in bacteria. J Cell Biochem 22:161-171

Yung BY-M, Kornberg A (1988) Membrane attachment activates DnaA protein, the initiation protein of chromosome replication in *Escherichia coli.* Proc Natl Acad Sci USA 85:7202-7205

ASSEMBLY OF SPHINGOLIPIDS INTO MEMBRANES OF THE YEAST, SACCHAROMYCES CEREVISIAE

P. Hechtberger, E. Zinser, F. Paltauf and G. Daum
Institut für Biochemie und Lebensmittelchemie
Technische Universität Graz
Petersgasse 12/2, A-8010 Graz
Austria

In the lower eukaryote, *Saccharomyces cerevisiae*, the inositol-containing ceramides inositolphosphate ceramide (IPC), mannosyl inositolphosphate ceramide (MIPC) and mannosyl diinositolphosphate ceramide (M(IP)$_2$C) were shown to be the major sphingolipids (Wagner & Zofcsik, 1966; Steiner et al., 1969; Smith & Lester, 1974).

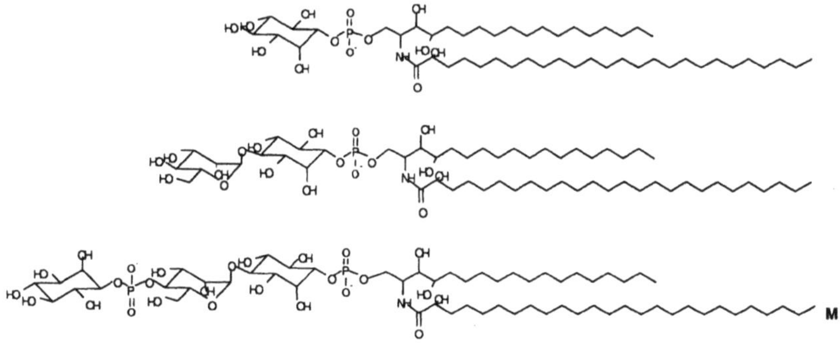

Fig. 1. Sphingolipids of the yeast, *Saccharomyces cerevisiae*. The position of the linkage between mannose and the second inositolphosphate group in M(IP)$_2$C is hypothetical.

Becker and Lester (1980) presented evidence that a yeast microsomal fraction contains an activity to introduce inositolphosphate from phosphatidylinositol into sphingolipids. Recently Puoti et al. (1991) reported that temperature-sensitive yeast secretory mutants (Schekman, 1985) blocked in early steps of the secretory pathway of proteins are unable to

NATO ASI Series, Vol. H 82
Biological Membranes:
Structure, Biogenesis and Dynamics
Edited by Jos A. F. Op den Kamp

synthesize $M(IP)_2C$, MIPC and a subclass of IPC at the non-permissive temperature. Patton and Lester (1991) reported, that similar to higher eukaryotes sphingolipids of the yeast are highly localized to the plasma membrane, although the authors did not present quantitative data.

An important prerequisite for a better understanding of the role of sphingolipids in yeast is the knowledge about their subcellular distribution, the single steps of the biosynthetic pathway, and the routes and mechanisms of subcellular transport. In order to quantitate the small amounts of inositol-containing sphingolipids in yeast organelle membranes we developed a sensitive assay procedure. Yeast secretory mutants were used to study the interplay between biosynthesis and intracellular transport of sphingolipids, and the secretory pathway of proteins.

MATERIALS AND METHODS

The wild type yeast strain, *Saccharomyces cerevisiae* X 2180, and temperature-sensitive secretory yeast mutants (a gift from R. Schekman, Berkeley, USA) were used throughout this study. Cells were grown either on YPD-medium (1 % yeast extract, 2 % peptone, 5 % glucose), or on inositol-free medium (Klig et al., 1985).

For [^{3}H]-inositol labelling secretory mutants and the corresponding wild type cells were grown overnight on inositol-free medium at the permissive temperature (24°C). For induction of the temperature block cells were transferred to new media, adjusted to an OD of 10, and shifted to the non-permissive temperature (37°C) for at least 30 min. Cells were pulse-labeled for 15 min in the presence of [^{3}H]-inositol (10 µCi in a total volume of 0.5 ml; spec. activity 20 Ci/mmol), and chased in the presence of 10 mM unlabeled inositol as indicated in legends to figures. For the extraction of inositol-containing lipids (phosphatidylinositol, sphingolipids) cells were harvested by centrifugation, suspended in chloroform/methanol/pyridine/water (60/30/1/6; by vol.), and frozen in liquid nitrogen. Then 1 ml glass beads were added, and lipids were extracted for 10 min at room temperature with vigorous shaking. Glass beads and cell debris were removed by centrifugation and reextracted, and the combined supernatants were collected and taken to dryness. For saponification lipids were resuspended in chloroform/methanol/water (16/16/5; by vol.) and incubated with an equal volume of 0.2 M methanolic NaOH at 30°C for 1 hour. Then 1 volume of 0.5 M EDTA was added, and samples were neutralized with 1 M acetic acid. Unsaponified lipids (among them sphingolipids) were extracted with 1 volume of chloroform. Samples were taken to dryness and resuspended in a small volume of chloroform/methanol/water (16/16/5; by vol.). Thin-layer chromatography of sphingolipids was carried out using Polygram Sil G plates (Macherey-Nagel) using chloroform/

methanol/0.25 M KCl (12/8/2; by vol.), or chloroform/methanol/4.2 M NH_3 (9/7/2; by vol.) as developing solvents. Prior to autoradiography plates were soaked in Amplify (Amersham) for 20 min at room temperature.

Standards of yeast sphingolipids were obtained by extraction of yeast spheroplasts with chloroform/methanol/pyridine/water (60/30/1/6; by vol.) overnight at 57°C. Saponification was carried out as described above. Unsaponified lipids were further purified by column chromatography on Silica gel 60 (Merck) using chloroform/methanol/water (16/8/1; by vol.) as an eluant. Individual mannose-containing sphingolipids, $M(IP)_2C$ and MIPC, were visualized by dipping thin-layer plates into a solution containing 15 ml conc. sulfuric acid, 5 ml water, 250 ml methanol and 700 mg orcinol, and heating for 10 min at 100°C. Quantitation was carried out by densitometric scanning of stained bands using a Shimadzu CS 930 Scanner.

Procedures for the isolation of yeast organelles, and for the measurement of markers of the respective subcellular fractions were the same as described by Zinser et al. (1991). Published procedures were used for the quantitation of phospholipids (Broekhuyse, 1968) and proteins (Lowry et al., 1951).

RESULTS AND DISCUSSION

As a first step to study the process of sphingolipid transport in yeast, we analyzed subcellular fractions for the occurrence of the two major yeast sphingolipids, mannosyl inositolphosphate ceramide (MIPC) and mannosyl diinositolphosphate ceramide ($M(IP)_2C$). We developed a simple and sensitive method to quantitate these lipids by direct optical scanning on thin-layer plates after chromatographic separation and derivatization with orcinol. Authentic standards were isolated (see Materials and Methods) and used as references. In agreement with the work of Patton and Lester (1991) highest concentrations of both sphingolipids, $M(IP)_2C$ and MIPC, were found in highly enriched preparations of the plasma membrane (Zinser et al., 1991). Other organellar fractions are largely devoid of sphingolipids.

Internal membranes are thought to be the site of synthesis of yeast sphingolipids (Becker & Lester, 1980), although a precise localization of this process has not been possible as yet. Angus and Lester (1972) reported that the inositolphosphate portion of IPC originates from phosphatidylinositol. Subsequent mannosylation and incorporation of a second inositol-phosphate group leads to $M(IP)_2C$. Our data obtained in a pulse-chase experiment (Fig. 2) confirm this finding.

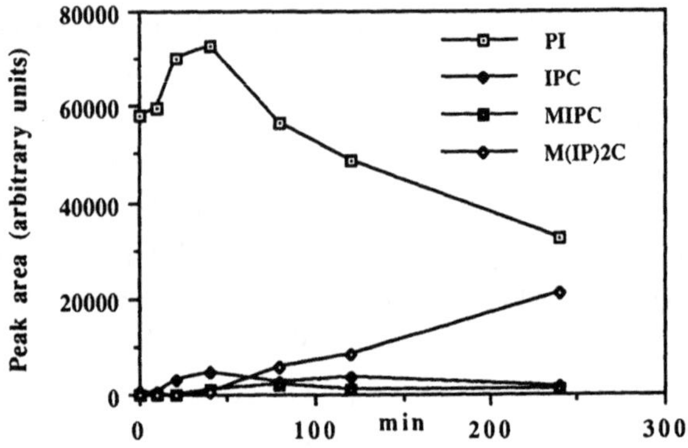

Fig. 2. Turnover of inositol-containing lipids in yeast. Yeast cells were labeled with [^{3}H]-inositol for 15 min, and chased in the presence of unlabeled inositol for 4 hours.

During the chase the amount of radioactivity in phosphatidylinositol (PI) decreases, whereas an increase of [^{3}H]-inositol in M(IP)$_2$C can be observed. The levels of radioactivity in IPC and MIPC are constant throughout the chase, pointing to their role as intermediates. Experiments with secretory mutants (see below, Fig. 2) clearly demonstrate that IPC is the first and MIPC the second product in the sequence of yeast sphingolipid biosynthesis.

The fact that the site(s) of synthesis (internal membranes) and the subcellular destination of sphingolipids (plasma membrane) are spatially separated, raises the question as to the intracellular transport of these lipids. One plausible mechanism would involve cotransport of sphingolipids and proteins along the secretory pathway. With yeast this hypothesis can easily be tested, because a whole set of temperature-sensitive secretory mutants (Schekman, 1985) with defects at all stages along the secretory pathway of proteins is available. Secretory mutants selected for this study are listed in Table 1.

If sphingolipids reach the plasma membrane by a process linked to protein secretion, temperature shift of secretory mutants from 24 to 37°C should not only inhibit protein secretion, but also affect the intracellular translocation of sphingolipids. For this purpose several secretory mutants were pulse-labeled with [^{3}H]-inositol, and the appearance of [^{3}H]-labeled phosphatidylinositol and sphingolipids was analyzed in a subsequent chase of 50 min. As can be seen from Fig. 3 four [^{3}H]-inositol labeled lipids appear in extracts of wild type cells, namely phosphatidylinositol (PI), inositolphosphate ceramide (IPC), mannosyl inositolphosphate ceramide (MIPC) and mannosyl diinositolphosphate ceramide (M(IP)$_2$C). The precursor-product relationship between these four lipids has been explained above. The

Table 1
Characteristics of *sec* mutants used in this study

Strain	Stage of defect in protein secretion	Gene product
sec53	Import into the ER	Phosphomannomutase[a)]
sec59	Import into the ER	Dolichol kinase[b)]
sec18	Transport from ER to Golgi	Yeast NSF-analog[c)]
sec21	Transport from ER to Golgi	105 kDa protein (membranous)[c)]
sec23	Transport from ER to Golgi	84 kDa protein (membranous)[c)]
sec7	Golgi	230 kDa protein (membr./sol.)[c)]
sec14	Golgi	PI transfer protein[c)]
sec1	Transport from Golgi to plasma membrane	83 kDa protein (soluble?)[d)]
sec4	Transport from Golgi to plasma membrane	GTP-binding protein[c)]
sec6	Transport from Golgi to plasma membrane	85 kDa protein (soluble ?)[c)]

[a)]Kepes & Schekman (1988); [b)]Heller et al. (1992); [c)]Pryer et al. (1992); [d)]Aalto et al. (1991)

temperature shift from 24 to 37°C has no influence on the labelling of sphingolipids in wild type cells. Several secretory mutants tested, namely *sec 59* , *7*, *14*, *4* and *6* do not show a defect in the labelling pattern of inositol-containing sphingolipids at the non-permissive temperature. On the other hand, the biosynthetic pathway of sphingolipids seems to be interrupted in some secretory mutants, when they are shifted to the non-permissive temperature of 37°C. This observation was made with one of the early secretory mutants, *sec 53*, which exhibits a defect in the import of proteins into the endoplasmic reticulum, and with several mutants, which are blocked at the transport of proteins between endoplasmic reticulum and Golgi at the non-permissive temperature (*sec 18*, *21* and *23*). Labelling patterns of *sec 53*, *18*, *21* and *23* mutants are similar, but reasons leading to these observations are different as will be outlined below. Finally *sec 1* has to be mentioned as an exception. In this mutant strain the final product in the biosynthetic pathway of sphingolipids, $M(IP)_2C$, is labeled neither under permissive nor under non-permissive conditions.

How can results obtained with secretory mutants be explained? We suggest the following model for the localization of biosynthetic steps of yeast sphingolipids, and for their transport from internal membranes to the plasma membrane (Fig. 4). The first step of synthesis, in which [^{3}H]-inositol is introduced into yeast sphingolipids, occurs in the endoplasmic reticulum. Early secretory mutants (*sec 53* and *59*) with a defect in the import of proteins into the endoplasmic reticulum (glycosylation) should not affect this process. This is

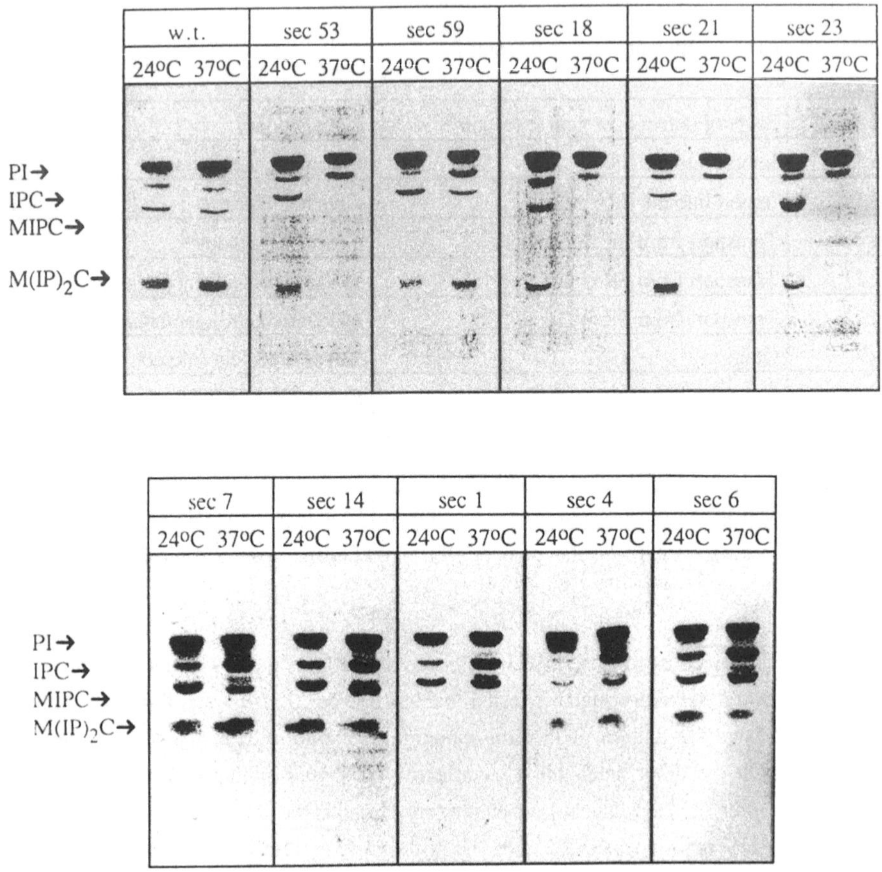

Fig. 3. Labelling (15 min pulse, 50 min chase) of yeast secretory mutants with [^{3}H]-inositol.

true for *sec 59*, but not for *sec 53*. This result can be explained by the fact, that the *SEC53* gene encodes phosphomannomutase (Kepes & Schekman, 1988), which is a key enzyme in the biosynthesis of GDP-mannose. The SEC59 gene product, on the other hand, is dolichol kinase (Heller et al., 1992), which synthesizes dolichyl phosphate and contributes to the mannosylation of proteins. The comparison of the labelling pattern of sphingolipids in the *sec 53* and *59* mutants clearly shows that GDP-mannose is the donor for the glycosylation of yeast sphingolipids. The SEC53 gene product itself does not directly influence the sphingolipid biosynthesis, but due to the lack of the product of the pathway blocked by the mutation mannosylation of IPC cannot occur at the non-permissive temperature.

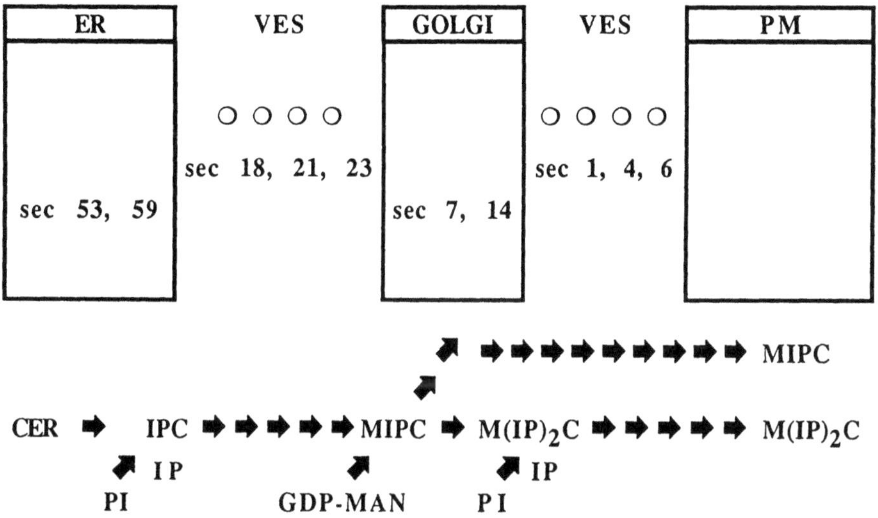

Fig. 4. Model for the biosynthesis and the intracelluar transport of yeast sphingolipids.

Protein transport from the endoplasmic reticulum to the Golgi involves vesicle flux (Pryer et al., 1992); the mutants *sec 18, 21* and *23* are blocked at this stage. In all these mutants IPC, the first inositol-containing substance on the biosynthetic pathway of sphingolipids, accumulates under non-permissive conditions. MIPC cannot be synthesized, because IPC, the substrate for the mannosylation, which obviously occurs in the Golgi, does not reach its site of metabolic conversion under non-permissive conditions. Therefore protein secretory transport vesicles must be involved in the transport of IPC from the endoplasmic reticulum to the Golgi.

In the Golgi mannosylation of IPC with GDP-mannose as a carbohydrate donor leads to the formation of MIPC. Further addition of inositolphosphate yields the final product, $M(IP)_2C$. It is not completely clear, if phosphatidylinositol is the direct donor of the inositolphosphate group or not. The scheme shown in Fig. 4 implies that biosynthesis of $M(IP)_2C$ is complete in the Golgi. This view is in good agreement with the fact that mutations in protein secretion at the stage of the late Golgi (*sec 7* and *14*) and in the late secretory pathway (*sec 4* and *6*) have no effect on the biosynthesis of $M(IP)_2C$. The exceptional characteristic of the *sec 1* mutation has to been pointed out. In this mutant the second inositolphosphate group is not attached to MIPC. The *SEC1* gene has been cloned (Aalto et al., 1991), but the function of the gene product is not known as yet.

The scheme outlined above leaves us with the question how $M(IP)_2C$ and MIPC are transported from the Golgi to the plasma membrane. Since the final product of the biosynthetic sequence, $M(IP)_2C$, is made in the Golgi, no further metabolic conversion can serve as a marker for the subcellular localization. In order to address this question we developed a different experimental strategy. We chose a secretory mutant, namely *sec 14*, in which the synthesis of $M(IP)_2C$ is complete, but the secretory pathway of proteins from the Golgi to the plasma membrane is blocked at 37°C. This mutant was labeled for 2 hours with $[^3H]$-inositol under permissive and non-permissive conditions, respectively. After this labelling period the plasma membrane and a microsomal fraction (containing the endoplasmic reticulum and the Golgi) were isolated and analyzed for inositol-labeled sphingolipids (Fig. 5).

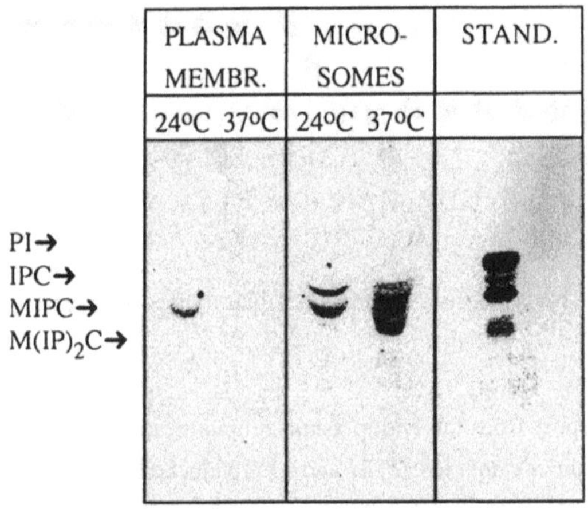

Fig. 5. Labelling pattern of inositol-containing sphingolipids in subcellular fractions of *Saccharomyces cerevisiae sec 14*. Lipids of the plasma membrane and of microsomes were saponified prior to thin-layer chromatographic separation.

In the plasma membrane of *sec14* cells labeled sphingolipids are detected when cells are grown at 24°C, but not when they are grown at the non-permissive temperature. This result indicates, that also late steps of sphingolipid transport to the plasma membrane are governed by the protein secretory machinery. Microsomes (including Golgi) of *sec 14* cells grown at 37°C, on the other hand, contain the whole set of labeled sphingolipids at large quantities, especially MIPC and $M(IP)_2C$. The amount of radioactivity in these sphingolipids is lower in control microsomes from cells cultivated at the permissive temperature. Taken together transfer along the secretory pathway of proteins seems to be the mechanism by which yeast sphingolipids reach the cell surface.

ACKNOWLEDGEMENTS

The excellent technical assistance of C. Hrastnik is greatfully acknowledged. This work was supported by the FWF in Austria (project S-5811 to G.D.)

REFERENCES

Aalto, M.K., Ruohonen, L., Hosono, K. and Keränen, S. (1991) Cloning and sequencing of the yeast Saccharomyces cerevisiae SEC1 gene localized in chromosome IV. Yeast 7, 643-650

Angus, W.W. and Lester, R.L. (1972) Turnover of inositol and phosphorus containing lipids in Saccharomyces cerevisiae; extracellular accumulation of glycerophosphorylinositol derived from phosphatidylinositol. Arch. Biochem. Biophys. 151, 483-495

Becker, G.W. and Lester, R.L. (1980) Biosynthesis of phosphoinositol-containing sphingolipids from phosphatidylinositol by a membrane preparation from Saccharomyces cerevisiae. J. Bacteriol. 142, 747-754

Broekhuyse, R.M. (1968) Phospholipids in tissues of the eye. Biochim. Biophys. Acta 152, 307-315

Heller, L., Orlean, P. and Adair, W.L., Jr. (1992) Saccharomyces cerevisiae sec 59 cells are deficient in dolichol kinase activity. Proc. Natl. Acad. Sci., USA, 89, 7013-7016

Kepes, F. and Schekman, R. (1988) The yeast SEC53 gene encodes phosphomannomutase. J. Biol. Chem. 263, 9155-9166

Klig, L.S., Homann, M.J., Carman, G.M. and Henry, S.A. (1985) Coordinate regulation of phospholipid biosynthesis in Saccharomyces cerevisiae: Pleiotropically constitutive opi1 mutant. J. Bacteriol. 162, 1135-1141

Lowry, O.H., Rosebrough, A.L., Farr, A.L. and Randall, R.J. (1951) Protein measurement with the Folin phenol reagent. J. Biol. Chem. 193, 265-275

Patton, J.L. and Lester, R.L. (1991) The phosphoinositol sphingolipids of Saccharomyces cerevisiae are highly localized in the plasma membrane. J. Bacteriol. 173, 3101-3108.

Pryer, N.K., Wuestehube, L.J. and Schekman, R. (1992) Vesicle-mediated protein sorting. Ann. Rev. Biochem. 61, 471-516

Puoti, A., Desponds, C. and Conzelmann, A. (1991) Biosynthesis of mannosylinositol-phosphoceramide in Saccharomyces cerevisiae is dependent on genes controlling the flow of secretory vesicles from the endoplasmic reticulum to the Golgi. J. Cell. Biol. 113, 515-525

Schekman, R. (1985) Protein localization and membrane traffic in yeast. Ann. Rev. Cell Biol. 1, 167-195

Smith, M.S. and Lester, R.L. (1974) Inositolphosphorylceramide, a novel substance and a chief member of a major group of yeast sphingolipids containing a single inositol phosphate. J. Biol. Chem. 249, 3395-3405

Steiner, S., Smith, S., Waechter, C.J. and Lester, R.L. (1969) Isolation and partial characterization of a major inositol-containing lipid in baker's yeast, mannosyl-diinositol diphosphoryl-ceramide. Proc. Natl. Acad. Sci., USA, 64, 1042-1048

Wagner, H. and Zofcsik, W. (1966) Sphingolipide und Glykolipide von Pilzen und höheren Pflanzen. II. Isolierung eines Sphingoglykolipids aus Candida utilis und Saccharomyces cerevisiae. Biochem. Z. 346, 343-350

Zinser, E., Sperka-Gottlieb, C.D.M., Fasch, E.-V., Kohlwein, S.D., Paltauf, F. and Daum, G. (1991) Phospholipid synthesis and lipid composition of subcellular membranes in the unicellular eucaryote Saccharomyces cerevisiae. J. Bacteriol. 173, 2026-2034

LOCALIZATION OF PHOSPHOLIPIDS IN PLASMA MEMBRANES OF MAMMALIAN CELLS

J.A.F. Op den Kamp, E. Middelkoop, R.J.Ph. Musters, J.A. Post, B. Roelofsen and A.J. Verkleij.
Institute of Biomembranes
Utrecht University
Padualaan 8
3584 CH Utrecht
The Netherlands

A transversal, asymmetric distribution of phospholipids is characteristic for most membrane systems investigated so far. Following a period in which an inventory was made of the asymmetry in various membrane systems, research was focused on questions regarding the biogenesis, the maintenance and the functional aspects of lipid asymmetry. As a result, a crucial factor involved in the maintenance of the asymmetry - the ATP dependent aminophospholipid translocase (flippase) - was detected. Both the existence of phospholipid asymmetry and the translocase were detected first in the erythrocyte membrane. However, if one wants to study dynamic characteristics of asymmetry and its maintenance, the biogenesis of this phenomenon, and in particular the influence that metabolic and physiological processes might have on the specific localization of membrane phospholipids, the erythrocyte provides a rather limited model system. Therefore, more recently emphasis was laid on studies with complex cells which can undergo severe physiological modifications. The cultured neonatal cardiomyocyte appeared to be a suitable model system and the study of its plasma membrane phospholipid asymmetry, its maintenance and changes therein upon metabolic alterations, can provide detailed information about the role that a specific phospholipid organization can play in proper cell functioning.

Phospholipid asymmetry in plasma membranes.

The best studied plasma membrane with respect to phospholipid localization is that of the erythrocyte and the existence of an asymmetric distribution of phospholipids was demonstrated first in this membrane. Initial information was obtained by the treatment of cells with probes designed to label amino-phospholipids and by treatment with phospholipases, which results later on were confirmed with a variety of studies using techniques that included exchange procedures, ESR and several combinations of them. Most procedures and tools, used to localize phospholipids, have been tested and applied on this system, viz. the human erythrocyte [Op den Kamp,1979; Roelofsen and Op den Kamp, 1994]. The erythrocyte membrane and the

NATO ASI Series, Vol. H 82
Biological Membranes:
Structure, Biogenesis and Dynamics
Edited by Jos A. F. Op den Kamp

plasma membrane of various cells investigated so far show a typical phospholipid asymmetry: the choline containing phosphatidylcholine (PC) and sphingomyelin (SM) are preferentially located in the outer membrane layer, whereas the aminogroup containing phosphatidylethanolamine (PE) and phosphatidylserine (PS) prefer the inner membrane layer. In addition to this polar headgroup asymmetry a specific localization of the various molecular species within one phospholipid class has been noted [Op den Kamp,1979; Roelofsen and Op den Kamp, 1994].

As indicated above, exchange procedures have been also applied to study phospholipid localizations by virtue of the existence of lipid transfer proteins. Several of these proteins have been isolated and purified and were shown to be useful tools in studies on membrane structure and function [Wirtz *et al*,1986]. The following paragraph presents an example of one of the applications.

The PC-specific transfer protein isolated from beef liver has the peculiar property of catalysing a truely one-for-one exchange. If we therefore together incubate two different PC containing membrane systems in the presence of this protein, an exchange of PC between the two systems occurs without any change in the respective PC contents. Using a radiolabelled PC of known specific radioactivity in the donor system, and simply measuring the final specific radioactivity of the PC in the acceptor system shows how much PC in the acceptor membrane is available for texchange [van Meer and Op den Kamp, 1982]. As illustrated in Fig.1, this procedure has been applied to determine the amount of accessible PC in the membrane of human and rat

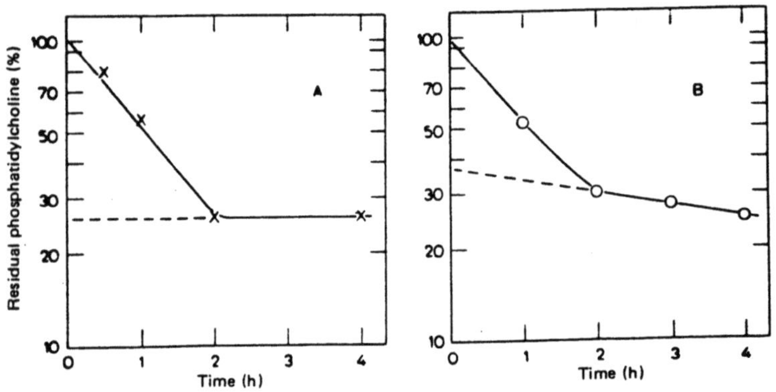

Fig.1 Human (A) and rat (B) erythrocytes were incubated with microsomes containing radiolabelled PC in the presence of PC transfer protein. The percentage erythrocyte PC not equilibrated with microsomal PC is plotted in a semi-logarithmic plot versus the incubation time. The solid lines connect the experimental points. The dashed lines represent the extrapolation of the slow phase to zero time [van Meer and Op den Kamp, 1982]

erythrocytes. Obviously, the accessible pool of PC represents the fraction of PC which is located in the outer leaflet of the membrane which, as can be deduced from the fact that the exchange profile reaches a plateau, does not appear to equilibrate rapidly with the PC pool present in the inner layer. The latter phenomenon indicates that the transbilayer movement (flip-flop) of this phospholipid in the erythrocyte membrane is a relatively slow process.

The same protein can be used to determine the actual -slow- rate of transbilayer movement more precisely and to investigate the influence of the fatty acid composition of the PC on this flip-flop process. To this end, incubations with erythrocytes are carried out in the presence of the PC transfer protein and vesicles containing radiolabelled PC of well defined fatty acid composition so as to introduce trace amounts of the radioactive molecules into the outer leaflet of the erythrocyte membrane. Following isolation of the cells, incubations are continued in the absence of donor vesicles and exchange protein for various time periods. Thereafter, cells are treated with a combination of phospholipase A_2 and sphingomyelinase C in order to hydrolyse all of the PC in the outer membrane layer. Phospholipid analysis subsequently reveals how much PC did escape the phospholipase treatment as it had migrated towards the inner membrane layer. Data obtained (Table 1) show that transbilayer movements are dependent to a large extent on the fatty acid composition of the phospholipid involved.

Table I

Rate of transbilayer movement of individual PC species in the human erythrocyte membrane at 37°C

PC species	Half-time (h)
1,2-Dipalmitoyl PC	26.3 ± 4.4
1,2-Dioleoyl PC	14.4 ± 3.5
1-Palmitoyl-2-linoleoyl PC	2.9 ± 1.7
1-Palmitoyl-2-arachidonoyl PC	9.7 ± 1.6

data are taken from Middelkoop *et al*, 1986

Maintenance of phospholipid asymmetry.

A similar approach as described above has been applied to measure flip-flop rates of phospholipids other than PC. For such measurements a non-specific phospholipid transfer protein was used . This way, trace amounts of radioactive PC, PE, PS and SM, when present in donor vesicles, are transferred in one single incubation into the outer layer of the erythrocyte membrane. Following subsequent incubation, washing of the erythrocytes and phospholipase treatment, the quantitation of residual radioactivity of each phospholipid class is carried out in order to determine the relative amount that had been migrated towards the inner membrane leaflet [Tilley et al 1986]. The procedure is outlined schematically in Fig. 2. and the results are depicted in Fig. 3.

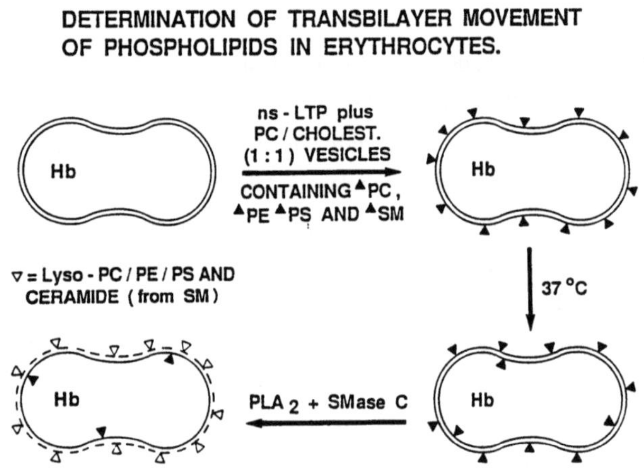

Fig. 2. Principle of an experiment to determine the rate of transbilayer movement of phospholipids in the erythrocyte membrane. In addition to the PC containing vesicles, indicated in the figure, it is possible to use any radioactive phospholipid in the donor vesicle system. Abbreviations; Hb, haemoglobin; PLEP, phospholipid exchange protein.

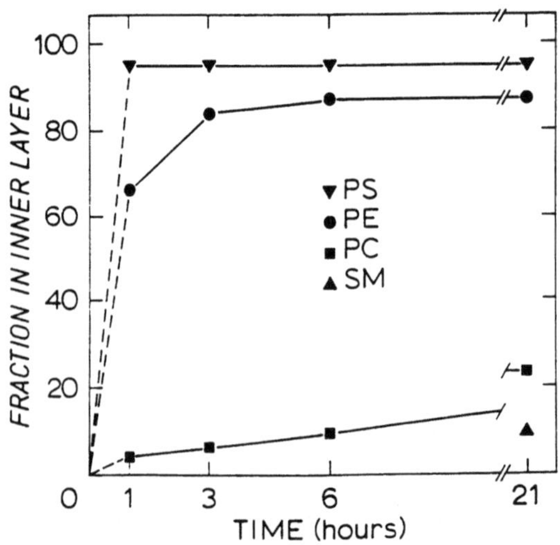

Fig. 3. Transbilayer migration of radiolabelled phospholipids, previously inserted into the outer membrane layer of human erythrocytes, using a non-specific lipid transfer protein.

It is obvious that large differences exist between different phospholipid classes. Whereas PC, and in particular SM, once incorporated in the outer membrane layer of the erythrocyte hardly undergo any transbilayer mobility, PE and PS move inward rapidly. An explanation for this phenomenon was offered by Devaux and co-workers who showed the existence of an ATP-dependent aminophospholipid translocase, or flippase, in the membrane of the red cell. This putative protein is held responsible for the maintenance of an asymmetric distribution of both aminophospholipids PE and PS [Seigneuret and Devaux, 1984].

A second factor involved in maintaining a specific phospholipid distribution between the two membrane leaflets is thought to be an interaction between the proteins present in the membrane skeleton underlying the lipid bilayer, and the polar head groups of the phospholipids present in this layer. Evidence to substantiate this view is originating from several experiments. To name a few: (i) spectrin and other skeletal proteins interact with PS containing monolayers and liposomes; (ii) modification of the membrane skeletal proteins results in accelerated transbilayer movement of phospholipids and enhanced accessibility of (amino)-phospholipids at the outer surface and (iii) uncoupling of the lipid bilayer from the membrane skeleton, as it occurs in sickled red cells, results in a loss of the constraints on lipid packing and transbilayer organization. The reader is referred to a recent review [Roelofsen and Op den Kamp, 1994] for an extensive description of the various observations.

Phospholipids in the myocardial sarcolemma.

During recent years, an increasing body of evidence has shown that the development of myocardial membrane dysfunction plays an important role in the pathogenesis of ischemic myocardial cell injury [Sen et al, 1987]. It is also known that ischemia induces a complex series of noxious alterations such as energy depletion, cellular acidosis, redistribution of cellular calcium ions and accumulation of metabolic intermediates (e.g. acyl-CoA esters, long-chain acyl carnitines and lysophospholipids), which may all contribute to the loss of stability and integrity of the sarcolemma in different ways. However, the exact mechanism(s) underlying irreversible ischemic membrane damage remain(s) to be established.

As the lipid bilayer plays an important role, not only in maintaining cell integrity but also in several other essential functions, investigations were initiated to unravel the behaviour of the lipid complement during ischemia. One of the surprising results showed that during the development of ischemic damage, alterations in the membranes of mitochondria and the sarcolemma occurred, resulting in the formation of spherical protein-free membranous structures, the so-called "blebs". It was postulated that a physico-chemical modification and reorganization of the membrane lipids was responsible for the final disorganization of the membrane [Verkleij and Post, 1987, Verkleij *et al*, 1990]. Considering the possible mechanisms responsible for a lateral membrane reorganization which is preceding the bleb formation, it could not be excluded that a change in the transversal phospholipid distribution is involved. We therefore checked, in detail, upon the phospholipid complement of the sarcolemma of the heart muscle cell. and used for that study isolated and neonatal cardiomyocytes kept under cell-culture conditions. These cultured heart cells have shown to be a useful model for studying effects of oxygen- and volume restrictions, thereby simulating ischemia [Vemuri *et al*, 1988, Musters *et al*, 1991]. First of all it was found that the sarcolemma bilayer exhibits a specific phospholipid distribution [Post et al, 1988], which is gradually lost after a prolonged period of ischemia [Musters *et al*,1994].

Techniques to elucidate the phospholipid distribution in the sarcolemma of these cells are, in principle, similar to those described above. However, additional difficulties arose from the fact that, in contrast to the erythrocyte, the cardiomyocyte contains a vast amount of intracellular membranes. In experiments in which probes do react exclusively with the phospholipids of the outer layer of the sarcolemma, a correct interpretation of the data requires detailed information on the phospholipid composition of the sarcolemma as well as on the percentage of total cellular phospholipid present in this membrane. The latter data can be obtained as follows. Intact cells are treated, under non-lytic conditions, with a sphingomyelinase C which hydrolyses only its substrate in the outer layer of the sarcolemma. Thereafter, part of the treated cells is extracted and analysed for its phospholipid content and the extent of hydrolysis. Another part of the same cell population is used to isolate pure sarcolemma (plasma membrane) which, in turn, is analysed as well. Then, calculations as first applied by Chap et al [1979], can be carried out in order to assess the amount of phospholipid present in the isolated membrane.
An essential prerequisite for the localization of the phospholipids in the plasma membrane of these complicated cells is, furthermore, the availability of a method to isolate plasma membranes fast and in a pure form. The "gas dissection " technique is ideally suited for this purpose. A description of this technique, as well as the procedures one has to follow in order to obtain the essential information on the phospholipids of the sarcolemma has been described previously [Post *et al* 1988]. The results are as follows. The sarcolemma contains 38 % of the total phospholipids of the cell. The major components are PC, PE, SM, PS and PI, and the distribution of these phospholipids over the two layers of the sarcolemma is as presented in Fig 4. It resembles the asymmetric distributions found in other plasma membrane systems, e.g. the choline containing phospholipids are located preferentially in the outer layer whereas PE and PS prefer the inner layer of the membrane.

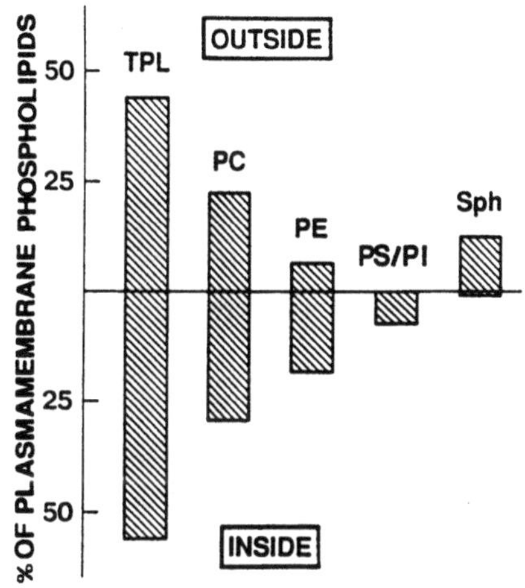

Fig.4 Phospholipid composition and transbilayer distribution in the sarcolemma of cultured rat heart cells.

Following these experiments on cardiomyocytes which were kept under oxygenated "control" conditions, experiments were carried out with cells kept under anoxic conditions. In addition, the limitations in perfusion rate, occurring in the *in vivo* situation, were mimicked by limiting the amount of -nutrient free- buffer in the culturing device. First of all, experiments were carried out on ^{3}H-[acetate]-pre-labelled cardiomyocytes and their 'gas-dissected' (non radio-labelled) sarcolemmal membranes in order to detect any breakdown in phospholipids under the "ischemic" conditions. After 60 min. of 'ischemia' no significant phospholipid hydrolysis could be detected and cellular high energy phosphate (ATP) levels had decreased to less than half of the control values. As could be expected from these results also the phospholipid composition of the sarcolemma did not undergo any appreciable modification during "ischemia".

Following various periods of incubation under these conditions up to one hour, the phospholipid distribution in the sarcolemma of these cells was investigated. Two approaches were applied, the labelling of aminophospholipids with the (at $4^{o}C$ non-permeant) probe trinitrobenzenesulfonic acid (TNBS) and the non-lytic treatment ot the cells with different phospholipases A_2. The results in both type of experiments appeared to be similar. The specific lipid topology of the outer sarcolemmal membrane leaflet is changing progressively during the anoxic incubations: a shifted transbilayer distribution of the hexagonal(II)phase promoting aminophospholipid phosphatidylethanolamine (PE) in favour of the outer sarcolemmal lipid monolayer is observed. Under normoxic conditions, 27 % of the total sarcolemmal PE can be hydrolysed at the outside of the cell by phospholipase A_2 under non-lytic conditions. After 60 min of simulated ischemia, this amount is found to be increased to about 37 %. Similar data are obtained using TNBS labelling as a tool to localise PE. This change clearly preceded the loss of integrity of the sarcolemma, which was monitored by the release of lactate dehydrogenase (LDH) as well as by scanning electron microscopy.

It is proposed that, in addition to the transversal phospholipid reorganization, uncontrolled transbilayer movement (flip-flop) of PE from the inner to the outer sarcolemmal membrane leaflet further destabilizes the lipid bilayer of the sarcolemma, and may provide a biochemical switch for the transition from reversible to irreversible membrane damage after a prolonged period of ischemia.

Investigations were carried out under auspices of the Netherlands Foundation for Chemical Research (SON) and with financial support from the Netherlands Organization for Scientific Research (NWO). J.A. Post is a recipient of a fellowship of the Royal Netherlands Academy of Arts and Sciences.

Literature

Chap, HJ, Zwaal, RFA and van Deenen LLM (1979) Action of highly purified phospholipases on blood platelets. Evidence for an asymmetric distribution of phospholipids in the surface membrane. Biochim Biophys Acta 467: 146-164.

Middelkoop, E, Lubin, BH, Op den Kamp JAF and Roelofsen B (1986) Flip-flop of individual molecular soecies of phosphatidylcholine in the human red cell membrane. Biochim Biophys Acta 855: 421-424

Musters,RJPH, Post JA, and Verkleij AJ, (1991), The isolated neonatal rat cardiomyocyte used in an in vitro model for 'ischemia'. I. A morpholgical study. Biochim Biophys Acta 1091: 270-277.

Musters,RJPH, Otten E, Biegelmann E, Bijvelt J, Keizer JJH, Post JA, Op den Kamp JAF and Verkleij AJ, (1994), Loss of phosphatidylethanolamine transbilayer asymmetry in the sarcolemma of the isolated neonatal rat cardiomyocyte during simulated ischemia. Circ Res, in press

Op den Kamp JAF(1979) Lipid asymmetry in membranes. Ann Rev Biochem 48:47-71

Roelofsen B and Op den Kamp JAF (1994) Plasma membrane phospholipid asymmetry and its maintenance: the human erythrocyte as a model. In: 'Current Topics in Membranes' vol. 41 in press.

Post JA, Langer GA, Op den Kamp JAF and Verkleij AJ(1988), Phospholipid asymmetry in cardiac sarcolemma. Analysis of intact cells and gas-dissected membranes.Biochim Biophys Acta 943: 256-266.

Seigneuret M and Devaux PF (1984) ATP-dependent asymmetric distribution of spin-labelled phospholipids in the erythrocyte membrane: relation to shape changes.Proc Natl Acad Sci US 81: 3751-3755.

Sen A, Buja LM, Willerson JT and Chien K (1987), Membrane phospholipid metabolism during myocardial ischaemia: past present and future. In:'Lipid metabolism in the normoxic and ischemic heart',Stam H and van de Vusse GJ (eds.), Springer Verlag New York pp.121-125.
Tilley L, Cribier S, Roelofsen B, Op den Kamp JAF and van Deenen LLM (1986) ATP-dependent translocation of aminophospholipids across the human erythrocyte membrane. FEBS lett. 194: 21-27.
van Meer G and Op den Kamp JAF (1982). Transbilayer movement of various phosphatidylcholine species in intact human erythrocytes. J Cell Biochem 19:193-204.
Vemuri R, Mersel M, Heller M and Pinson A, (1988), Studies on oxygen and volume restriction in cultured cardiac cells: possible rearrangement of sarcolemmal lipid moieties during anoxia and ischemia-like states. Mol Cell Biochem 79: 39-46.
Verkleij AJ and Post JA, (1987), Physicochemical properties and organization of lipids in membranes. Its possible role in myocardial injury. Basic Res Cardiol 82: 85-92
Verkleij AJ, Post JA and Schneijdenberg CTWM, (1990), Is acidosis the clue to the loss of structure and functioning of the sarcolemma? Cell Biol Intl Rep 14 No.4: p.335-341
Wirtz KWA, Op den Kamp JAF and Roelofsen B (1986) Phosphatidylcholine transfer protein: properties and applications in membrane proteins. In: Progress in Protein-Lipid Interactions, vol.2 (Watts A and de Pont JJHHM eds.) Elsevier, Amsterdam, pp 221-266

SOME ASPECTS OF INTRACELLULAR LIPID TRAFFIC

Vytas A. Bankaitis
Department of Cell Biology
University of Alabama at Birmingham
School of Medicine
Birmingham, Alabama 35294-0005
U.S.A.

A universal feature of eukaryotic cells is a set of intracellular organelles that are dedicated to the execution of specific cellular functions. These organelles are distinguished by their unique complement of resident proteins and by the quantitative characteristics of the lipid compositions of their membranes. The sum of numerous analyses indicate that the lipid compositions of organellar membranes differ from each other with respect to: (i) the relative proportions of the lipid species measured and, in some cases, to the presence of a particular lipid in only a few biological membranes, and (ii) the lipid asymmetry observed between the external and internal leaflets of a particular membrane. Moreover, as the endoplasmic reticulum represents the major site of cellular lipid synthesis in eukaryotic cells, this organelle can be considered as the point at which the membrane bilayer is set up in the cell. These collective observations indicate that organelle biogenesis is not predominantly driven by biosynthetic processes inherent to each organelle. Rather, specific lipid transport and sorting phenomena must play a determining role in the assembly of intracellular organelles. In this lecture, I will discuss some of the mechanisms presently under consideration for lipid transport and sorting between organelles, and discuss new work that may unify aspects of lipid and protein trafficking in eukaryotic cells. This lecture cannot give an exhaustive review of lipid trafficking. Excellent reviews to that effect have recently been published (Pagano, 1990; Voelker, 1991).

For the purpose of this lecture, I find it useful to categorize lipid traffic into two general mechanisms: vesicular and nonvesicular. The vesicular mechanism I subdivide into conventional and nonconventional modes, whereas the nonvesicular mechanism includes: (i) transport of lipid monomers between physically distinct membranes, and (ii) the transport of lipids between membranes by a "collision" event. This conceptual scheme is illustrated in Figure 1. Examples highlighting each of these mechanisms will be discussed.

NATO ASI Series, Vol. H 82
Biological Membranes:
Structure, Biogenesis and Dynamics
Edited by Jos A. F. Op den Kamp

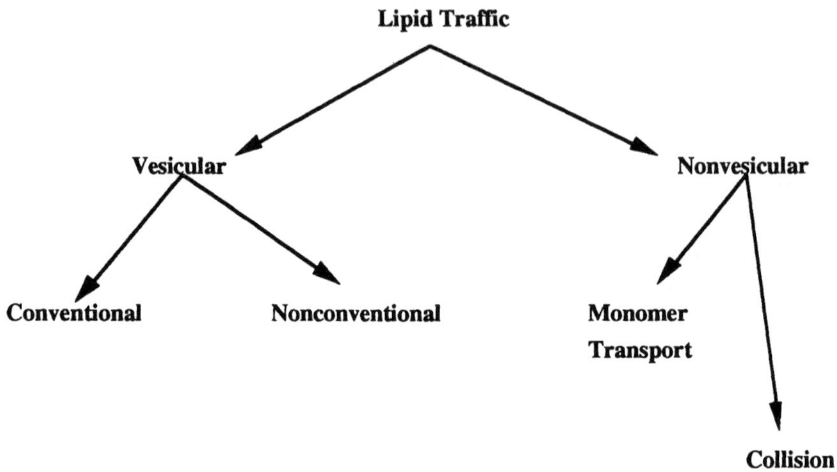

Figure 1. Conceptual mechanisms for lipid transport.

I. Conventional vesicular trafficking of lipids.

I define conventional vesicular trafficking of lipids as the transport of lipids in the same transport vesicles that are involved in protein trafficking through the secretory/endocytic pathway. Quantitatively, this must be a very (if not the most) significant pathway for bulk lipid transport in the cell. It is also an obvious mechanism. Nevertheless, it is instructive to examine several cases where this type of trafficking has been studied. Wattenberg (1990) reconstituted glycolipid transport between Golgi compartments in a cell-free system. The basis of the transport assay was the transport of lactosylceramide from a donor Golgi preparation to acceptor Golgi membranes that had the capability to convert lactosylceramide to ganglioside GM3 by a sialylation reaction. Thus, incorporation of labeled sialic acid into GM3 recorded the desired inter-Golgi lipid transport event. By comparing the biochemical requirements, the pharmacology and the kinetic parameters of glycolipid transport to those of secretory glycoprotein transport, Wattenberg concluded that glycolipid and glycoprotein transport through the Golgi employ similar, if not identical, mechanisms.

In another study, Conzelmann and colleagues demonstrated that the synthesis of a particular class of sphingolipid in yeast, the mannosylinositolphosphoceramide $M(IP)_2C$ (Puoti *et al.*, 1991), is sensitive to secretory defects that prohibit protein transport from the endoplasmic reticulum to the Golgi complex. One current interpretation of the data is that some lipid precursor of $M(IP)_2C$, either sphingolipid or phosphatidylinositol (PI), must be delivered from the ER to the Golgi via a vesicular carrier that also is involved in glycoprotein transport.

Finally, the use of fluorescently-labeled short acyl chain (i.e. C_6-NBD) derivatives of both sphingolipids and glycerophospholipids has permitted a facile examination of lipid trafficking in living cells (reviewed by Pagano, 1990). Incorporation of NBD-labeled PC or various NBD-glycolipids into the outer leaflet of the plasma membrane recapitulates the classic endocytic pathway.

II. Nonconventional vesicular trafficking of lipids.

I define this mode of lipid trafficking to involve vesicular carriers of lipid that play no clear role in bulk glycoprotein transport through the secretory pathway. This represents an operational definition that is not without some degree of ambiguity as such a mechanism may not be easily distinguished from other nonvesicular transport mechanisms. Nevertheless, work on the problem of cholesterol transport from the ER to the plasma membrane, primarily in the lab of Bob Simoni, suggests a nonconventional vesicular mode of cholesterol transport from the ER to the Chinese hamster ovary cell plasma membrane. These studies employ pulse-radiolabeling of cells to label either cholesterol or a marker secretory protein followed by rapid isolation of plasma membrane. Appearance of label in the PM fraction records the arrival of cholesterol at the cell surface. Again, the kinetic and biochemical properties of cholesterol and marker protein transport to the PM are compared. The $t_{1/2}$ of cholesterol transport to the PM was approximately 10 min (Kaplan and Simoni, 1985a), a value within the bounds of estimates of bulk membrane flow from the ER to the plasma membrane (Wieland *et al.*, 1987). This cholesterol transport required metabolic energy, exhibited a dramatic cold-sensitivity as transport ceased at 15°C, and labeled cholesterol accumulated in the ER and in a lipid-rich fraction upon cold challenge. These features were consistent with a vesicular mode of cholesterol transport as transport of VSV G protein exhibited similar characteristics, and VSV G accumulated in a lipid-rich fraction that was biochemically similar to that observed in cells challenged with the 15°C condition. This lipid-rich fraction may represent the intermediate compartment between ER and Golgi (Urbani and Simoni, 1990). Although these data initially suggest a conventional route of cholesterol transport to the PM, VSV G trafficking from this lipid-rich fraction is pharmacologically distinguishable from that of cholesterol (Kaplan and Simoni, 1985a; Urbani and Simoni, 1990). Whereas cholesterol transport to the PM is resistant to agents that block conventional vesicular traffic (i.e. brefeldin A and monensin), VSV G transport is sensitive. On the basis of these findings, a substantial fraction of cholesterol synthesized in the ER is considered to traffic to the PM via a nonconventional vesicular route -- one that may bypass the Golgi complex. A summary of this concept is presented in Figure 2.

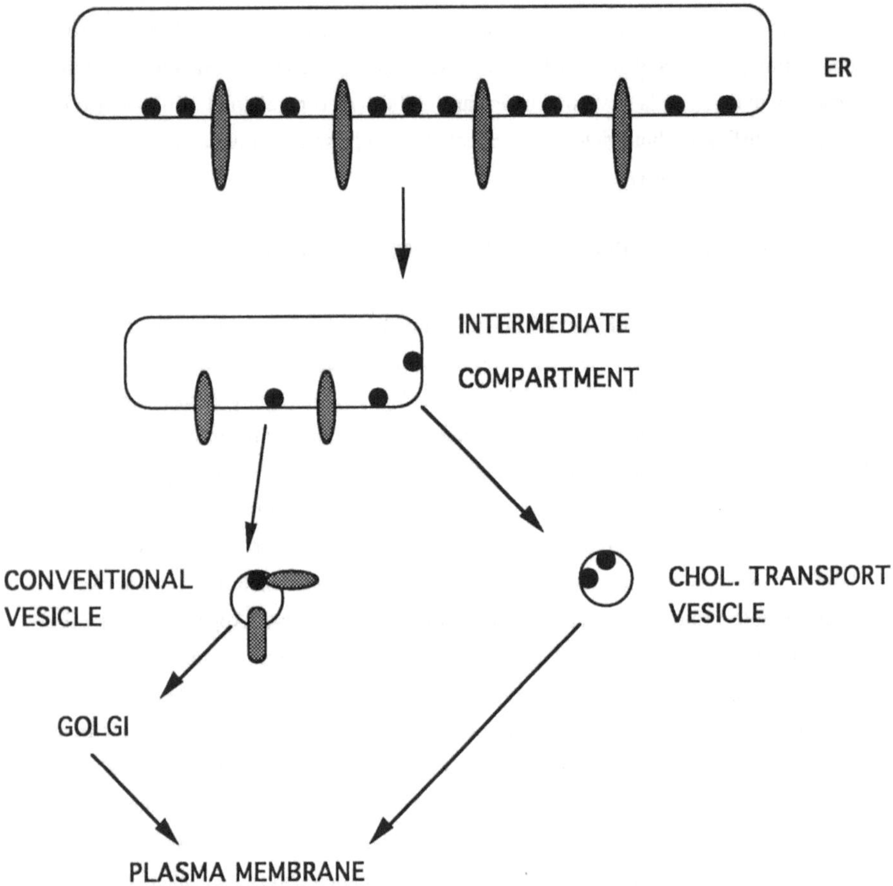

Figure 2. An example of nonconventional vesicular lipid transport. VSV G prorein is symbolized by the stippled spikes whereas cholesterol is indicated by the solid circles.

III. Nonvesicular lipid transport: the bangosome concept.

Nonvesicular lipid transport is a broad category into which fall lipid transport processes whose kinetics and biochemical properties are not characteristic of properties generally expected of processes that involve vesicle formation and/or fusion. Several examples are described here. An intriguing example of what appears to be a nonvesicular mode of lipid traffic involves the cell biology associated with the decarboxylation of phosphatidylserine to phosphatidylethanolamine. The PS decarboxylase is a mitochondrial enzyme that appears to be localized to the zones of adhesion between the mitochondrial inner and outer membranes. Thus, some fraction of the PS pool must be transported from the ER to the mitochondrion where it is converted to PE. Some fraction of the PE must then traffic back to the ER. This pathway provides a ready assay for transport of PS to the mitochondrion. One can radiolabel PS with serine and monitor the formation of labeled PE by decarboxylation. In principle, this assay is directly analagous to those that employ stage-specific processing events to record the transport of proteins between intracellular compartments. This assay has been successfully applied both *in vivo* and *in vitro*.

The $t_{1/2}$ of PS transport to the mitochondrion has been estimated to be approximately 7hrs in baby hamster kidney cells, and this transport of PS to the mitochondrion is abolished by depletion of intracellular ATP both in mammalian cells and in yeast (Voelker, 1985). The reconstitution of PS transport to the mitochondrion in saponin permeabilized cells has provided a particularly powerful system with which to study this process. In the permeabilized cell system, the $t_{1/2}$ of PS transport is about 3hrs, is dependent upon ATP hydrolysis, does not require ongoing PS synthesis, at the least does not require a high concentration of cytosol and may not require cytosolic proteins at all, and is resistant to a dilution of approximately 50-fold: a property not immediately consistent with transport via a freely diffusible vehicle such as a vesicle or a phospholipid transfer protein (Voelker, 1990; see below). PS transport into isolated mitochondria has also been reconstituted (Voelker, 1989), and this reaction shows intriguing differences when compared to the *in vivo* and permeabilized cell systems. The most significant discrepancy is that PS transport in the most purified system is insensitive to depletion of mitochondrial ATP, to various uncouplers of oxidative phosphorylation, or to dissipation of the mitochondrial electrochemical gradient. Moreover, efforts to reconstitute this process *in trans* have thus far failed. The various data have been reconciled in a model where PS transport consists of an ATP-dependent formation of a dedicated mitochondrial -- ER complex that now renders PS transport to the mitochondrial outer membrane possible. Formation of this complex may represent the step that cannot yet be reconstituted *in trans*, and may not require cytosolic factors for its formation. This formation of complex may be by collision -- hence the bangosome. Subsequent transfer of PS to the mitochondrial outer membrane may be an ATP-independent step that might utilize a PS/PE transfer protein that executes a net transfer of PS from the ER to the mitochondrial outer membrane (Figure 3). The bangosome concept may not uniquely apply to PS transport from the ER to mitochondria. Lipid transfer between *trans*-Golgi elements may also employ a variation of this mode of transport (Cooper *et al.*, 1990).

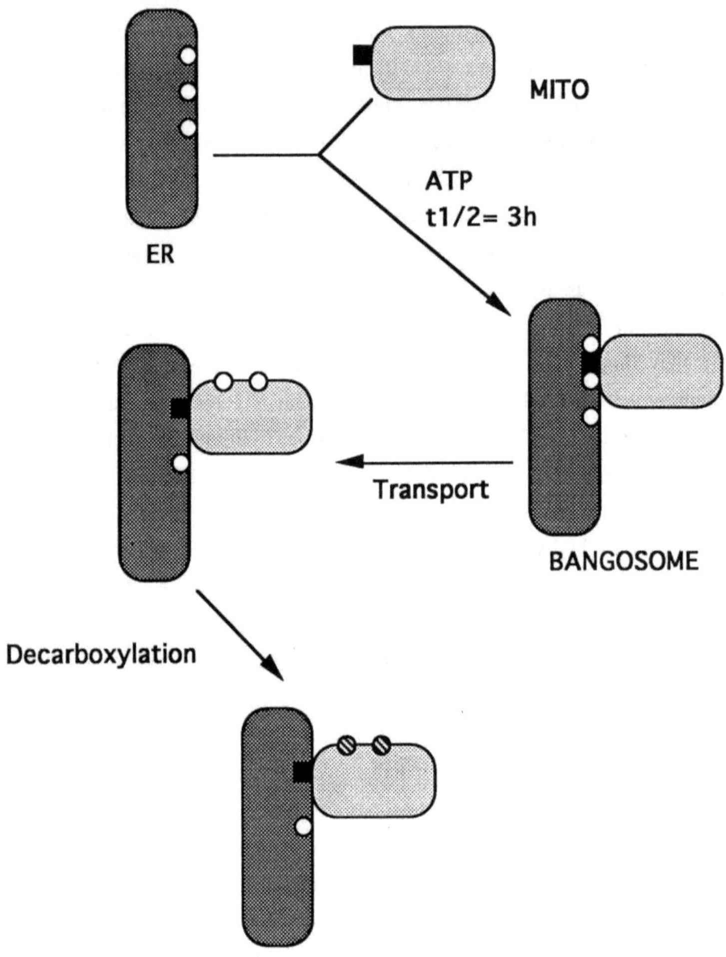

Figure 3. The bangosome model for PS transport to the mitochondrion. PS monomers are depicted as open circles and PE monomers derived from PS are represented by hatched circles. The PS/PE transfer protein is represented by the solid square.

In addition to PS transport from the ER to mitochondria, there are several other examples of lipid transport that suggest nonvesicular modes. Using an *in vivo* pulse-radiolabeling regimen in conjunction with a rapid PM purification, Kaplan and Simoni

(1985b) found that newly synthesized PC trafficked from the ER to the PM with a $t_{1/2}$ of 1min -- a rate much faster than can seemingly be accounted for by vesicular traffic. This transport was unaffected by low temperature challenge (15°C), agents that deplete cellular ATP, or agents that otherwise block vesicular traffic through the secretory pathway. Similar conclusions were drawn from experiments measuring the transport of PC from the ER to mitochondria and transport of nascent PE from the ER to the PM, respectively (Kobayashi and Pagano, 1989; Vance *et al.*, 1991). What is the mechanistic basis for such seemingly nonvesicular lipid transport that also does not appear to utilize interorganelle collision events to facilitate lipid transfer? It has long been speculated that an enigmatic class of cytosolic proteins, the phospholipid transfer proteins (PL-TPs), could represent bulk carriers of lipid monomers through the cytosol since these are capable of carrying out this reaction in an energy-independent manner *in vitro*(reviewed by Wirtz, 1991). These proteins will be covered in more detail in my research lecture.

IV. Caveolae: Lipid microdomains that execute protein sorting reactions?

Until recently, glycoprotein and lipid sorting have been treated as separate issues. An important conceptual step toward unifying aspects of these two processes was provided by Kai Simons and Gerritt van Meer who suggested that glycolipid microdomains may participate in the sorting of proteins to the apical PM of epithelial cells (Simons and van Meer, 1988). This model was in part driven by the finding that the outer leaflet of the apical plasma membrane is highly enriched in glycolipid. It is proposed that sphingolipids assemble into microdomains in the lumenal leaflet of the trans-Golgi membrane by virtue of interlipid hydrogen binding. These microdomains then, in some manner, serve as affinity surfaces for proteins destined for the apical PM thus providing the basis of apical sorting. Some biochemical evidence to support this model has recently become available. Caveolae, are nonclathrin coated plasma membrane invaginations that are highly enriched in glycolipids, thus fulfilling a principle criterion of a glycolipid microdomain. The predominant component of the caveolar coat is a 21kD protein called caveolin or VIP21 that may recycle between PM and Golgi (Dupree et al., 1993; Rothberg et al., 1992). Moreover, at least some GPI-anchor proteins assemble into detergent resistant vesicles that do not contain basolateral protein markers but have the characteristics of VIP21-containing vesicles (Brown and Rose 1992; Dupree *et al.*, 1993). It seems likely that caveolae may indeed represent the glycolipid microdomains proposed by Simons and van Meer (1988) to execute apical sorting.

V. References.

Brown, DA, and Rose, JK (1992) Sorting of GPI-anchored proteins to glycolipid-enriched membrane subdomains during transport to the apical cell surface. Cell 68: 533-534.

Cooper, MS, Cornell-Bell, AH, Chernjavsky, A, Dani, JW, and Smith, SJ (1990) Tubulovesicular processes emerge from the trans-Golgi cisternae, extend along microtubules, and interlink adjacent trans-Golgi elements into a reticulum. Cell 61: 135-145.

Dupree, P, Parton, RG, Raposo, G, Kurzchalia, TV, and Simons, K (1993) Caveolae and sorting in the trans-Golgi network of epithelial cells. EMBO J. 12: 1597-1605.

Kaplan, MR, and Simoni, RD (1985a) Transport of cholesterol from the endoplasmic reticulum to the plasma membrane. J. Cell Biol. 101: 446-453.

Kaplan, MR, and Simoni, RD (1985b) Intracellular transport of phosphatidylcholine to the plasma membrane. J. Cell Biol. 101: 441-445.

Kobayashi, T., and Pagano, RE (1989) Lipid transport during mitosis: Alternative pathways for delivery of newly synthesized lipids to the cell surface. J. Biol. Chem. 264: 5966-5973.

Pagano, RE (1990) Lipid traffic in eukaryotic cells: mechanisms for intracellular transport and organelle-specific enrichment of lipids. Curr. Op. in Cell Biol. 2: 652-663.

Puoti, A, Desponds, C, and Conzelmann, A (1991) Biosynthesis of mannosylinositolphosphoceramide in *Saccharomyces cerevisiae* is dependent on genes controlling the flow of secretory vesicles from the endoplasmic reticulum to the Golgi. J. Cell Biol. 113: 515-525.

Rothberg, KG, Heuser, JE, Donzell, WC, Ying, Y-S, Glenney, JR, and Anderson, RGW (1992) Caveolin, a protein component of caveolae membrane coats. Cell 68: 673-682.

Simons, K., and van Meer, G (1988) Lipid sorting in epithelial cells. Biochemistry 27: 6197-6202.

Urbani, L, and Simoni, RD (1990) Cholesterol and vesicular stomatitis virus take separate routes from the endoplasmic reticulum to the cell surface. J. Biol. Chem. 265: 1919-1923.

Vance, JE, Aasman, EJ, and Szarka, R (1991) Brefeldin A does not inhibit the movement of phosphatidylethanolamine from its site of synthesis to the cell surface. J. Biol. Chem. 266: 8241-8247.

Voelker, DR (1985) Disruption of phosphatidylserine translocation to the mitochondria in baby hamster kidney cells. J. Biol. Chem. 260: 14671-14676.

Voelker, DR (1989) Reconstitution of phosphatidylserine import into rat liver mitochondria. J. Biol. Chem. 264: 8019-8025.

Voelker, DR (1990) Characterization of phosphatidylserine synthesis and translocation in permeabilized animal cells. J. Biol. Chem. 260: 14340-14346.

Voelker, DR (1991) Organelle biogenesis and intracellular lipid transport in eukaryotes. Microbiol. Rev. 55: 543-560.

Wattenberg, B (1990) Glycolipid and glycoprotein transport through the Golgi complex are similar biochemically and kinetically. Reconstitution of glycolipid transport in a cell-free system. J. Cell Biol. 111: 421-428.

Wieland, FT, Gleason, ML, Serafini, TA, and Rothman, JE (1987) The rate of bulk flow from the endoplasmic reticulum to the cell surface. Cell 50: 289-300

Wirtz, KWA (1991) Phospholipid transfer proteins. Annu. Rev. Bioch. 60: 73-99

Intracellular Localization and Mechanisms of Regulation of Phosphatidylinositol Transfer Protein in Swiss Mouse 3T3 Cells

G.T. Snoek and K.W.A. Wirtz
Centre for Biomembranes and Lipid Enzymology
Utrecht University
Padualaan 8
3584 CH Utrecht
The Netherlands

Introduction

Proteins which are able to catalyze the transfer of phospholipid molecules between membranes have been purified and characterized from a wide spectrum of cells and tissues [reviewed by K.W.A. Wirtz, 1991]. Phospholipid transfer proteins with different specificities toward polar headgroups of phospholipids include the phosphatidylcholine transfer protein (PC-TP) which specifically binds and transfers PC [Kamp *et al.* 1973; Lumb *et al.*, 1976] and the phosphatidylinositol transfer protein (PI-TP) which binds and transfers PI, and to a lesser extent PC [Helmkamp *et al.*, 1974; Demel *et al.*, 1977; DiCorletto *et al.*, 1974]. In addition, a non-specific lipid transfer protein (identical to sterol carrier protein 2) has been identified which catalyzes the transfer of phospholipids and cholesterol between membranes [Bloj and Zilversmit, 1977; Crain and Zilversmit, 1980]. The characteristics of these proteins have been studied *in vitro*. However, the physiological function of the phospholipid transfer proteins is not yet clear. In mammalian cells most phospholipids are synthesized on the endoplasmic reticulum. This implies that specific mechanisms of transport must operate to redistribute phospholipids from the site of synthesis to the proper localization in the cell. It has been suggested that the phospholipid transfer proteins are involved in these transport processes. Our investigation is focussed on mammalian PI-TP, which is believed to be involved in the intracellular transport processes, particularly pertaining to those membranes that have an active PI metabolism [Wirtz *et al.*, 1987; van Paridon *et al.*, 1987; Cleves *et al.*, 1991]. Presumably, transfer of PI to these membranes occurs by PI-TP directly or by flow of membrane vesicles, the PI content of which has been modulated by PI-TP. Recently, the interest in the physiological role of PI-TP has strongly increased with the observation that PI-TP in yeast is identical to the SEC14 protein. This protein is involved in secretory processes at the level of

NATO ASI Series, Vol. H 82
Biological Membranes:
Structure, Biogenesis and Dynamics
Edited by Jos A. F. Op den Kamp

the late Golgi compartment [Bankaitis *et al.*, 1989; Salama *et al.*, 1990]. Localization studies showed that PI-TP in yeast is associated with Golgi structures. Furthermore, it was shown that deletion of the gene encoding for PI-TP in yeast is lethal for the organism [Aitken *et al.*, 1990]. These studies have been interpreted to indicate that PI-TP has the capacity to control the phospholipid composition of the yeast Golgi membranes which is critical to the secretory competence of these membranes. However, the primary structure of mammalian PI-TP has no similarity with yeast SEC14p [Dickeson *et al.*, 1989; Salama *et al.*, 1990]. Therefore, it is possible that the cellular functions of both proteins are different.

Cellular localization of PI-TP in Swiss mouse 3T3 fibroblasts

The cellular localization of PI-TP was studied by indirect immunofluorescence [Snoek *et al.*, 1992]. In exponentially growing (non-confluent cultures in bicarbonate-buffered DMEM +7.5% FCS) 3T3 cells, PI-TP was localized in the cytoplasm, associated with structures

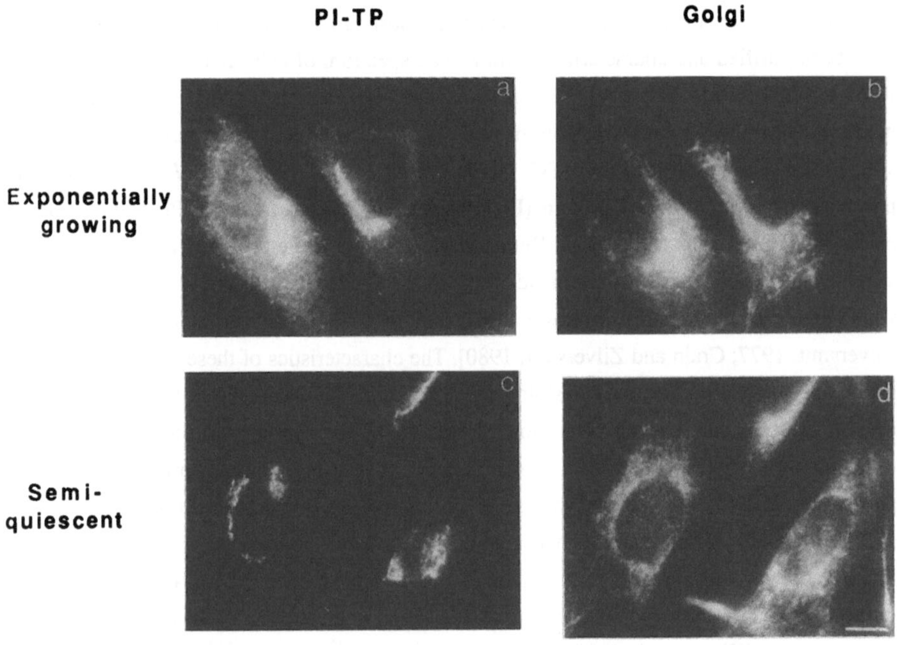

Fig. 1. Indirect immunofluorescence microscopic images of exponentially growing (panels a and b) and semi-quiescent (serum-starved, panels c and d) Swiss mouse 3T3 fibroblast cells. Panels a and c: specific labeling of PI-TP visualized by GAR-FITC; panels b and d: Golgi structures visualized by Ricin-TRITC. Bar: 5 μm. Experimental methods are described in Snoek *et al.*, 1992, 1993.

surrounding the nucleus and inside the nucleus (fig. 1a). The structures around the nucleus have been shown to coincide with Golgi structures which are labelled with the Golgi-specific lectin Ricin (fig. 1b). Association of PI-TP with Golgi membranes has been indicated by a second method using Brefeldin A [Snoek *et al.*, 1992]. The concentration of PI-TP in semi-quiescent (serum-starved) cells was shown to be decreased when compared to exponentially growing cells (fig. 1c). Only Golgi structures were labelled as shown by Ricin labelling (fig. 1d). No or very little nuclear or cytoplasmic labelling was found. In quiescent (density-arrested) 3T3 cells the labelling of PI-TP was even further decreased [Snoek *et al*, 1993].

Redistribution of PI-TP upon stimulation of semi-quiescent 3T3 cells

In order to determine the localization of PI-TP during the transition from quiescence to growth, quiescent 3T3 cells have been incubated with phorbol 12-myristate,13-acetate (PMA) and bombesin. Bombesin binds to a specific receptor which activates the phospholipase C-dependent breakdown of PIP_2 [Cook *et al.*, 1990; Cook and Wakelam, 1991], whereas PMA activates protein kinase C [Nishizuka, 1986]. In semi-quiescent cells, the association of PI-TP with the Golgi system is very distinct (fig. 2a,b). In response to PMA, a clear redistribution of PI-TP can be detected within 10 min (fig. 2c). Rounding of the cells was observed after 3 h of PMA treatment (fig. 2d). In the case of bombesin, no change in the localization of PI-TP was apparent after 10 min. However, after 3 h of stimulation, changes in the localization were observed (fig. 2f). These observations strongly suggest that those compounds that activate protein kinase C, affect the localization of PI-TP in quiescent 3T3 cells. Therefore, we investigated whether PI-TP is phosphorylated upon stimulation of cellular protein kinase C.

Phosphorylation of PI-TP *in vitro* and *in vivo*

Analysis of the amino acid sequence of rat brain PI-TP (using the PC-Gene program) demonstrated the presence of 5 serine and threonine residues which could function as potential phosphorylation sites for protein kinase C [Dickeson *et al.*, 1989; Snoek *et al.*, 1993]. When PI-TP from bovine brain is incubated with purified protein kinase C from rat brain in the presence of Ca^{2+} and phosphatidylserine, PI-TP becomes phosphorylated (fig. 3a, lane 8). This phosphorylation is not observed in the presence of EGTA (lane 7). The Ca^{2+}-phosphatidylserine-dependent autophosphorylation of rat brain protein kinase C is shown in lane 6. Phosphorylation of PI-TP *in vivo* was investigated in semi-quiescent 3T3 cells which were prelabelled with [^{32}P]phosphate. PI-TP was immunoprecipitated from control cells and from cells stimulated with PMA or bombesin for 30 min. or 3 h. In control cells some phosphorylation of PI-TP could be observed (fig. 3b, lane 1). After stimulation with PMA for 30 min, the phosphorylation of PI-TP was increased (fig. 3b, lane 2) and this increased level

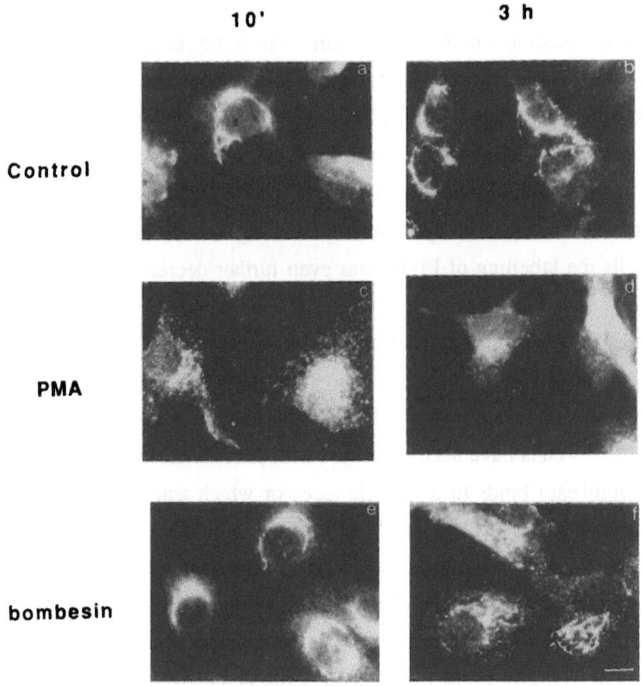

Fig. 2. Immunofluorescence microscopic images of PI-TP in semi-quiescent and stimulated Swiss mouse 3T3 fibroblast cells. Semi-quiescent cells were incubated for 10 min (panels a, c and e) or 3 h (panels b, d and f) with DBH (panels a, b), PMA (50 ng/ml; panels c, d) or bombesin (10 nM, panels e, f). Bar: 5 μm. Experimental methods are described in Snoek *et al.*, 1992, 1993.

of phosphorylation was maintained for at least 3 h (fig. 3b, lane 3). However, upon stimulation with bombesin, no increase in phosphorylation of PI-TP could be detected, neither after 30 min nor after 3 h (not shown, Snoek *et al.*, 1993).

Since PI-TP in non-stimulated, semi-quiescent 3T3 cells does contain some ^{32}P-label already, we have attempted to determine the proportion of phosphorylated and non-phosphorylated PI-TP in these cells. To this end, the cytoplasmic fractions of 3T3 cells were analyzed on isoelectric focusing gels, followed by immunoblotting. Previously, it was shown by isoelectric focusing of purified PI-TP that two forms can be distinguished, that is PI-TP I (isoelectric point 5.5) containing one molecule of PI and PI-TP II (isoelectric point 5.7) containing one molecule of PC. Figure 3c (lane 1) shows that PI-TP I is present in semi-quiescent 3T3 cells; PI-TP II is not detected under these conditions. However, PI-TP II has been identified in

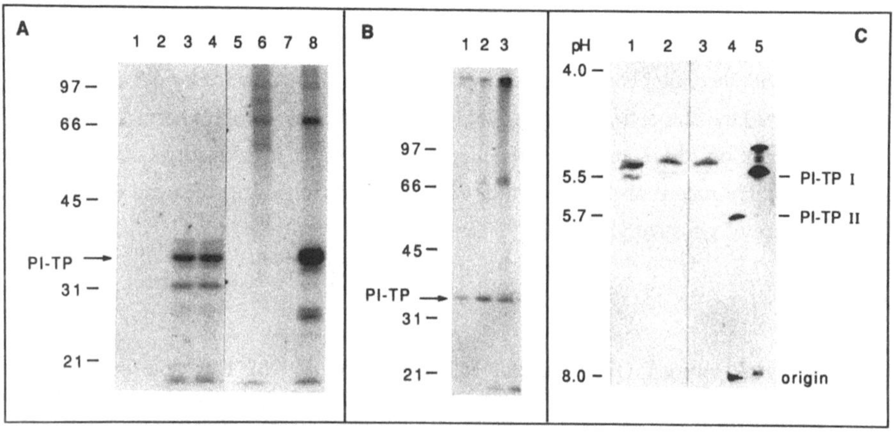

Fig. 3. Phosphorylation of PI-TP *in vitro* and *in vivo*. (a) Phosphorylation of purified PI-TP by protein kinase C. PI-TP was purified as described by van Paridon *et al.* (1987). Protein kinase C was purified as described by Huang *et al.* (1986). PI-TP (3 μg) and protein kinase C (0.5 μg) were incubated as described in the presence of EGTA (lanes 1, 3, 5 and 7) or in the presence of Ca^{2+}, phosphatidylserine and diacylglycerol (lanes 2, 4, 6 and 8). Lanes 1, 2, 5 and 6: protein kinase C; lanes 3, 4, 7 and 8: protein kinase C and PI-TP. Lanes 1-4: Coomassie Brilliant Blue staining of the SDS-PAGE gel; lanes 5-8: autoradiograph of lanes 1-4. (b) *In vivo* phosphorylation of PI-TP. Swiss mouse 3T3 cells (semi-quiescent) were metabolically labelled with [^{32}P]phosphate. PI-TP was immunoprecipitated from the cell lysate as described [Snoek *et al.*, 1993]. An autoradiograph of an SDS-PAGE gel is shown. Lane 1: control cells; lanes 2 and 3: cells incubated with PMA (50 ng/ml) for 30 and 180 min respectively. (c) Western blot of an isoelectric focusing gel of a 3T3 cell lysate. Cell culture, isoelectric focusing and Western blotting procedures are described in [Snoek *et al.*, 1992]. Lane 1: control cells; lanes 2 and 3: cells incubated with PMA (50 ng/ml) for 30 and 180 min, respectively; lane 4: PI-TP II (1 μg); lane 5, PI-TP I (3 μg).

exponentially growing 3T3 cells (data not shown). In addition to PI-TP I, a second, more acidic form of PI-TP is observed in semi-quiescent 3T3 cells (fig. 3c, lane 1). Since the relative intensity of the two bands changes when the cells are stimulated with PMA for 30 or 180 min (fig. 3c, lanes 2, 3), we propose that the more acidic form of PI-TP is a phosphorylated form. Another argument in favor of this contention is the observation that incubation of the cytoplasmic fraction of 3T3 cells with acid or alkaline phosphatase before isoelectric focusing leads to an apparent shift of the acid form of PI-TP I to the more alkaline (non-phosphorylated) form (I.S.C. de Wit, unpublished results). Although purified PI-TP is readily phosphorylated by protein kinase C *in vitro*, it is not yet known whether protein kinase C is responsible for the phosphorylation of PI-TP *in vivo* or whether other protein kinases are (also) involved [Thomas, 1992].

We think it likely that a specific transfer of PI from the endoplasmic reticulum to the plasma membrane may occur by vesicle flow and/or by PI-TP. In this transfer process phosphorylation of PI-TP could be a powerful mechanism to regulate its function either by affecting its association with the Golgi complex (thereby controlling vesicle flow) or by directly affecting its phospholipid transfer activity. It is, as yet, unknown whether the phosphorylation of PI-TP has any effect on these proposed functions. Our current investigations are aimed at characterizing the nature of the interaction of PI-TP with Golgi structures and at disclosing the importance of phosphorylation for PI-TP function.

References

Aitken JF, van Heusden GPH, Temkin M, Dowhan W (1990) The gene encoding the phosphatidylinositol transfer protein is essential for cell growth. J Biol Chem 265:4711-4717

Bankaitis VA, Malehorn DE, Emr SD, Greene R (1989) The *Saccharomyces cerevisiae* SEC14 gene encodes a cytosolic factor that is required for transport of secretory proteins from the yeast Golgi complex. J Cell Biol 108:1271-1281

Bloj B, Zilversmit DB (1977) Rat liver proteins capable of transferring phosphatidylethanolamine. Purification and transfer activity of other phospholipids and cholesterol. J Biol Chem 252:1613-1619

Cleves AE, McGee T, Bankaitis VA (1991) Phospholipid transfer proteins: a biological debut. TI Cell Biol 1:30-34

Cook SJ, Palmer S, Plevin R, Wakelam MJO (1990) Mass measurement of inositol 1,4,5-triphosphate and *sn*-1,2,-diacylglycerol in bombesin stimulated Swiss 3T3 mouse fibroblasts. Biochem J 265:617-620

Cook SJ, Wakelam MJO (1991) Hydrolysis of phosphatidylcholine by phospholipase D is a common response to mitogens which stimulate inositol lipid hydrolysis in Swiss 3T3 fibroblasts. Biochem Biophys Acta 1092: 265-272

Crain RC, Zilversmit DB (1980) Two nonspecific phospholipid exchange proteins from beef liver. I. Purification and characterization. Biochemistry 19:1433-1439

Demel RA, Kalsbeek R, Wirtz KWA, van Deenen LLM (1977) The protein-mediated net transfer of phosphatidylinositol in model systems. Biochim Biophys Acta 466:10-22

Dickeson SK, Lim CN, Schuyler GT, Dalton TP, Helmkamp GM, Yarbrough LR (1989) Isolation and sequence of cDNA clones encoding rat phosphatidylinositol transfer protein. J Biol Chem 264:16557-16564

DiCorletto PE, Warach JB, Zilversmit DB (1979) Purification and characterization of two phospholipid exchange proteins from bovine heart. J Biol Chem 254:7795-7802

Helmkamp GM, Harvey MS, Wirtz KAW, van Deenen LLM (1974) Phospholipid exchange between membranes. Purification of bovine brain proteins that preferentially catalyze the transfer of phosphatidyl inositol. J Biol Chem 249:6382-6389

Huang KP, Nakabayashi H, Huang FL (1986) Isozymic forms of rat brain Ca^{2+}-activated and phospholipid dependent protein kinase. Proc Natl Acad Sci USA 83:8535-8539

Kamp HH, Wirtz KWA, van Deenen LLM (1973) Some properties of phosphatidyl choline exchange protein purified from beef liver. Biochem. Biophys. Acta 318:313-325

Lumb RH, Kloosterman AD, Wirtz KWA, van Deenen LLM (1976) Some properties of phospholipid exchange proteins from rat liver. Eur J Biochem 69:15-22

Nishizuka Y (1986) Studies and perspectives of protein kinase C. Science 233:305-312

Salama SR, Cleves AE, Malehorn DE, Whitters EA, Bankaitis VA (1990) Cloning and characterization of *Kluyveromyces lactis* SEC14, a gene whose product stimulates Golgi secretory function in *Saccharomyces cerevisiae*. J Bacteriol 172:4510-4521

Snoek GT, de Wit ISC, van Mourik JHG, Wirtz KWA (1992) The phosphatidylinositol transfer protein in 3T3 mouse fibroblast cells is associated with the Golgi system. J Cell Biochem 49:339-348

Snoek GT, Westerman J, Wouters FS, Wirtz KWA (1993) Phosphorylation and redistribution of the phosphatidylinositol transfer protein in phorbol-12myristate,13-acetate- and bombesin stimulated Swiss mouse 3T3 fibroblasts. Biochem J 291:649-656

Thomas G (1992) MAP-kinase by any other name smells just as sweet. Cell 68:3-6

van Paridon PA, Gadella TFJ, Somerharju PJ, Wirtz KWA (1987) On the relationship between the dual specificity of the bovine brain phosphatidylinositol transfer protein and membrane phosphatidylinositol levels. Biochim Biophys Acta 903:68-77

Wirtz KWA (1991) Phospholipid transfer proteins. Annu Rev Biochem 60:73-99

Wirtz KWA, Helmkamp Jr GM, Demel RA (1987) The phosphatidylinositol exchange protein from bovine brain. In: "Protides of the Biological Fluids" (Peeters H, eds) Pergamon Press, Oxford and New York, pp. 25-31

PHOSPHATIDYLINOSITOL TRANSFER PROTEIN FUNCTION IN YEAST

Vytas A. Bankaitis
Department of Cell Biology
University of Alabama at Birmingham
School of Medicine
Birmingham, Alabama 35294-0005
U.S.A.

All eukaryotic cells contain a set of cytosolic proteins that act as diffusible carriers for the energy-independent transport of phospholipid monomers between membrane bilayers *in vitro* (Rueckert and Schmidt, 1990; Wirtz, 1991). These phospholipid transfer proteins (PL-TPs) are classified into three general categories on the basis of their ligand specificity; i.e. which phospholipids (PLs) serve as substrates for these proteins in the cell-free PL-transfer reaction. These categories include the nonspecific PL-TPs, the phosphatidylcholine transfer proteins, and the phosphatidylinositol (PI)/phosphatidylcholine (PC) transfer proteins. Members of the last class of PL-TPs are frequently referred to as the phosphatidylinositol (PI) transfer proteins in recognition of the fact that PI is clearly the preferred ligand of these PI-TPs *in vitro*. The mechanisms by which PL-TPs catalyze phospholipid transfer *in vitro* have been the subject of detailed biochemical analysis. Although there has been much discussion about the possible involvement of PL-TPs in various aspects of intracellular trafficking of lipids in cells, very little information has been forthcoming with respect to the question of what precise cellular functions PL-TPs fulfill *in vivo*. The primary reasons for this lack of progress have been several-fold. First, it is essentially impossible, even with today's technology, to measure phospholipid transfer *in vivo*. Thus, one is left with an *in vitro* assay that sets the general premise from which models for *in vivo* function have been extrapolated. While this conceptual arrangement is reasonable in cases where one is analyzing authentic enzymatic activities, there has been considerable discussion as to whether PL-TPs truly fit the criteria for consideration as enzymes that catalyze PL transfer *in vitro*. Helmkamp (1986) has argued that it is appropriate to consider PL-TPs as true biological catalysts on the basis of their substrate specificity, their clear ability to enhance the rate of PL-transfer between membrane bilayers *in vitro*, and inhibition. On the other hand, several properties of PL-

NATO ASI Series, Vol. H 82
Biological Membranes:
Structure, Biogenesis and Dynamics
Edited by Jos A. F. Op den Kamp

TPs have been difficult to reconcile with such a view. These have included the ability of PL-TPs to utilize most any natural or synthetic membrane bilayer as PL donors or acceptors in the transfer reaction, and the propensity of these proteins to catalyze efficient PL-exchange reactions, but not net PL-transfer reactions, *in vitro*. It is not unreasonable to question whether the cell could gainfully employ such seemingly promiscuous, and possibly nonvectorial, catalysts in PL trafficking reactions. Second, elucidation of PL-TP function *in vivo* requires the acid test of what are the consequences of loss of a particular PL-TP function, if any, to a cell or an organism. This lack of a suitable *in vivo* system with which to analyze the *in vivo* function of a particular PL-TP has been especially significant. Although the crucial *in vivo* approaches are just now beginning to bear fruit, the fact remains that the true *in vivo* function of any PL-TP is not yet known in any metazoan system, and that in only one case has there been a direct demonstration that the biological function of a PL-TP somehow interfaces with the metabolism of one of its ligands (see below). Nevertheless, the conservation of antigenic and biochemical properties of PI-TPs from mammals to *Drosophila* has been suggested to be indicative of some important cellular function for the PI-TP in higher eukaryotes (Dickeson *et al.*, 1989).

The yeast *Saccharomyces cerevisiae* exhibits a PI-TP whose molecular mass and catalytic properties are very similar to those of the mammalian PI-TP (Szolderits *et al.*, 1989). Yet, the mammalian and yeast proteins share no primary sequence homology (Bankaitis *et al.*, 1989; Dickeson *et al.*, 1989). The first opportunity for a detailed *in vivo* analysis of PL-TP function was identified by our finding, in collaboration with Bill Dowhan's laboratory, that the *Saccharomyces cerevisiae SEC14* gene product (SEC14p), whose function is essential for protein transport from the yeast Golgi complex (Bankaitis *et al.*, 1989; Novick *et al.*, 1980), is the yeast PI-TP (Bankaitis *et al.*, 1990). It must be noted that this finding was made possible by the work of Randy Schekman's laboratory that originally identified *SEC14* (Novick *et al.*, 1980), and Bill Dowhan's laboratory that determined the primary sequence of the yeast PI-TP amino-terminus (Aitken *et al.*, 1990). This lecture will focus on what we know about the *in vivo* function of SEC14p in yeast, and what future directions we will pursue. In particular, I wish to address four questions: **(i)** does the PI-TP activity of SEC14p reflect some essential functional property of this protein, **(ii)** what is the basis for the seemingly specific Golgi requirement for SEC14p activity, **(iii)** what is the biological function of SEC14p, and **(iv)** what sorts of mechanisms for SEC14p function *in vivo* can be considered that are consistent with all of the data. At the end, I plan to close with some comments identifying what types of *in vivo* biological reactions we might reasonably expect mammalian PI-TPs to be involved with.

I. Relationship between SEC14p function and PI-TP activity.

As discussed above, the relationship between the *in vivo* function of a PL-TP and its *in vitro* PL-transfer properties is unclear. The best evidence to date in support of the idea that there exists some important relationship between these two activities is provided by the finding that the *sec14-1*^{ts} allele encodes a thermolabile PI-TP (Bankaitis *et al.*, 1990). This finding, although consistent with such a concept, does not constitute a proof as the *sec14-1*^{ts} gene product (Sec14p^{ts}) may itself exhibit some general, but reversible, denaturation at restrictive temperatures (Novick *et al.*, 1980; our unpublished data). Such a property could result in the exhibition by Sec14p^{ts} of thermolabile PI-TP activity, even though this activity may not represent an essential functional activity of this protein *in vivo*. A more direct test of the relationship in question is provided by the availability of a mammalian (i.e. rat) PI-TP cDNA recovered by Dickeson *et al.* (1989). Although the rat PI-TP and the yeast SEC14p share very similar catalytic properties from the standpoint of substrate specificity and substrate preference in the *in vitro* phospholipid transfer reaction (Szolderits *et al.*, 1989), these polypeptides share no detectable primary sequence homology (Bankaitis *et al.*, 1989; Dickeson *et al.*, 1989). This conservation of biochemical function, in the absence of any structural similarity at the primary sequence level, provides an opportunity to address the question of whether the sole essential function of SEC14p *in vivo* was in some fashion represented by its PI-TP activity *in vitro*. If this were strictly the case, one might predict that expression of a catalytically active rat PI-TP in yeast should complement the growth and secretory defects associated with *sec14* defects. In agreement with this general prediction, expression of active rat PI-TP in yeast effects a specific complementation of the growth and Golgi secretory defects associated with the *sec14-1*^{ts} allele. However, rat PI-TP-mediated complementation of *sec14* mutations is not absolute as rat PI-TP expression fails to rescue *sec14* null mutations. There are several general possibilities as to why rat PI-TP fails to completely substitute for SEC14p *in vivo*. Perhaps rat PI-TP operates in yeast by an *in vivo* mechanism unrelated to that employed by SEC14p. For example, one protein may function as a PI-TP *in vivo* whereas the *in vitro* PI-TP activity of the other might indirectly reflect a specific phospholipid binding activity that is utilized in some other fashion *in vivo*. Alternatively, rat PI-TP may function in a mechanistically similar fashion to SEC14p, yet be an incomplete surrogate by virtue of its inability to execute some other essential activity of SEC14p. Some clues to this effect are discussed below. Nevertheless, these collective data provide the strongest demonstration yet that the *in vitro* PI-TP activity of SEC14p reflects some functional property of this protein *in vivo*.

II. The basis for the Golgi-specific function of SEC14p.

The promiscuity of PL-TPs with respect to PL donor and acceptor membranes *in vitro* is not immediately consistent with the the idea that PL-TPs exhibit membrane specificity *in vivo*. Yet, *sec14-1^{ts}* mutants exhibit a specific block in secretory glycoprotein transport from what we consider to be a late yeast Golgi compartment (Franzusoff *et al.*, 1989; Cleves *et al.*, 1991). A combination of double-label immunofluorescence and subcellular fractionation experiments indicate that SEC14p is found both as a cytosolic and peripheral Golgi membrane protein (Cleves *et al.*, 1991). As Golgi membranes represent a minor intracellular membrane system in yeast, these data reveal a specificity of membrane targeting of SEC14p *in vivo* that is not apparent in the SEC14p-mediated PL-TP reaction *in vitro*. Moreover, these data demonstrate a direct association of the SEC14p with the Golgi complex, i.e. the organelle that becomes dysfunctional under SEC14p deficient conditions. From these findings we infer that SEC14p is directly involved in stimulating Golgi secretory function *in vivo*, and that the molecular basis for the Golgi-specific function of SEC14p is likely determined by its efficient targeting to yeast Golgi membranes. How is SEC14p targeted to yeast Golgi membranes? Two general models can be entertained. First, one can envision that Golgi membranes exhibit some unique PL domain that recruits and/or retains SEC14p by virtue of its PL-binding properties. Alternatively, it is possible that Golgi membranes harbor a SEC14p receptor that imposes targeting specificity. By direct analogy to the logic described above, we ask whether rat PI-TP exhibited Golgi membrane tropism in yeast. Double-label immunofluorescence experiments, coupled with subcellular fractionation data, indicate that active rat PI-TP does not efficiently target to yeast Golgi membranes. The demonstration that rat PI-TP is much less effective at associating with yeast Golgi membranes than SEC14p suggests that SEC14p targeting to yeast Golgi membranes is not a simple function of some unique Golgi phospholipid domain recruiting and/or retaining SEC14p by virtue of its specific phospholipid binding properties. If this particular targeting mechanism were operative in cells we would have expected rat PI-TP to localize in a similar manner to SEC14p since these PI-TPs share very similar phospholipid binding/transfer properties. Some other mechanism (receptor-mediated?) contributing to the localization of SEC14p to Golgi membranes is implied. Experimental support for this possibility has been obtained. Fusion of the amino-terminal 129 SEC14p residues to the rat PI-TP amino-terminus yields a hybrid protein that is inactive as a PI-TP, but targets to Golgi membranes in a manner similar to SEC14p (see Figure 1).

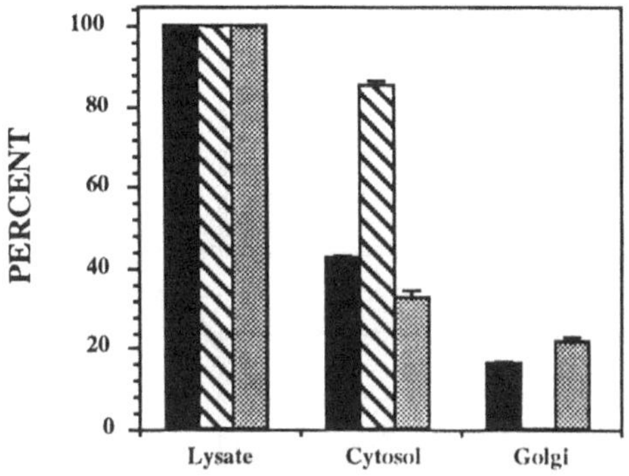

Figure 1. The amino-terminal 129 SEC14p residues target a heterologous PI-TP to yeast Golgi membranes. The fractionation profiles of SEC14p (solid bars), rat PI-TP (cross-hatched bars), and a SEC14p129-rat PI-TP fusion protein (stippled bars) are summarized. The rat PI-TP and SEC14p-rat PI-TP fusion proteins were fractionated in a *sec14* null strain carrying a *cki* bypass SEC14p mutation so as to eliminate any potential competition between endogenous SEC14p and the test protein for Golgi targeting. Data are presented as percent of total test protein in starting lysate.

III. The biological function of SEC14p.

To date, the most penetrating clues relating to SEC14p function *in vivo* have been obtained from analyses of yeast mutants that no longer require SEC14p in order to survive and efficiently execute Golgi secretory function (Cleves *et al.*, 1989; Cleves *et al.*, 1991). These studies revealed that one mechanism for bypass of SEC14p function is inactivation of the CDP-choline pathway, a phosphatidylcholine (PC) biosynthetic pathway that consists of three reactions resulting in the incorporation of choline into PC (Kennedy and Weiss, 1956).

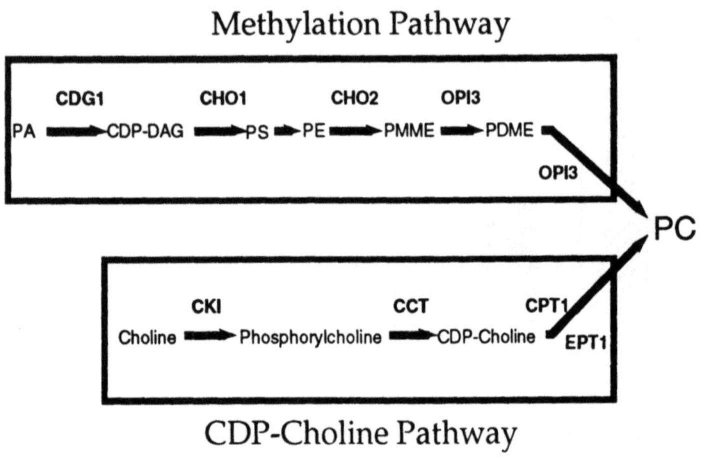

Figure 2. Phosphatidylcholine biosynthesis in yeast. The two pathways for phosphatidylcholine synthesis in yeast are shown. Genes encoding the structural enzymes of each pathway are indicated at the site of action of their respective gene products. Abbreviations: phosphatidic acid (PA), CDP-diacylglycerol (CDP-DAG), diacylglycerol (DAG), phosphatidylserine (PS), phosphatidylethanolamine (PE), phosphatidylmonomethylethanolamine (PMME), phosphatidyldimethylethanolamine (PDME), and phosphatidylcholine (PC).

The finding that the cellular SEC14p requirement is obviated by inactivation of a specific avenue for PC biosynthesis provided the first demonstration of a direct physiological relationship between an *in vitro* ligand of a PL-TP and the function of that PL-TP *in vivo*. On the basis of such genetic data, we have proposed that SEC14p functions to maintain an appropriately elevated PI/PC ratio in Golgi membranes, a ratio that is somehow critical to the secretory competence of these membranes (Cleves *et al.*, 1991). We have obtained biochemical data that demonstrate the *in vivo* function of SEC14p to be the maintenance of an appropriately reduced PC content in yeast Golgi membranes. This insight comes from experiments where we measured Golgi PL composition as a function of SEC14p activity under conditions where Golgi secretory function is uncoupled from the usual SEC14p requirement. We also find that PC synthesis via the CDP-choline pathway is a significant contributor to the rate of PC biosynthesis in cells, irrespective of whether this pathway contributes to net cellular PC synthesis or whether it is relegated to a choline salvage activity that cannot contribute to net cellular PC synthesis. The finding that disruption of CDP-choline pathway function eliminates the dramatic increase in Golgi PC that is a direct result of SEC14p dysfunction provides a satisfying biochemical rationale for why mutational inactivation of the CDP-choline pathway results in bypass of the normally essential SEC14p requirement for Golgi secretory function and cell viability. It remains to be determined why Golgi PC content is an important factor in whether Golgi membranes will be competent for sustaining Golgi secretory function, or not.

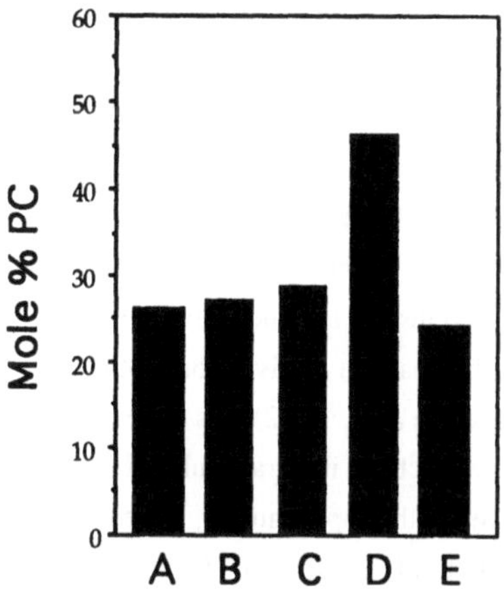

Figure 3. SEC14p controls the PC content of yeast Golgi membranes. The PC content of yeast Golgi membranes is presented as a function of SEC14p activity under conditions permissive for Golgi secretory function: (A) *SEC14*-- 25°C; (B) *sec14ts*-- 25°C; (C) *SEC14, bsd2*--37°C; (D) *sec14ts, bsd2*--37°C; (E) *sec14ts, cki-284::HIS3*--37°C. Note that SEC14p dysfunction results in a dramatic increase in Golgi PC (D) relative to SEC14p-proficient conditions (A,B,C). The *cki* disruption is epistatic both to the increase in Golgi PC caused by loss of SEC14p function, and to all of the effects associated with the *sec14ts* block.

IV. Potential mechanisms for SEC14p function.

How might SEC14p function interface with PC synthesis via the CDP-choline pathway? The genetic data can be interpreted to indicate that PC synthesis via the CDP-choline pathway is inherently toxic to Golgi secretory function and, consequently, incompatible with cell viability. SEC14p would serve to detoxify the deleterious effect of CDP-choline pathway activity *in vivo*. Until now, we have considered SEC14p function in terms of its *in vitro* PI/PC transfer activity (Cleves *et al.*, 1991). By this view, SEC14p could effect net removal of PC from yeast Golgi membranes via its PC transfer activity, an interpretation that is entirely consistent with the data indicating SEC14p to function in maintaining an appropriately reduced Golgi PC content, and that SEC14p is found both as a cytosolic species and as a peripheral Golgi protein (Bankaitis *et al.*, 1989; Cleves *et al.*, 1991b). The difficulties with this mechanism for SEC14p function are that it fails to offer an explanation for why PC is the relevant substrate *in vivo* when PI is the preferred SEC14p ligand *in vitro*, and that this model provides no insight into the relevance of the cellular SEC14p requirement in the face of CDP-choline pathway activity, but not PE methylation pathway activity (Cleves *et al.*, 1991, McGee *et al.*, 1992). As there exists no *in vivo* evidence to indicate that any PL-TP genuinely functions in intracellular PL transport *per se*, the *in vitro* PL transfer activity must be viewed with proper caution. Indeed, it is clear that SEC14p does not play an essential role in bulk PL transport *in vivo* (Cleves *et al.*, 1991), as had been proposed (Wieland *et al.*, 1987; Rothman, 1990).

A new possibility for how SEC14p functions *in vivo* is suggested by our finding that overproduction of SEC14p results in a specific reduction in cellular PC synthesis via the CDP-choline pathway. This alternative mechanism invokes a sensor function for SEC14p rather than a genuine PL transfer function. In such a model, the various genetic and biochemical data are considered to indicate that SEC14p serves to regulate CDP-choline activity in yeast Golgi membranes as a function of its PL-liganded state. For instance, the PC-bound form of SEC14p could potentially serve as a specific allosteric repressor of Golgi PC synthesis via the CDP-choline pathway. This model also accounts for all of the available data concerning SEC14p function *in vivo*. Finally, we find the sensor model for SEC14p function *in vivo* to be attractive not only because it is consistent with all of the *in vivo* data, but also because it reconciles a curious property of SEC14p-mediated PL transfer activity *in vitro* with SEC14p function *in vivo*. It is known that many PL-TPs, SEC14p included, efficiently catalyze PL exchange reactions *in vitro* but are not efficient catalysts of net PL transfer reactions (Szolderits *et al.*, 1989; Wirtz, 1991; Rueckert and

Schmidt, 1991). The sensor model predicts that the functional reaction mode for SEC14p *in vivo* will be simple PL exchange. Thus, the increase in Golgi PC that occurs as a direct consequence of SEC14p dysfunction is predicted to reflect loss of a local SEC14p-dependent downregulation of PC synthesis via the CDP-choline pathway and not loss of some preferential SEC14p-mediated transport of PC from the Golgi.

V. References

Aitken, JF, van Heusden, GPH, Temkin, M, and Dowhan, W (1990) The gene encoding the phosphatidylinositol transfer protein is essential for cell growth. J. Biol. Chem. 265: 4711-4717

Bankaitis, VA, Malehorn, DE, Emr, SD, and Greene, R (1989) The *Saccharomyces cerevisiae SEC14* gene encodes a cytosolic factor that is required for transport of secretory proteins from the yeast Golgi complex. J. Cell Biol. 108: 1271-1281

Cleves, AE, Novick, PJ, and Bankaitis, VA (1989) Mutations in the *SAC1* gene suppress defects in yeast Golgi and yeast actin function. J. Cell Biol. 109: 2939-2950

Cleves, AE, McGee, TP, Whitters, EA, Champion, KM, Aitken, JR, Dowhan, W, Goebl, M, and Bankaitis, VA (1991) Mutations in the CDP-choline pathway for phospholipid biosynthesis bypass the requirement for an essential phospholipid transfer protein. Cell 64: 789-800

Dickeson, SK, Lim, CN, Schuyler, GT, Dalton, TP, Helmkamp, GM, Jr and Yarbrough, LR (1989) Isolation and sequence of cDNA clones encoding rat phosphatidylinositol transfer protein. J. Biol. Chem. 264: 16557-16564

Helmkamp, GM, Jr (1986) Phospholipid transfer proteins: mechanisms of action. J. Bioenergetics and Biomembranes 18: 71-91

Franzusoff, A, and Schekman, R (1989) Functional compartments of the yeast Golgi apparatus as defined by the *sec7* mutation. EMBO J. 8: 2695-2702

MgGee, TP, Fung, MKY, and Bankaitis, VA (1992) A reply to interpreting the effects of blocking phosphatidylcholine biosynthesis. Trends in Cell Biol. 2: 71-72

Rothman, JE (1990) Phospholipid transfer market. Nature 347: 519-520

Rueckert, DG, and Schmidt, K (1990) Lipid transfer proteins. Chemistry and Physics of Lipids 56: 1-20

Skinner, HB, Alb, JG,Jr, Whitters, EA, Helmkamp, GM,Jr, and Bankaitis, VA (1993) Complementation of a *Saccharomyces cerevisiae* Golgi defect by a mammalian phospholipid transfer protein. EMBO J. (submitted)

Szolderits, G, Hermetter, A, Paltauf, F and Daum, G (1989). Biochim. Biophys. Acta 986: 301-309

Whitters, EA, Cleves, AE, McGee, TP, Skinner, HB, and Bankaitis, VA (1993) SAC1p is an integral membrane protein that influences the cellular requirement for phospholipid transfer protein function and inositol in yeast. J. Cell Biol. (In Press)

Wieland, FT, Gleason, ML, Serafini, TA, and Rothman, JE (1987) The rate of bulk flow from the endoplasmic reticulum to the cell surface. Cell 50: 289-300

Wirtz, KWA (1991) Phospholipid transfer proteins. Annu. Rev. Bioch. 60: 73-99

Conversion of the amphiphilic 115 kDa Form of Glycosyl-Phosphatidylinositol-specific Phospholipase D to an active, hydrophilic 47 kDa Form

Marius C. Hoener, Jian Liao and Urs Brodbeck
Institute of Biochemistry and Molecular Biology
University of Bern
CH-3012 Bern
Switzerland

INTRODUCTION

More than 100 cell surface proteins have been shown to be anchored to cell membranes by a glycosyl-phosphatidylinositol (GPI) anchor. GPI-anchored proteins include cell surface enzymes, complement regulatory proteins, cell-adhesion molecules and receptors, both forms of the prion protein, lymphoid antigens, as well as many parasitic cell-surface antigens (Cross, 1990). The GPI-anchoring is thought to be a possible targeting signal in the sorting of apically located surface proteins of polarized cells (Rodriguez-Boulan and Powell, 1992), and it may function to prolong the cell-surface half-life of proteins (Lemansky et al., 1990). It has been suggested by Rothberg et al. (1990) that certain GPI-anchored receptors can undergo a novel, clathrin-independent form of endocytosis and recycling. It was found recently that GPI-anchored proteins are associated intracellularly with vesicles enriched with glycosphingolipids which are also sorted preferentially to the apical cell surface (Brown and Rose, 1992). A further function of the GPI-anchor could be the regulated release of the protein moiety by GPI-specific phospholipases. Additionally, the increased lateral mobility of GPI-anchored proteins within the membrane may promote cell-cell and protein-protein interaction by facilitating the recruitment of these proteins into membrane domains engaged in adhesion.

In mammals, the only purified and well characterized GPI-specific phospholipase is the D-type enzyme. GPI-specific phospholipase D (GPI-PLD) which hydrolyzes the inositol-phosphate linkage of GPI-anchored proteins has originally been discovered in serum (Davitz et al., 1987; Low and Prasad, 1988) and later in whole brain (Hoener et al., 1990; Hoener and Brodbeck, 1993), neurons (Sesko and Low, 1991), cerebrospinal fluid (Hoener and Brodbeck,

---

Abbreviations: GPI, glycosyl-phosphatidylinostiol; GPI-PLD, glycosyl-phosphatidylinositol-specific phospholipase D; HDL, high-density lipoproteins.

NATO ASI Series, Vol. H 82
Biological Membranes:
Structure, Biogenesis and Dynamics
Edited by Jos A. F. Op den Kamp

1992), milk (Hoener and Brodbeck, 1992), in the islets of Langerhans (Metz et al., 1991), liver (Scallon et al., 1991; Heller et al., 1992), mast cells of liver, adrenal gland, and lung (Metz et al., 1992; Stadelmann et al., 1993). GPI-PLD has been purified and characterized from serum (Davitz et al., 1989; Huang et al., 1990; Hoener and Brodbeck, 1992), and its primary sequence is known from bovine liver cDNA (Scallon et al., 1991). Two distinct GPI-PLD cDNA were also found in human liver and pancreas (Tsang et al., 1992)

In serum, GPI-PLD is associated with high-density lipoproteins (HDL). The purified enzyme shows amphiphilic properties and forms high molecular aggregates as seen on gel filtration and density gradient centrifugation (Hoener and Brodbeck, 1992). The major component of the HDL fraction, apolipoprotein A-I, is able to associate with GPI-PLD, thereby disaggregating the high molecular forms (Hoener et al., 1993). Triton X-100 was also able to dissociate aggregated GPI-PLD. However, it inhibited enzyme activity at detergent concentrations above the critical micellar concentration. In this report, we show that purified GPI-PLD from bovine serum exists in different forms, about 8% of which consist of a proteolytically degraded yet active 47 kDa enzyme. In vitro, the purified 115 kDa GPI-PLD from bovine serum could also be proteolytically degraded by a lysosomal fraction and by cathepsin D to a 47 kDa protein with identical N-terminus as the full-size enzyme, thereby retaining GPI-anchor-hydrolyzing activity. It is thus possible that in vivo, GPI-PLD may undergo a similar proteolytic processing in lysosomes or in another acidic cell compartement.

RESULTS

Purification of GPI-PLD from bovine serum by separation in Triton X-114 followed by column chromatography on DEAE-cellulose, octyl-Sepharose, and concanavalin-A-Sepharose resulted in a 20% pure enzyme as estimated by SDS/PAGE (Hoener and Brodbeck, 1992). For further purification, GPI-PLD was subjected to chromatography on hydroxyapatite which resolved the enzyme into a number of peaks (Fig. 1). Fractions were pooled as indicated by the bars in Fig. 1. Pool I eluted at 5 mM sodium phosphate pH 6.4 and contained about 12% of the total enzyme with a specific activity of 2460 units per mg of protein. The highest specific activity (around 7000 units per mg of protein) was found in pools II and III (47% of total activity) while pools IV to VI (40% of total activity) showed lower specific activities (Table 1).

SDS/PAGE in reducing conditions of pool II revealed one band migrating to an apparent molecular mass of about 115 kDa (Fig. 2A). Similarly, one predominant band was seen in pools III and IV whereas pools I, V, and VI showed a number of protein bands. Western blot using mAb 117-1 (IgG_1, k chain) revealed that pool I not only contained immunoreactive material appearing at 115 kDa but also at lower molecular masses (Fig. 2B). The most prominent band appeared at 47 kDa while minor bands were clearly visible at 92, 81, 49, and 40 kDa. Some of these additional bands were also detected in the other pools. The same results were obtained with SDS/PAGE under non reducing conditions (data not shown).

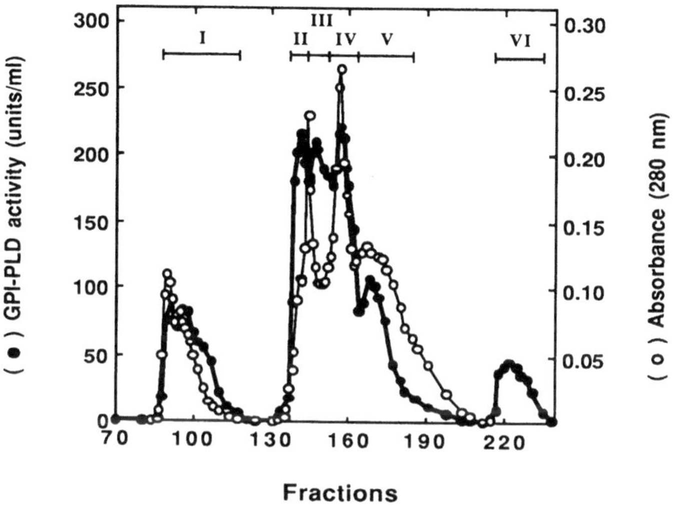

Fig. 1. *Elution profile of GPI-PLD from a hydroxyapatite column.*

Partially purified GPI-PLD from bovine serum after chromatography on concanavalin A (200 ml) was chromatographed on a hydroxyapatite column (4 x 16 cm). Fractions of 5 ml were collected. The enzyme in pool I was eluted with 300 ml 5 mM sodium phosphate pH 6.4, pools II to V were eluted with 450 ml 100 mM sodium phosphate pH 6.4, and pool VI was eluted with 250 ml 100 mM sodium phosphate pH 6.4 containing 0.1% Triton X-100. Absorbance at 280 nm (O) and GPI-PLD activity (●) were measured.

Table 1. *Chromatography of GPI-PLD from bovine serum on hydroxyapatite.*

Fraction	Volume (ml)	Total protein (mg)	Total activity (units[a])	Specific activity (units/mg)
Pool I	2.5	2.56	6300	2460
Pool II	48	1.45	10800	7450
Pool III	54	1.95	13800	7080
Pool IV	60	2.81	10900	3880
Pool V	138	4.84	4800	990
Pool VI	85	7.37	5000	680

[a] 1 unit equals 1 nmol of membrane form of acetylcholinesterase converted per min at 37 °C

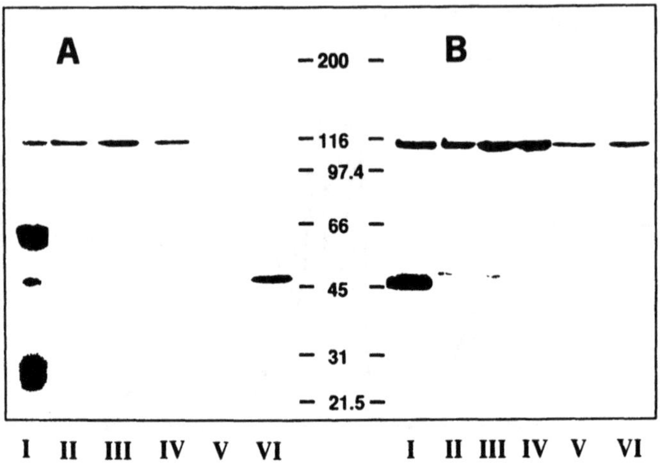

Fig. 2. *SDS/PAGE of GPI-PLD of pools I to VI after hydroxyapatite chromatography.*

Electrophoresis was carried out in reducing condition on a 5-15% gradient polyacrylamide gel. (A) Protein pattern (Coomassie staining) and (B) Western blot with mAb 117-I after SDS/PAGE of GPI-PLD of pools I-VI (lanes I-VI) after hydroxyapatite chromatography (see Fig. 1). Aliquots of 80 µl of pools II-VI were put on lanes II-VI. Pool I was concentrated 70 times, and an aliquot of 80 µl was put on lane I. The gel was calibrated with myosin (200 kDa), ß-galactosidase (116 kDa), phosphorylase b (97.4 kDa), bovine serum albumin (66 kDa), ovalbumin (45 kDa), carbonic anhydrase (31 kDa), and trypsin inhibitor (21.5 kDa).

Peptide sequencing of the band appearing at 47 kDa gave an identical N-terminal sequence for the first 10 amino acids to that of the band at 115 kDa except in position 7 where leu instead of ile was determined. The appearance of the 47 kDa band in which the first 6 amino acids of the N-terminus were identical to the sequence of the band of 115 kDa suggested that the band at 47 kDa arose from the 115 kDa band by proteolytic cleavage from the C-terminus. Consequently, we investigated the possibility that this conversion could also be mediated by proteases in vitro. As a source of proteolytic enzymes, we prepared a crude lysosomal fraction from bovine liver by subcellular fractionation in isotonic sucrose. Increasing amounts of the lysosomal fraction were added to GPI-PLD (pool III enzyme) and incubated at 37 °C for 1 h. Then, the enzyme was subjected to SDS/PAGE, and the products were visualized by Western blot with mAb 117-1. As seen from Fig. 3, lanes B to D, increasing amounts of the lysosomal fraction yielded increasing amounts of immunoreactive material with a major band at around 49 kDa. In addition, several faint bands at molecular masses between the parent enzyme (115 kDa) and the main reaction products as well as below were seen. Proteolytic conversion of the high molecular mass form of GPI-PLD could also be obtained by the lysosomal protease cathepsin D which yielded two main bands appearing at 47 and 49 kDa (Fig. 3, lanes E and F). While incubation of GPI-PLD with the lysosomal fraction did not decrease GPI-PLD activity, incubation with cathepsin D caused a loss of enzyme activity. Incubation of GPI-PLD with bovine serum did not degrade the purified 115 kDa enzyme.

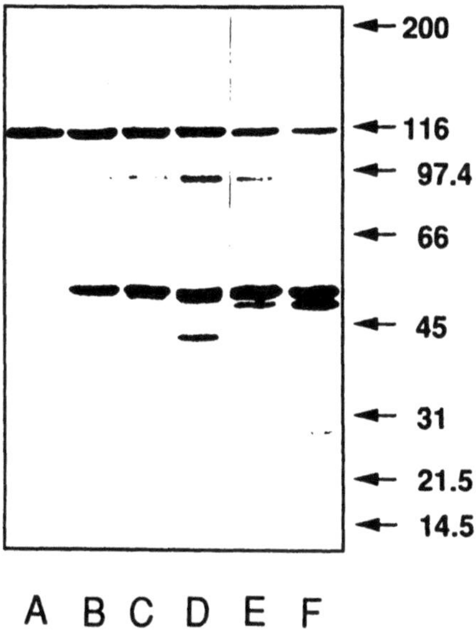

Fig. 3. *SDS/PAGE of GPI-PLD from bovine serum after incubation with increasing amounts of a lysosomal fraction or cathepsin D.*

Electrophoresis was carried out in reducing condition on a 5-15% gradient polyacrylamide gel. GPI-PLD of pool III (about 3 μg) was incubated with lysosomal fraction at 37 °C for 1 h, or with cathepsin D in 20 mM sodium phosphate pH 5.0 containing 0.02% Triton X-100 at 37 °C for 20 h. Lane A GPI-PLD without, lanes B to D GPI-PLD in presence of 0.25, 1, and 5 μl of lysosomal fraction, respectively, and lanes E and F GPI-PLD in presence of 1.1 munits and 5.7 munits cathepsin D, respectively. Molecular mass markers are as described in Fig. 2.

As previously shown (Hoener and Brodbeck, 1992), the partially purified enzyme formed multiple aggregates which, in sucrose density gradient centrifugation, sedimented in absence of detergent with sedimentation coefficients up to 14.5 S while in presence of detergent, the enzyme sedimented at about 6 S. The sedimentation behavior of GPI-PLD contained in pools I to VI was, therefore, investigated. As shown in Fig. 4A, GPI-PLD contained in pool I yielded three active enzyme forms (a, b, and c) which, in absence of detergent, sedimented at 14.9 S, 11.4 S, and 7.0 S, respectively. In presence of 0.1% Triton X-100 in the gradient, all three enzyme forms converged to one form sedimenting at 5.7 S. Western blot with mAb 117-1 demonstrated that pool a contained only the 115 kDa form and pool c the 47 kDa form of GPI-PLD, while pool b consisted of a mixture of both forms (inset Fig. 4A).

GPI-PLD of pools II to VI aggregated during density gradient centrifugation in the absence of detergent (Fig. 4B). The enzyme of pool II sedimented as a broad peak with a maximum appearing at 14.5 S. The enzyme in pools III and IV gave two activity peaks with s-values of about 13.9 S and 10.8 S, while enzyme in pools V and VI sedimented with a main peak at about 10.2 S. In the presence of 0.1% Triton X-100 in the gradient, GPI-PLD of all pools sedimented as one form at about 6.0 S.

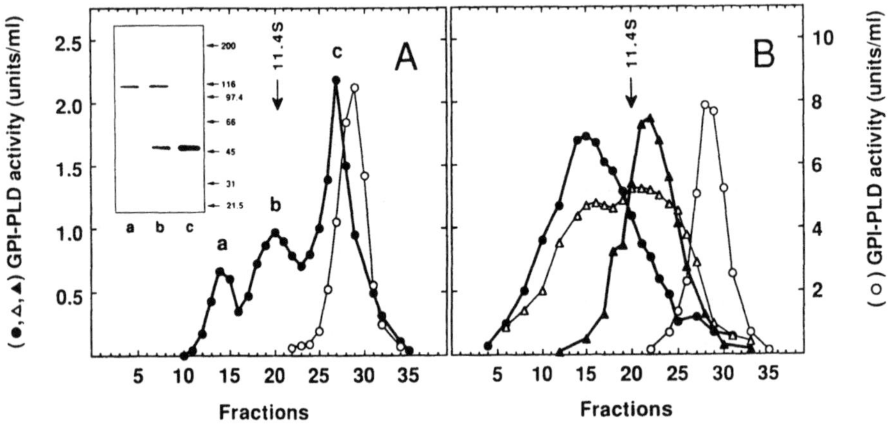

Fig. 4. *Sedimentation analysis of GPI-PLD contained in different pools after hydroxyapatite chromatography.*

Sucrose density gradients were carried out as described by Hoener and Brodbeck (1992). (A) Sedimentation profiles of GPI-PLD in pool I of Fig. 1 in the absence (●) or presence of 0.1% Triton X-100 (○). Fractions 12-16, 17-23, and 25-31 of the gradient devoid of Triton X-100 were combined as pools a, b, and c, respectively, and subjected to SDS/PAGE followed by Western blotting with mAb 117-1 (insert). The gel was calibrated as in Fig. 2. (B) Pools II (●), III (△), and V (▲) were run on a gradient in the absence of Triton X-100. Pools IV and VI displayed the same sedimentation behavior as pools III and V, respectively (unpublished results). In presence of 0.1% Triton X-100, all pools showed the same sedimentation behavior (○). GPI-PLD activity was assayed under standard assay conditions. The arrow indicates the position of catalase (11.4 S).

DISCUSSION

GPI-PLD from bovine serum has an apparent molecular mass of 115 kDa and contains major hydrophobic sequences in the C-terminal half of the primary structure which are responsible for its observed association in serum with HDL (Hoener and Brodbeck, 1992). Our present results demonstrate that the enzyme purified from bovine serum exists in multiple molecular forms which are partly resolved by chromatography on hydroxyapatite (Fig. 1). The pools differ in the specific activity of GPI-PLD as well as in the sedimentation behavior on density gradients. The multiple bands recognized by mAb 117-1 on Western blots indicate that the heterogeneity of the enzyme may be in part due to proteolytic cleavage of the 115 kDa enzyme. The most prominent species detected was a product with an apparent molecular mass of 47 kDa which was catalytically active. Results from density gradient centrifugation clearly showed that the 47 kDa entity no longer aggregated in absence of detergent, while the uncleaved enzyme fully displayed its aggregating properties (Fig. 4A). Since the N-terminal sequence of the 47 kDa breakdown product was for the first 6 amino acids identical to the native enzyme of 115 kDa molecular mass, proteolytic cleavage must have taken place from the C-terminal end of the enzyme, thereby removing the major hydrophobic sequences. SDS/PAGE (protein staining) of GPI-PLD contained in pool I revealed major bands around 66 kDa and 28 kDa. It was thus possible that the 115 kDa enzyme was cleaved into two peptides, the immunoreactive one of 47 kDa and the other, non-immunoreactive of 66 kDa. Sequencing of the bands around 66 kDa as well as those around 28 kDa (Fig. 2A, lane 1) gave, however, no sequences corresponding to C-terminal stretches of GPI-PLD. This result clearly showed that the peptides migrating to 66 and 28 kDa did not arise from the 115 kDa form of GPI-PLD but represented impurities in pool I. Since no C-terminal peptides could be detected by SDS/PAGE in the samples containing 47 kDa form, proteolytic cleavage must have taken place by a stepwise degradation from the C-terminus. This notion is sustained by the additional bands seen with mAb 117-1 at molecular masses in between the native 115 kDa form and the main proteolytic 47 kDa form (Fig. 2B, 3).

Interestingly, mAb 117-1 reacted more strongly with the 47 kDa band than with the uncleaved enzyme. We noted (unpublished data) that mAb 117-1 reacted with two CNBr-fragments which both originate in position 331 (numbering of residues according to the known sequence of GPI-PLD from bovine liver, see Scallon et al., 1991). On the other hand, it did not recognize fragments near the N- nor the C-terminus indicating that the immunoreactive determinant resides in a hydrophilic stretch near the C-terminal end of the 47 kDa protein. GPI-PLD occurs in relatively high amounts in bovine serum (40 mg/l; Hoener and Brodbeck, 1992), whereas comparatively low amounts are measurable in individual organs (Hoener and Brodbeck, 1992; Heller et al, 1992; Metz et al., 1992). Since GPI-PLD could be proteolyzed by the addition of a subcellular fraction containing lysosomal enzymes as well as by the addition of cathepsin D, it is possible that GPI-PLD may be similarly processed in vivo. It is thus conveivable that GPI-PLD may undergo an endocytotic pathway which would direct the enzyme to secondary lysosomes, where the proteolytic cleavage of GPI-anchored proteins may take place. In that respect, GPI-PLD may resemble lysosomal enzymes such as arylsulfatase A (Fujii et al., 1992), β-hexosaminidase (Mahuran et al., 1988), or cathepsin D (Yonezawa et al., 1988). Alternatively, a clathrin-independent pathway might also be envisaged.

A possible reason for the existence of multiple molecular forms and for the determined wobble in position 7 of serum GPI-PLD may lie in the synthesis of the enzyme in different organs. In human, a major source of the serum enzyme is thought to be the pancreas (Tsang et al., 1992) which synthesizes a form of GPI-PLD different from liver (J.P. Kochan, personal communication). Another source of the enzyme may be liver mast cells in which the presence of GPI-PLD was demonstrated by immunohistochemical staining (Metz et al., 1992).

ACKNOWLEDGMENTS

We thank Dr. P. Bütikofer for valuable advice, and Drs B. Nørgaard-Pedersen and C. Koch for their help in the production of mAbs. This work was supported by the Swiss National Science Foundation grant No. 31-29875.90.

REFERENCES

Brown DA and Rose JK (1992) Sorting of GPI-anchored proteins to glycolipid-enriched membrane subdomains during transport to the apical cell surface. Cell 68: 533-544

Cross GAM (1990) Glycolipid anchoring of plasma membrane proteins. Annu. Rev. Cell Biol. 6: 1-39

Davitz MA, Hereld D, Shak S, Krakow J, Englund PT and Nussenzweig V (1987) A glycan-phosphatidylinositol-specific phospholipase D in human serum. Science 238: 81-84

Davitz MA, Hom J and Schenkman S (1989) Purification of a glycosyl-phosphatidylinositol-specific phospholipase D from human plasma. J. Biol. Chem. 264: 13760-13764

Fujii T, Kobayashi T, Honke K, Gasa S, Shimizu T and Makita A (1992) Proteolytical processing of human lysosomal arylsulfatase A. Biochim. Biophys. Acta 1122: 93-98

Heller M, Bieri S and Brodbeck U (1992) A novel form of glycosylphosphatidylinositol-anchor converting activity with a specificity of a phospholipase D in mammalian liver membranes. Biochim. Biophys. Acta 1109: 109-116

Hoener MC and Brodbeck U (1992) Phosphatidylinositol-glycan-specific phospholipase D is an amphiphilic glycoprotein that in serum is associated with high-density lipoproteins. Eur. J. Biochem. 206: 747-757

Hoener MC and Brodbeck U (1993) Phosphatidylinositol glycan-anchor-specific phospholipase D from mammalian brain. Meth. Neurosci. 18: 3-13

Hoener MC, Stieger S and Brodbeck U (1990) Isolation and characterization of a phosphatidylinositol-glycan-anchor-specific phospholipase D from bovine brain. Eur. J. Biochem. 190: 593-601

Hoener MC, Bolli R and Brodbeck U (1993) Glycosyl-phosphatidylinositol-specific phospholipase D: interaction with and stimulation by apolipoprotein A-I. FEBS Lett. 327: 203-206

Huang KS, Li S, Fung WJC, Hulmes JD, Reik L, Pan YCE and Low MG (1990) Purification and characterization of glycosyl-phosphatidylinositol-specific phospholipase D. J. Biol. Chem. 265: 17738-17745

Lemansky P, Fatemi SH, Gorican B, Meyale S, Rossero R and Tartakoff AM (1990) Dynamics and longevity of the glycolipid-anchored membrane protein, Thy-1. J. Cell Biol. 110: 1525-1531

Low MG and Prasad ARS (1988) A phospholipase D specific for the phosphatidylinositol anchor of cell-surface proteins is abundant in plasma. Proc. Natl. Acad. Sci. 85: 980-984

Mahuran DJ, Neote K, Klavins MH, Leung A and Gravel RA (1988) Proteolytic processing of pro-α and pro-β precursors from human β-hexosaminidase. J. Biol. Chem. 263: 4612-4618

Metz CN, Thomas P and Davitz MA (1992) Immunolocalization of a glycosylphosphatidylinositol-specific phospholipase D in mast cells found in normal tissue and neuro-fibromatosis lesions. Am. J. Pathol. 140: 1275-1281

Metz CN, Zhang Y, Guo Y, Tsang TC, Kochan JP, Altszuler N and Davitz MA (1991) Production of the glycosylphosphatidylinositol-specific phospholipase D by the islets of Langerhans. J. Biol. Chem. 266: 17733-17736

Rodriguez-Boulan E and Powell SK (1992) Polarity of epithelial and neuronal cells. Annu. Rev. Cell Biol 8: 395-427

Rothberg KG, Ying Y, Kolhouse JF, Kamen BA and Anderson RGW (1990) The glycophospholipid-linked folate receptor internalizes folate without entering the clathrin-coated pit endocytic pathway. J. Cell Biol. 110: 637-649

Scallon BJ, Fung WJC, Tsang TC, Li S, Kado-Fong H, Huang KS and Kochan JP (1991) Primary structure and functional activity of a phosphatidylinositol-glycan-specific phospholipase D. Science 252: 446-448

Sesko AM and Low MG (1991) Immunolocalization of glycosyl phosphatidylinositol-specific phospholipase D (GPI-PLD) in mammalian tissues. FASEB J. 5: A839

Stadelmann B, Zurbriggen A and Brodbeck U (1993) Distribution of glycosylphosphatidylinositol-specific phospholipase D mRNA in bovine tissue sections. Cell Tissue Res.: in press

Tsang TC, Fung W-J, Levine J, Davitz MA, Burns DK, Huang K-S and Kochan JP (1992) Isolation and expression of two human GPI-PLD cDNAs. FASEB J.: A1922

Yonezawa S, Takahashi T, Wang X-J, Wong RNS, Hartsuck JA and Tang J (1988) Structures at the proteolytic processing region of cathepsin D. J. Biol. Chem. 263: 16504-16511

MAGNETIC RESONANCE STUDIES OF PROTEIN-LIPID INTERACTIONS*

Anthony Watts,
Department of Biochemistry,
University of Oxford,
OXFORD, OX1 3QU,
UK

Proteins and lipids co-exist in biomembranes as a functioning and dynamic entity. Both spin-label electron spin resonance and solid state nuclear magnetic resonance methods have been used to quantitate the mutual interactions between proteins and lipids in both reconstituted and natural membranes. The information available from these methods includes the stoichiometry of the lipid-protein interaction which can be used to make estimates about the protein hydrophobic surface area, and hence protein size. Both the exchange rates of lipids into and out of the protein-lipid interface, as well as any selectivity of lipid-protein associations, has been determined for a number (>15) of integral membrane proteins. Such physical properties do have functional significance in many cases.

Two spectroscopic methods have been used extensively for studying lipid protein interaction in membranes, namely NMR and ESR. Each method has sensitivity to different motional ranges (Watts, 1987) but they do give complimentary information on systems where both methods have been applied. Spin-label ESR is able to sense fast (τ_c ~ ns) motions and gives rise to distinct spectral components attributed to lipids either directly at the protein interface, or in the bulk membrane phase away from the protein. Because of the sensitivity to slower (μs) molecular motions of deuterium and phosphorous-31 NMR, an averaged spectrum is observed since lipids are exchanging quickly between all environments of the membrane. The information that is gained from each methods is consistent when the biochemistry of the system is controlled. In one case studied, that of the M13 coat, trapped lipids in protein aggregates are observed by each method, although the lipid interfacing the protein aggregates, which can be distinguished by both methods, exchanges quickly on the NMR time-scale and slowly on the ESR time-scale (Figure 1). This chapter will review some of the recent advances in the field of lipid-

NATO ASI Series, Vol. H 82
Biological Membranes:
Structure, Biogenesis and Dynamics
Edited by Jos A. F. Op den Kamp

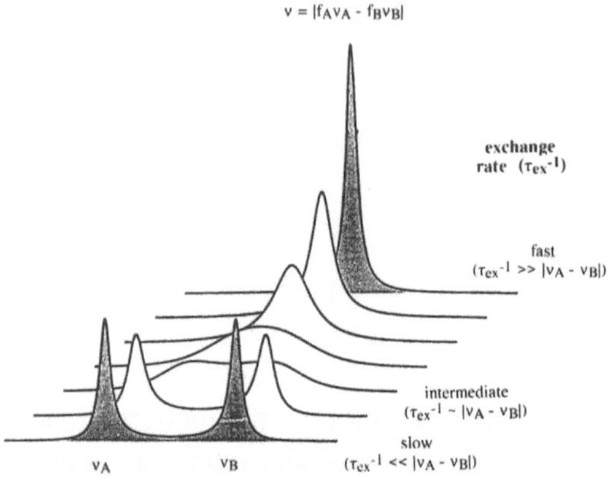

Figure 1. *Diagram showing chemical exchange (at a rate τ_{ex}^{-1}) from environment A (giving a magnetic resonance spectral line frequency, ν_A) to environment B (frequency, ν_B), for fast intermediate and slow exchange defined as shown.*

protein interactions as studied by magnetic resonance methods, and the reader is referred to a number of reviews for earlier work (Marsh & Watts, 1982; Watts & de Pont, 1985,1986; Watts, 1987; Marsh and Watts, 1988; Watts & van Gorkom, 1991; Watts & Spooner, 1992; Watts, 1993), and references therein.

Spin-label ESR

The spin-label ESR method is well suited to studying natural membranes. The probe nitroxide label is covalently attached to a bilayer lipid (Fig. 2) and is introduced into the membrane system, either a reconstituted membrane or natural purified membrane whose composition is known, at a level of 1 spin-labelled lipid to 150 - 200 endogenous membrane lipids. The sensitivity is such that 1 - 2mgs of membrane lipid are required for each experiment. The conventional ESR spectrum for the label is then recorded using computational averaging methods. The recorded spectra from membranes containing large, membrane spanning integral proteins usually consists of two spectral components. These can be deconvoluted into their constituent parts by using subtraction methods (Fig. 3). For this, one spectral component needs to be known and obtained from some other means, possibly by computer simulation or from the same spin-label in a protein-free membrane. When deconvoluted, the information which is available is:

- the proportion of spectral intensity in each component, with a sum of 100%;

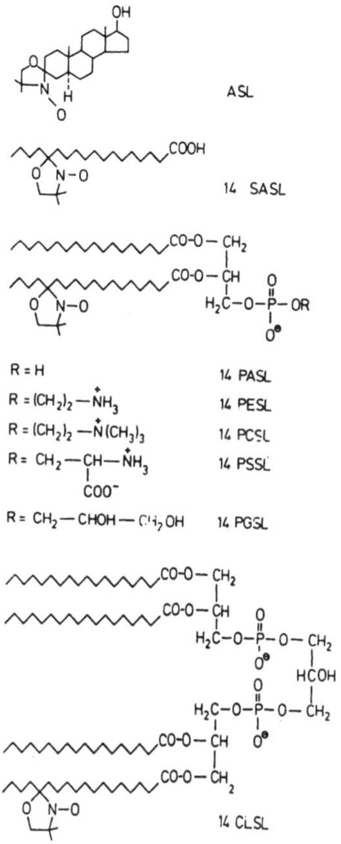

Figure 2. Chemical formulae of some lipid spin-labels used to study protein-lipid interactions in membranes.

- the dynamic characteristics of each of the two spin-label environments giving rise to the spectral components.

It can be shown, from experiments with protein-free bilayers, or reconstitution experiments where the amount of protein in the bilayer can be varied, that one of the spin-label ESR spectral components is induced by the presence of the protein. The other, narrower component is very similar to that recorded for the spin-label in lipid bilayers in the absence of protein. It has been shown now in a number of examples that the spin-labels are able to detect the presence of large proteins in the bilayer. The lipids which are at any instant at the lipid protein interface, are reduced in their rate of motion such that they give rise to spin-label ESR spectra typical for the label undergoing slow motion (τ_c < ns), on the ESR time-scale.

Since two spectral components are monitored, two site slow chemical exchange is occurring on the ESR anisotropy averaging time-scale (Fig. 2). Within the ESR dynamic window, then, two populations of lipid exist, one with slow acyl chain motion at the protein lipid interface, and one with faster motion in the bulk bilayer phase. Lateral diffusion between the two motionally distinct environments is about an order of magnitude slower than for lipid-lipid exchange ($\tau_c \sim 10^7$s) and it is this fortuitous aspect of membrane dynamics which permits the spin-label ESR methods to be so useful for such studies.

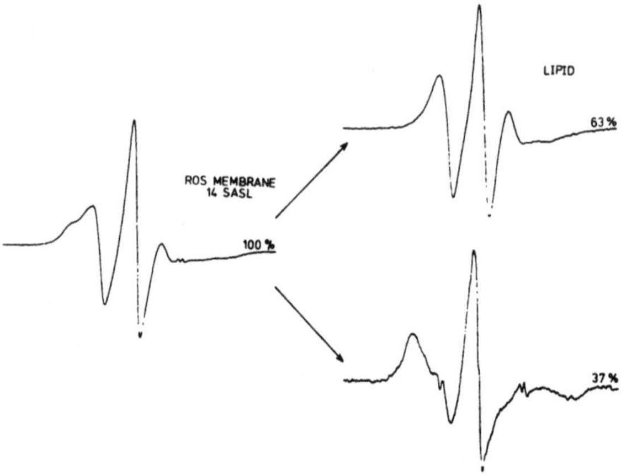

Figure 3. *A two-component spin-label ESR spectrum (left) from a stearic acid spin-labelled at the C14 position along the acyl chain, intercalated into bovine rod outer segment membranes containing with a lipid:rhodopsin mole ratio of 65:1. The spectrum is deconvoluted into a narrow component similar to that from the same label in protein-free bilayers (upper right) and a component (lower right) typical for a motionally restricted (on the ESR time-scale) spin-label (Watts et al, 1979).*

The stoichiometry of lipid-protein interactions (ie: the number of lipids interacting with the protein interface) is calculated from the spectral deconvolution. The fraction of lipid interacting with the protein as a proportion of the total lipid in the membrane per protein (determined chemically) can then yield a number of lipids which are momentarily at the protein lipid interface. The functional significance of the protein-lipid stoichiometry is that a number of integral proteins, for which the function can be determined, have been shown to require a minimal number of lipids to retain function; reducing this number substantially reduces their activity. This number of lipids per protein below which activity is reduced, corresponds well with *N*, the number of lipids at the protein-lipid interface monitored by the spin-label ESR approach. For example, cytochrome *c* oxidase requires at least 50 - 60 lipids to function as an oxidase of cytochrome *c* (Knowles *et al.*, 1979), and the Ca^{++}-Mg^{++}-ATPase requires at least 30 lipids around its interface to hydrolyse ATP. In spin-label ESR studies, it is found that similar stoichiometries of lipids are measured at the protein-lipid interface and in the motionally restricted environment of the membrane, on the ESR time-scale.

In structural terms, since the size of a lipid is known from X-ray studies on bilayers (~ 0.96nm chain width), it is now possible to calculate the hydrophobic perimeter of an integral protein. Thus if *N* lipids are interacting with the protein from the spin-label ESR methods (as is the case for bovine rhodopsin), then the protein has a hydrophobic diameter given by [(0.96 x

$N)/2\pi$] nm. This determination thus gives a value which is independent of the molecular weight or total size of the protein, and is determined only by the part of the protein which is inside the bilayer hydrophobic core which is interacting with the lipid chains. This kind of determination has been used for a number of proteins, including bovine rhodopsin, cytochrome *c* oxidase, bacteriorhodopsin, the acetyl choline receptor, etc., and subsequent independent methods of determining protein size (diffraction methods, hydrodynamics, etc) has confirmed protein size as determined by the spin-label method.

Although the spin-label ESR spectra are two component in nature, this does not imply that lipid-protein exchange does not occur. In fact, temperature studies show that with some systems, the spectral components, both narrow and broad, show changes which are characteristic for lipids increasing in their exchange rate between the two environments of the membrane. This is particularly true for systems in which the protein is not aggregated and in a low state of oligomerization, such as rhodopsin. For proteins which are highly aggregated, or in a very large complex, then exchange is slower on the ESR time-scale and the spectral components, especially the broader component, show little temperature dependence. Since the spin-labelled lipids are introduced into the membranes as freely diffusing lipids, if they sense an environment such as the protein-lipid interface, even if this is in between an oligomeric protein complex, then exchange into that environment must have been possible, as is exchange out of that environment. Exchange of lipids is then governed, presumably, by the rate of protein-protein association, a process which will release lipids slowly (longer than ms) into the bulk phase.

When an integral protein is exposed to a range of different lipid types, it appears that it may select specific lipids as neighbours rather than others. Spin-labels have been used to quantitate this selectivity. The ideal way to do this seems to be to take a single protein-lipid complex, with one lipid type, usually PC, and split it into aliquots. Each aliquot is labelled with a spin-labelled lipid of a particular type. Since most lipid types can now be synthesized (Fig 2), and the only difference in the spin-label is the head group moiety. Any difference in the quantitation of the number of lipids interacting with the protein, deduced from spectral subtractions, can therefore be ascribed to simply a head group effect. It has been shown that cytochrome *c* oxidase does display a selectivity for cardiolipin as well as phosphatidic acid, and a number of other proteins for negatively charged lipids, such as PS and PG. Two types of selectivity may occur, simple charge mediated and a more specific selectivity related to particular binding sites on the protein at the polar-apolar interface, or in a functional significant site of the protein (see Marsh and Watts, 1988 for a review).

Experiments in which the order of the lipid chains at the protein-lipid interface have been investigated, have shown that the lipids at the protein-lipid interface are more disordered than in the bilayer away from the protein (Marsh & Watts, 1988). This is not surprising since proteins have a rather rough and "squishy" surface to present to the bilayer in which the lipids have a fair degree of order (Bloom, 1979).

If the protein-lipid interface is a region which is relatively disordered compared to the rest of the bilayer, then it might be suspected that preferential partitioning of lipophilic molecules into such a region may occur. Indeed it has been shown that a range of general anaesthetics do partition into the lipid interface of the nicotinic acetyl choline receptor (nAChR) in receptor-rich membranes (Fraser *et al.* 1990). Spin-labels do sense the protein-lipid interface, as with other large integral proteins (Fig. 4). It has been found that the degree of displacement of lipids in the receptor-rich membranes, as determined from the spin-label ESR experiments, depends upon the amount of general anaesthetic in the membrane (Fig. 5). The degree of partitioning into this region has a functional relevance in that the displacement of lipid correlated well with the desensitization of the receptor by the same molecules, when normalized for the differential partitioning behaviour of the anaesthetics. To support this, the thermal stability of the protein, when measure by differential scanning calorimetry (DSC), is found to be reduced

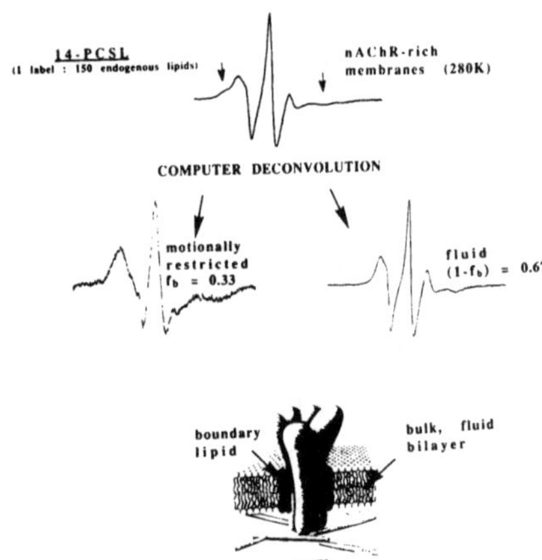

Figure 4. *Spin-label ESR spectra for the C14 labelled phosphatidylcholine probing acetylcholine receptor-rich membranes (upper spectrum) and the result of the spectral deconvolution. Lower diagram shows the suggested shape of the protein and the boundary layer and bulk phase being probed by the labels to give the two spectral components (Fraser et al., 1990).*

considerably in the bilayer in the presence with the reduction in stability (from a measure of the denaturation temperature) being directly correlated with amount of hexanol in the bilayer. It is probable that partitioning to the lipid-protein interface of lipophiles, to destabilize a protein conformation and hence reduce or inhibit activity, may be a general phenomenon for many protein.

In summary, the detection of protein-lipid interaction by spin-label ESR methods has provided a method for determining the stoichiometry, selectivity, exchange dynamics, perturbation and functional implications of the protein-lipid interactions in biological membranes. The methods has good sensitivity and fortuitously the time-scale of the method is perfect for such studies of membranes by virtue of the fact that the anisotropy averaging of the sin-label ESR properties by lipid exchange is in the most sensitive range (ns - μs) for lipid-lipid exchange in membranes.

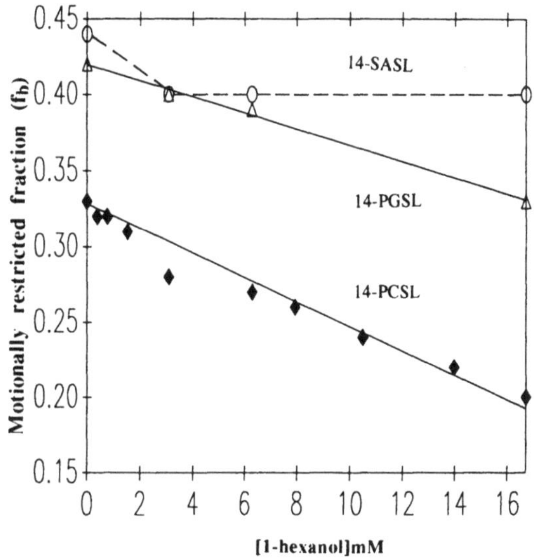

Figure 5. *Dependence of the fraction of lipids motionally restricted by the acetyl choline receptor in receptor-rich membranes as a function of hexanol concentration in the membrane. The fatty acid spin-label (14-SASL) is partly displaced due to strong interaction at a non-boundary site, and the other two labels displaced with increasing hexanol.*

Solid state NMR

The major advantage of using solid state NMR methods for studying lipid-protein interactions, is that the nucleus being detected can be a non-perturbing probe, for example, a deuteron instead of a proton, or the indigenous phosphorus-31 nucleus. With the deuterium NMR methods, some chemical synthesis is necessary and, unless some biosynthetic ways of incorporating the nucleus into the membranes, as done by Seelig and co-workers, then there are certain restrictions as to the membrane system that can be studied. The more usual approach has been to make reconstituted bilayers containing a protein of interest and then use that for study of lipid-protein interactions. Although quantitative titrations of lipid and protein can be carried out, it may not be possible to produce complexes of the required high protein density, even though the methods can detect peptides and protein at up to 1000 lipid for one protein. In addition, to demonstrate the functional relevance of the system, requires quantitation of the activity of the protein in the complex and often proteins may not be active at either high protein density, or in the lipid chosen for the study. Thus it can be assumed that studies should ideally by obtained from fully functional protein containing systems, and wherever possible, the functional implications of the protein related to the structural information gained.

Our approach has been to study protein-lipid interaction which occur at the membrane surface (Watts, 1988; Watts & van Gorkom, 1991; Watts & Spooner, 1991). For this, we have deuterated the head groups of a number of phospholipids and studied the NMR of these complexes (Fig. 6). The perturbation of the bilayer surface is then related to the protein interaction. Both peripheral and integral proteins have been studied, although we have also studied peptides which are know to bind to membrane surfaces (Sixl and Watts, 1985).

*__Figure 6.__ Chemical formulae of phospholipids which have been selectively deuterated (at positions marked with *) in their head-groups for studies of protein-lipid interactions by deuterium (and phosphorus) NMR.*

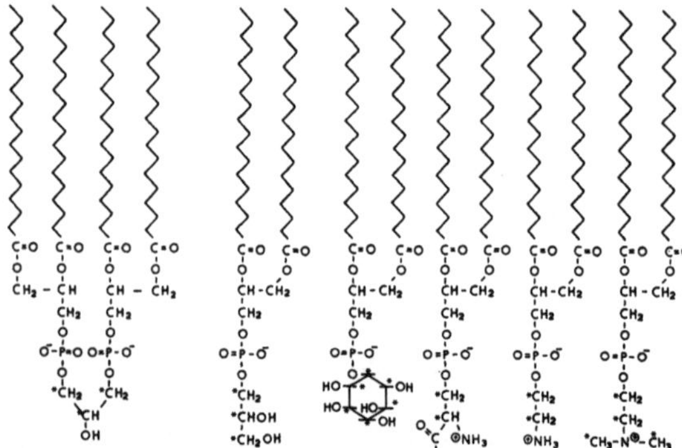

Myelin basic protein (MBP) holds the myelin sheath together through protein lipid interactions. The protein has a molecular weight of 18.4kDa and is found to be cleaved in patients with MS into two fragments of 12.6 and 5.8kDa, which can also be achieved *in vitro* with protease (Fig. 7).

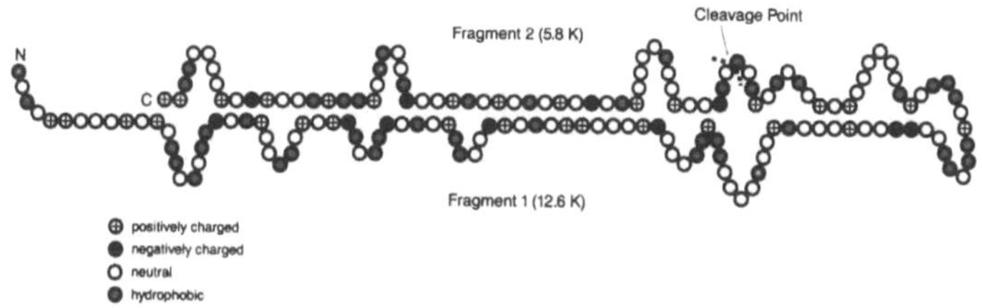

Figure 7. *Myelin basic protein in a string representation, showing the positive, negative, neutral and hydrophobic residues, as well as the cleavage point in the protein. The protein penetrates the myelin at the sites shown by the curves in the sequence.*

The interaction with PC bilayers by this protein is minimal at the bilayer surface, but when the bilayer contains a charged lipid, such as PG, the protein binds strongly and perturbs the bilayer lipid packing (Sixl *et al.*, 1984). Only one component NMR spectra are recorded, implying that there are no two long-lived environments for the lipids, one protein associated and one not protein-associated in fast exchange (Fig. 1), as there is in spin-label ESR experiments with large proteins (see above). The titration of protein with lipid shows that one protein interacting with 1000 lipids is detected in the methods. By assuming a fast exchange analysis for the protein-lipid association (Sixl *et al*, 1984), it was shown that the number of lipids interacting with MBP or its two fragments in separate experiments, is related directly to the number of charges on the whole protein or its fragments (Hayer-Hartl *et al.*, 1993) (Fig. 8). Indeed, the analysis shows that one lipid probably interacts with one charged residue on the protein in a stoichiometric way. The proteins tend to phase separate the charged lipids from the mixed PC/PG complex, as with charged peptides such as polymyxin (Sixl & Watts, 1985), although lipids can still exchange freely into and out of the complex. A similar analysis with the changes in the phosphorus-31 chemical shift anisotropy (csa) which can be measured in the same complexes, shows a similar behaviour, even though this is a completely different nucleus, thereby confirming the mode of interaction is general throughout the bilayer surface. The analysis also confirms that the lipid is relatively disordered, or at least complete averaging of the deuterium quad-

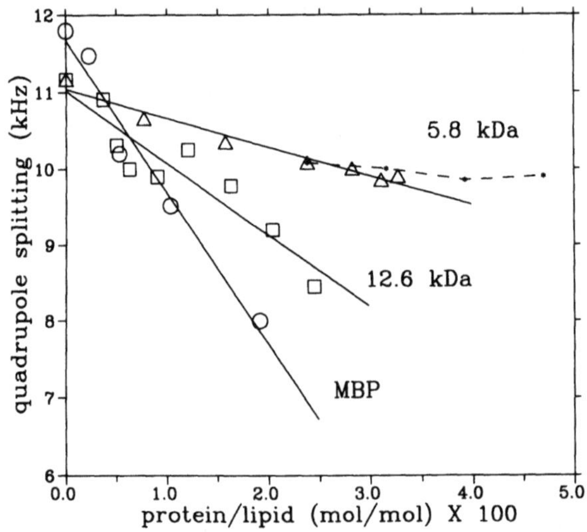

Figure 8. The change in quadrupole splittings for the α-CD_2 group of head-group deuterated phosphatidylglycerol in bilayers as a function of protein-lipid ratio for peripherally bound intact myelin basic protein and its two fragments. A fast exchange of lipids into and out of the protein-lipid interface, on the deuterium NMR time-scale, was assumed and in so doing, the slopes of the dependences (MBP: fragment 1: fragment 2, 1.0:0.65:0.35) relate to the ratio of the number of charges on the proteins (MBP: fragment 1: fragment 2, 1.0:0.47:0.19) (Hayer-Hartl et al., 1993).

rupolar anisotropy can occur for lipids at the protein interface. It is assumed therefore, that the protein itself is high unstructured, that is, does not have smooth surfaces, or that bilayer itself is distorted up to the protein hydrophobic area (Fig. 9). It may be that the bilayer forms a meniscus or "knecks" into the protein hydrophobic area, forming a local curved region in the bilayer around which the lipids can diffuse, thus averaging out the anisotropy through lateral diffusion.

The integral protein bacteriorhodopsin (BR), has been incorporated into lipid bilayers with a view to investigate the perturbation of bilayer surfaces by the protein residues which reside at the protein-lipid interface. The PC head group, when deuterated, has been shown to be a "membrane surface" voltmeter, where the sensitivity of the α- and β-methylenes are able to reveal the sign and magnitude of the charge at the bilayer surface (Gale & Watts, 1992). For BR, we showed that there are some 4 - 5 negative charges residing at the bilayer surface as monitored by deuterated PC. This correlates well with the known structure of the protein and probable disposition of residues from model building. It is not clear of course, which charges are detected, or what is

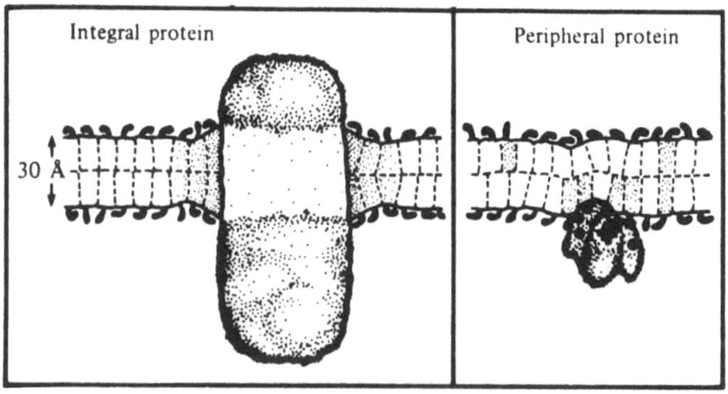

Figure 9. Possible ways in which a bilayer surface can be perturbed by an integral and peripheral protein (adapted from Israelachivili et al., 1980)

the asymmetry of the charge distribution. Although no attempts were made to quantitate their interactions, the perturbation of the bilayer surface by rhodopsin (Ryba *et al.*, 1986) and by band 3 (Dempsey *et al.*, 1986) has been examined. In each case, fast exchange of the lipids into the protein lipid interface has been observed. Also, the deuterium NMR spectra were reduced to essentially isotropic components when at the protein interface. This implies that either the lipid head groups are highly disordered or that the protein-bilayer match is not perfect and some bilayer curvature or distortion occurs, thereby averaging the deuterium anisotropy through motional factors, such as local curvature.

In another example, that of the M13 bacteriophage coat protein (Mr ~ 5,000) and composed of a single transmembrane pass, the protein has been reconstituted into PC bilayers and studied by both spin-label ESR and deuterium NMR (van Gorkom, 1991). This system may be rather unusual, but it an interesting model to which the consistency of NMR and ESR to study protein lipid-interactions can be applied. By both magnetic resonance methods, it appears that motionally restricted lipid is induced by the protein. This is the first example where such a situation has been observed, although other proteins may show the same phenomenon. When careful examination of the NMR and ESR data is made, it is found that the stoichiometry of the lipid-protein interaction is different for each method. This can be rationalized if the motionally restricted lipid observed by both methods, is composed of two types, both trapped and interfacial lipid. The NMR method can only detect the trapped lipids and hence shows the lower stoichiometry, whilst the ESR method detects the trapped lipid as well as the interfacial lipid, which is a total of about 25% more than the trapped lipid per protein monomer, assuming that the aggregate is a linear arrangement of protein monomers, as suggested. Thus, the lipid which exchanges from the interface of the protein-trapped lipid complex, is detected by ESR as in slow exchange with the bilayer, but in fast exchange in the NMR experiment (Fig. 10). Further more, the life-time of a lipid-protein aggregate must be long when compared with the deuterium NMR time-scale for anisotropy averaging, which puts a slower limit on the life-time of an aggregate of ms. The complexes are completely homogeneous when viewed by electron microscopy and density gradient centrifugation. In addition, the aggregates are induced at high protein concentrations since the protein behaves as a monomer at low protein concentration (van Gorkom, 1991).

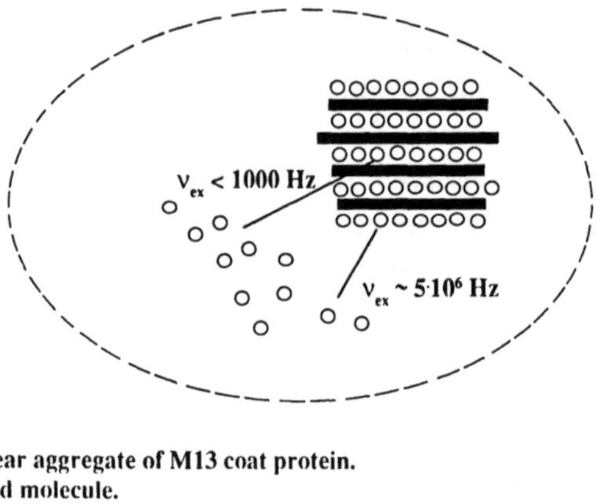

Figure 10. *The suggested arrangement of the M13 coat protein in bilayers showing how the trapped lipids* ***and*** *interfacial boundary lipids are both detected in the spin-label ESR experiment since both lipid types exchange slowly on the ESR time-scale ($\nu_{ex} < 10^7$Hz for slow exchange), whereas in the deuterium NMR experiment, only the trapped lipids are detected since the interfacial lipids are in fast exchange ($\nu_{ex} > 10^5$Hz for fast exchange) on the NMR time-scale (van Gorkom, 1991).*

In summary, deuterium NMR has been used to study protein-lipid interactions in complexes made by reconstitution methods. The mode of interaction seems to be driven through hydrophobic interactions as well as some electrostatic contributions. It is not easy to deconvolute these, but as the structure of more membrane proteins become available, and the types and kinds of residues identified which interact with the lipids assigned, then specific details of how the bilayer matches the protein interface may become available to explain the spectral changes observed in the NMR experiment.

CONCLUSIONS

Both spin-label ESR and solid state (^{2}H and ^{31}P) NMR methods appear to give similar information of the nature of lipid-protein interactions in bilayers. The reasons for previous apparent inconsistencies, lie in the different experimental time-scales for the methods, with NMR being sensitive to slower motions than the spin-label ESR (Watts, 1881). Therefore, a slow lipid motion on the ESR time-scale can be seen as a fast motion by deuterium NMR. To date, only a limited number of examples of the application of both methods to the same system have been attempted. Also, the analysis for fast exchange as required for the NMR approach, is not readily capable of identifying selectivity of the interactions. Although the methods is very sensitive, quantitation to give selectivity has not been shown, except that interaction occurs, or not.

ACKNOWLEDGEMENTS

The work described here has been supported by SERC (UK), MRC (UK), EC, Wellcome Foundation, NIAAA (USA) and the MS Soc of GB.

REFERENCES

Bloom M, (1979) Squishy proteins in fluid membranes. Can J. Phys. 57: 2227-2230

Dempsey CE, Ryba NJP and Watts A (1986) Evidence from deuterium Nuclear Magnetic Resonance for the Temperature-dependent reversible Self-association of Erythrocyte Band 3 in Dimyristoyl Phosphatidylcholine Bilayers. Biochemistry 25: 2180-2187

Fraser D, Louro S, Horvath LI, Miller K and Watts A (1990) A study of the effect of general anesthetics on lipid-protein interactions in acetylcholine receptor enriched membranes from Torpedo nobiliana using nitroxide spin-labels. Biochemistry: 29, 2664-2669.

Gale P & Watts A (1992) Effect of bacteriorhodopsin on the orientation of the headgroup of 1,2-dimyristoyl-sn-glycero-3-phosphocholine in bilayers - a ^{31}P and ^{2}H-NMR study. BBA 1106: 317-324.

Hayer-Hartl M, Brophy PJ, Marsh D and Watts A (1993) Interaction of two complementary fragments of the bovine spinal cord myelin basic protein with phosphatidylglycerol bilayers studied by 2H and 31P NMR spectroscopy. Biochemistry (in press)

Israelchivilli, JN, Marcelja, S, Horn RG (1980) Physical principles of membrane organization, Qu. Rev. Biophysics 13: 121-200

Knowles PF, Watts A and Marsh D (1979) Spin-label studies of lipid immobilization in DMPC-substituted cytochrome oxidase. Biochemistry 18: 4480-4487.

Marsh D and Watts A (1988) Association of lipids with membrane proteins. In Recent Advances n Membrane Fluidity (Aloia R, ed) 2: 163-200, Alan R Liss, Inc., New York.

Ryba NJP, Dempsey CE and Watts A (1986) Protein-lipid interactions at membrane surfaces: A deuterium and phosphorus NMR studies of the interaction between bovine rhodopsin and the bilayer head-groups of DMPC. Biochemistry 25: 4818-4825

Sixl F and Watts A (1985) Deuterium and phosphorus NMR studies on the binding of polymyxin-B to lipid bilayers-water interfaces. Biochemistry 24: 7906-7910.

Sixl F, Brophy PJ and Watts A (1984) Selective protein-lipid interactions at membranes surfaces: a deuterium and phosphorus NMR study of the association of myelin basic protein with the bilayer head groups of DMPC and DMPG. Biochemistry 23: 2032-2039.

van Gorkom L (1991) D. Phil. Thesis, University of Oxford, UK

Watts A, Volotovski ID and Marsh D (1979) Rhodopsin-lipid association in bovine rod outer segment membranes. Biochemistry 18:5006-5013

Watts A and de Pont JJHHM (eds) (1985) Progress in Protein-Lipid Interactions Vol. 1, Elsevier, Amsterdam.

Watts A (1981) Protein-lipid Interactions. Nature 294: 512.

Watts A (1987) NMR methods to characterize lipid–protein interactions at membrane surfaces. J. of Bioenergetics and Biomembranes 19: 625-653.

Watts A (1991) Magnetic Resonance Studies of Lipid-Protein Interfaces and Lipophilic Molecule Partitioning. In Molecular and cellular mechanisms of alcohol and anesthetics (E Rubin, KW Miller and SH Roth, eds). Annals of the New York Academy of Science 625: 653-669

Watts A (ed) (1993) Protein Lipid Interactions. New Comprehensive Biochemistry series Vol. 25 Elsevier, Amsterdam.

Watts A and de Pont JJHHM (eds) (1986) Progress in Protein-Lipid Interactions Vol. 2, Elsevier, Amsterdam.

LIPID-PROTEIN INTERACTIONS WITH THE HYDROPHOBIC SP-B AND SP-C LUNG SURFACTANT PROTEINS IN DIPALMITOYLPHOSPHATIDYLCHOLINE BILAYERS

Jesús Pérez-Gil[1], Cristina Casals[1] and Derek Marsh[2]

[1]Departamento de Bioquímica y Biología Molecular I, Fac. Químicas
Universidad Complutense, 28040 Madrid, SPAIN.
[2]Abteilung Spektroskopie, Max-Planck-Institut für biophysikalische Chemie
Göttingen, GERMANY.

Pulmonary surfactant is a lipid/protein system which has the essential role of stabilizing the lung airway spaces during successive respiratory cycles. It mainly consists of 90% lipids and 10% of some specific proteins. The biophysical activity of surfactant is assumed to reside in the behaviour of the main phospholipid component, dipalmitoyl-phosphatidylcholine (DPPC), as a monomolecular film in the alveolar air-water interface. However, the presence of several specific proteins has been shown to be essential in the development of the complex surfactant cycle *in vivo*. Surfactant is secreted by epithelial type II cells and must be transported through the aqueous lining layer until it reaches the air-liquid interface where it must spread as the monolayer form. Two proteins (SP-B and SP-C) having unusual structural properties -including an extreme hydrophobicity- have been demonstrated to promote the adsorption of surfactant lipids from the hypophase to the interface. In this context, exogeneous surfactant materials consisting of suspensions of lipids plus these hydrophobic surfactant proteins are already being used with success in treating neonatal respiratory distress syndrome.

We are still far from understanding the nature of the interactions between lipid and protein components in surfactant that lead to such a specialized biophysical function of surfactant *in vivo*. The main objective of the present work was to use electron spin resonance (ESR) spectroscopic techniques to obtain further information on lipid-protein interactions ocurring in systems consisting of the isolated SP-B and SP-C proteins reconstituted in bilayers of the main surfactant phospholipid, DPPC.

NATO ASI Series, Vol. H 82
Biological Membranes:
Structure, Biogenesis and Dynamics
Edited by Jos A. F. Op den Kamp

EXPERIMENTAL

Dipalmitoylphosphatidylcholine (DPPC) was from Avanti Polar Lipids (Birmingham, AL). Spin-labelled phosphatidylcholines, with a nitroxide group at different positions of the sn-2 acyl chain were synthesized as described in Marsh and Watts (1982). The spin labels were stored at -20°C, in chloroform/methanol (2:1, v/v) solutions at a concentration of 1 mg/mL.

Surfactant proteins SP-B and SP-C were isolated from pig lungs as described elsewhere (Perez-Gil et al., 1993). Their purity was routinely checked by electrophoresis, determination of amino acid composition and N-terminal sequencing analysis. The proteins were quantitated by amino acid analysis. Isolated SP-C was found to retain its two palmitoylated cysteine residues, analyzed as previously described (Perez-Gil et al., 1992a).

To reconstitute the proteins, appropriate amounts of DPPC and the selected spin-labelled phosphatidylcholine (PCSL) (1 mol %) were dissolved in chloroform/methanol 2:1 v/v solution, and mixtured with the desired amount of protein (stored also as a chloroform/methanol 2:1 -v/v- solution). Samples were then evaporated under a N_2 stream and dried under vacuum in a dessicator overnight. Lipid and protein/lipid samples were hydrated with 100 μL of buffer (50 mM HEPES, NaCl 150 mM, EDTA 5 mM, pH 7) at 50°C for 1 h, with occasional vortexing. The reconstituted material was then pelleted by centrifugation at 3000 rpm in a bench centrifuge using the 100-μL capillaries for ESR spectroscopy. ESR spectra were recorded on a Varian E-12 Century Line 9-GHz Spectrometer equipped with a nitrogen gas flow temperature regulation system. The sealed capillaries (1 mm diameter) were placed in a quartz tube containing silicone oil for thermal stability. Spectra were digitized using an IBM PC interfaced to the spectrometer. Instrumental settings were 10-mW microwave power, 1.25-G modulation amplitude, 100-kHz modulation frequency, 0.25s time constant, 4-min scan time, 100-G scan range, and 3245-G center field. Several scans, typically 3-5, were accumulated to improve the signal to noise ratio. After the ESR experiments, the samples were again suspended in the above buffer and aliquots were taken to assay the protein and lipid contents for determination of the lipid/protein ratios of the recombinants.

The secondary structure of the proteins reconstituted in DPPC vesicles was analyzed by recording the far-UV circular dichroism spectra of samples in a Jobin Ivon Mark III dichrograph fitted with a 250 W Xenon lamp. Cells of 0.1 cm optical path were routinely used and the spectra were recorded at 0.2 nm/s scanning speed. Four different spectra were recorded independently and averaged for each sample. Ellipiticity was calculated taking a mean molecular weight per residue of 110 and 106 for SP-B and SP-C, respectively.

Biophysical activity of the reconstituted proteins was assayed by injecting the lipid/protein suspensions in the subphase of a King-Clements adsorption surface balance, as previously described (Perez-Gil et al., 1992b).

RESULTS AND DISCUSSION

SP-B and SP-C, when present in reconstituted bilayers of DPPC, promote the adsorption of lipids from the hypophase to the air-water interface. It can be seen in

Figure 1 that mixtures of DPPC and SP-B or DPPC and SP-C (10% protein/lipid ratio, w/w, in either case) adsorb more rapidly than do DPPC vesicles in the absence of protein. This result has been interpreted as reflecting *in vitro* a biophysical function for the hydrophobic surfactant proteins *in vivo*, i.e. improving the movement of surfactant lipids through the aqueous alveolar lining layer to reach the interface where they must spread efficiently as a monolayer.

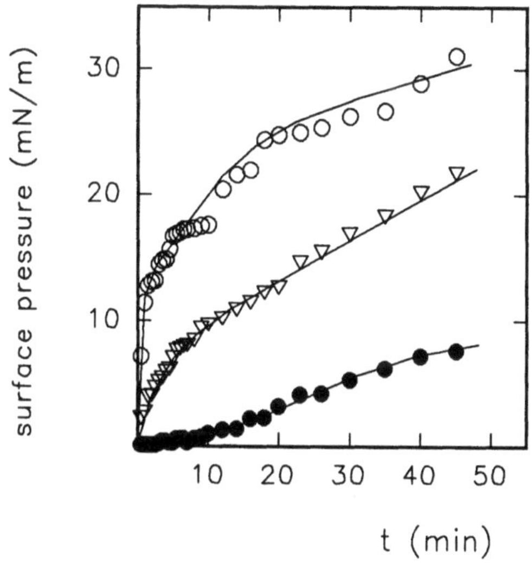

Figure 1. Adsorption kinetics of dispersions of DPPC in the absence (●) and in the presence of 10% (w/w) of either SP-B (○) or SP-C (▽). Each point represents the average of 3 independent experiments. In a lipid assay, 100 μg of DPPC suspension were injected into a total subphase volume of 6 mL of buffer HEPES 50mM, NaCl 150 mM, $CaCl_2$ 5 mM, pH 7, thermostatted at 25°C.

The enhancement of interfacial adsorption promoted by SP-B or SP-C in DPPC, shown in Figure 1, is qualitatively and quantitatively similar to the biophysical activity previously reported for similar samples (Perez-Gil et al., 1992b; Morrow et al., 1993). SP-B seems to be more active in promoting interfacial adsorption than is SP-C.

Circular dichroism spectroscopy yields information on the secondary structure that SP-B and SP-C adopt when reconstituted in a lipid environment. Figure 2 shows the CD spectra of porcine SP-B and SP-C reconstituted in DPPC bilayers. The spectra are also similar to those previously reported (Perez-Gil et al., 1993; Morrow et al., 1993) and indicate a mainly alpha-helical structure for both proteins: SP-B is estimated to possess 45% alpha-helical content while SP-C is calculated to have as much as 70% of its structure in an alpha-helical conformation. This is consistent with the secondary structure determined for these proteins by FTIR spectroscopy (Vandenbussche et al., 1992a,b). Our samples are, therefore, good models for further study of the lipid-protein interactions related to the biophysical activity of hydrophobic surfactant proteins.

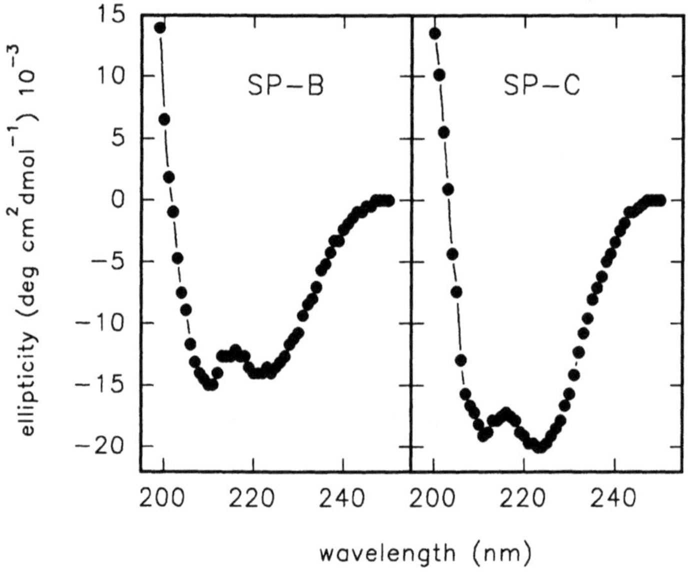

Figure 2. Far-UV CD spectrum of SP-B and SP-C reconstituted in DPPC bilayers. Dots represent the experimental data resulting from averaging four spectra.

Electron spin resonance (ESR) spectroscopy has proved to be a uniquely useful tool for the study of lipid-protein interactions in several systems (for a review, see Knowles and Marsh, 1991). Figure 3 shows the effect of the presence of SP-B or SP-C in fluid-phase bilayers of DPPC on the ESR spectrum of phosphatidylcholine spin-labelled at two different positions in the sn-2 acyl chain; one close to the lipid headgroup (5-PCSL) and other closer to the middle part of the bilayer (12-PCSL). The ESR spectra in the presence of either protein have larger outer hyperfine splittings and total spectral anisotropy than in their absence. This fact would indicate that SP-B and SP-C cause a reduction in the acyl chain mobility of DPPC molecules. The effect is greater for SP-C than it is for SP-B. The difference could originate from a different location of the two proteins in the bilayer. It has been proposed that SP-B does not penetrate far into the phospholipid bilayer and has therefore little effect in the lipid acyl chains (Vandenbussche et al., 1992b; Morrow et al., 1993) while SP-C could consist mainly of a transmembranal alpha-helix, which could have pronounced consequences on the bilayer or monolayer lipid packing (Simatos et al., 1990; Perez-Gil et al., 1992a).

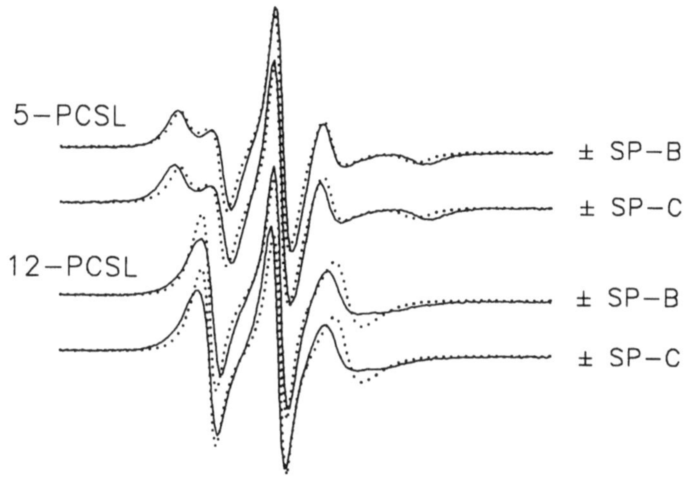

Figure 3. ESR spectra of phosphatidylcholine spin labelled at the 5th (5-PCSL) and the 12th (12-PCSL) positions of the sn-2 acyl chain in DPPC dispersions alone (dotted line) and in dispersions containing 20% (w/w) of either porcine SP-B or SP-C (solid line). Buffer 50 mM HEPES, NaCl 150 mM, $CaCl_2$ 5 mM, pH 7. Total spectral width is 100 G and T=45°C.

Two distinguishable components were never observed in the ESR spectra at any hydrophobic surfactant protein/lipid ratio up to 60% (w/w). Such complex spectra have been reported for other membrane proteins and explained by the coexistence of a restricted lipid population interacting directly with the intramembranous section of the protein. The effects of SP-B and SP-C on the ESR spectrum of DPPC bilayers containing PC spin labels were strongly dependent on temperature. As can be seen in Figure 4, both

SP-B and SP-C clearly affected the gel-to-liquid-crystalline thermotropic phase transition of DPPC bilayers. The transition was broadened and attenuated by the presence of the proteins, the latter effect being mainly manifested in the fluid phase of the lipids. Both proteins caused ordering of the fluid phase without affecting very much the mobility of the gel phase, on the conventional ESR timescale. The effect was again larger for SP-C than for SP-B. These results are consistent with scanning calorimetric data published by others (Simatos et al., 1990; Shiffer et al., 1993).

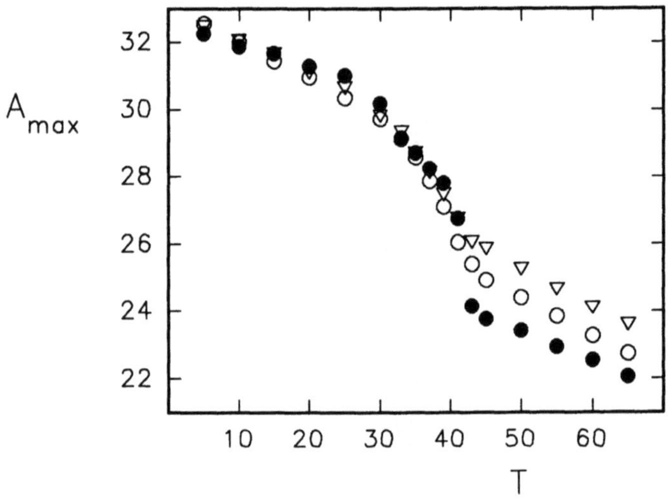

Figure 4. Temperature dependence of the outer hyperfine splitting constant, Amax, of a 5-PCSL phosphatidylcholine spin label in DPPC dispersions without (●) and containing 20% SP-B (w/w) (○) or 20% (w/w) SP-C (▽).

Although SP-B and SP-C do not seem to affect very much the mobility of the acyl chains of DPPC in the gel phase, both proteins showed a dramatic effect on the mixing behaviour of the spin labels in the DPPC gel phase. Figure 5 shows that very small amounts of either SP-B or SP-C, on the order of 5-10% (w/w), drastically reduced the spin-spin broadening effect which is significant in the absence of protein. Such spin-spin broadening arises from the partial exclusion of the spin probes from the tightly packed gel domains of DPPC, which results in locally higher concentrations of spin probes with consequent dipole-dipole and exchange interactions between probe molecules. The presence of low percentages of SP-B and SP-C increases the solubility of the spin probes in the gel domains and, therefore, decreases the spin-spin broadening component. Such an effect can be followed from the decrease in the linewidths of the ESR spectrum caused by the presence of the proteins (Figure 5). Other studies using epifluorescence microscopy to analyze the effect of SP-C in spread monolayers of DPPC also concluded that the protein improves the solubility of a fluorescent lipid probe in the gel phase domains (Perez-Gil et al., 1992a).

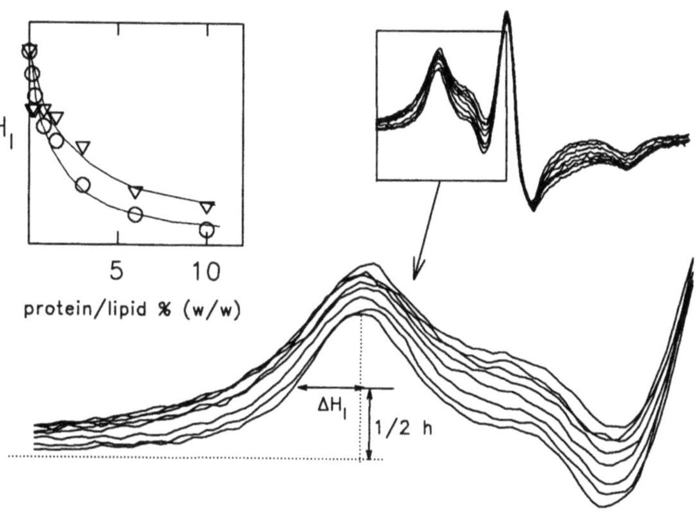

Figure 5. Effect of SP-B and SP-C on the spin-spin broadening of the ESR spectra from a 14-PCSL probe (2% molar ratio) in bilayers of DPPC in the gel phase (T=10°C). Plotted are the ESR spectra in the presence of (from upper to lower) 0, 0.25, 0.5, 1.0, 2.0, 3.0, 6.0 and 10.0% (w/w) of SP-B. Insert: Linewidth of the low field outermost peak, ΔH_l, versus the weight percent of SP-B (O) or SP-C (∇).

This effect, detected by the solubility of the lipid probe in the gel phase, could have a significant parallelism with the functional role of the hydrophobic proteins in the surfactant system *in vivo*. This is firstly because such properties have been shown *in vitro* at near-physiological protein/lipid ratios (on the order of 3%, w/w) and secondly because the functionally competent surfactant system, under physiological conditions, i.e. as a

DPPC-enriched monolayer at 37°C, is thought to be in the gel state. The presence of SP-B and/or SP-C in surfactant bilayers or monolayers could give the necessary flexibility to such gel phases that is required to develop its dynamic behaviour without reducing very much the highly ordered structure which, in turn, is necessary to achieve the minimal surface tension values of the monolayer.

ACKNOWLEDGEMENTS

This work has been partially supported by the Grant PM89-0037 from the Direccion General de Investigacion Cientifica y Tecnica from Spain. J. P.-G. was a recipient of a Short Term EMBO Fellowship.

REFERENCES

Hawgood, S.; Shiffer, K (1991). Structure and properties of the surfactant-associated proteins. Annu. Rev. Physiol. 53, 375-394.

Knowles, P.F., Marsh, D (1991). Magnetic resonance of membranes. Biochem. J. 274, 625-641.

Marsh, D.; Watts, A (1982). In "Lipid-protein Interactions" (Jost, P.C. & Griffith, O.H., Eds.), vol.2, pp 53-126. Wiley-Interscience, NY.

Morrow, M.A.; Perez-Gil, J.; Simatos, G.; Boland, C.; Stewart, J.; Absolom, D.; Sarin, V.; Keough, K.M.W. (1993). Pulmonary surfactant-associated protein SP-B has little effect on acyl chains in dipalmitoylphosphatidyl-choline dispersions. Biochemistry 32, 4397-4402.

Perez-Gil, J.; Tucker, J.; Simatos, G.A.; Keough, K.M.W. (1992a). Interfacial adsorption of simple lipid mixtures combined with hydrophobic surfactant protein from pig lung. Biochem. Cell Biol. 70, 332-338.

Perez-Gil, J.; Nag, K.; Taneva, S.; Keough, K.M.W. (1992b) Pulmonary surfactant protein SP-C causes packing rearrangements of dipalmitoylphosphatidylcholine in spread monolayers. Biophys. J. 63, 197-204.

Perez-Gil, J.; Cruz, A.; Casals, C. (1993) Solubility of hydrophobic surfactant proteins in organic solvent/water mixtures. Structural studies on SP-B and SP-C in aqueous organic solvents and lipids. Biochim. Biophys. Acta (in press).

Shiffer, K.; Hawgood, S.; Haagsman, H.P.; Benson, B.; Clements, J.A.; Goerke, J. (1993) Lung surfactant proteins SP-B and SP-C alter the thermodynamic properties of phospholipid membranes: a differential calorimetry study. Biochemistry 32, 590-597.

Simatos, G.A.; Forward, K.B.; Morrow, M.R.; Keough, K.M.W. (1990). Interaction between dimyristoylphosphatidylcholine and low molecular weight pulmonary surfactant protein SP-C. Biochemistry 29, 5807-5814.

Vandenbussche, G.; Clercx, A.; Curstedt, C.; Johansson, J.; Jornvall, H.; Ruysschaert, J.M. (1992a) Structure and orientation of the surfactant-associated protein C in a lipid bilayer. Eur. J. Biochem. 203, 201-209.

Vandenbussche, G.; Clercx, A.; Clercx, M.; Curstedt, T.; Johansson, J.; Jornvall, H.; Ruysschaert, J. -M. (1992b). Secondary structure and orientation of the surfactant protein SP-B in a lipid environment. A Fourier Transform Infrared spectroscopy study. Biochemistry 31, 9169-9176.

PROTEIN-MEMBRANE INTERACTIONS IN THE COMPLEX BIOLOGICAL MILIEU

A. Chonn* , S. C. Semple and P. R. Cullis
The University of British Columbia
Department of Biochemistry
Vancouver, British Columbia
Canada

Alterations in the lipid composition of biological membranes can have dramatic effects on their ability to interact with soluble proteins. In a series of studies employing large unilamellar vesicles (LUVs) produced by an extrusion technique, we have characterized the influence of membrane components on protein-membrane interactions. Much of our understanding of how and why proteins interact with seemingly inert membrane surfaces stems mainly from studies involving one or two protein component systems. These studies, however, do not accurately reflect the interactions that occur in the complex biological milieu (reviewed by Horbett and Brash, 1987). There have been very few studies reported on the interactions of proteins with liposomal systems incubated with whole blood. There are two main reasons for this. First, the large majority of studies on the association of plasma proteins with liposomes in vitro have been performed employing multilamellar systems. Due to the variable lamellarity of liposomes of different lipid compositions, quantification of the amount of various proteins associated per liposome has not been possible. Second, convenient techniques have not been available for the isolation of liposomes, particularly LUVs, from blood components. We have recently described a rapid method for the isolation of well-defined large unilamellar liposomes from the blood of liposome-treated mice (Chonn et al., 1991). With such a procedure now available, we have started to biochemically and immunologically characterize the amount and type of proteins associated with liposomes exposed to the complex biological milieu.

Liposomes, prepared by an extrusion technique (Hope et al., 1985; Nayar et al., 1991) and having an average size of 100 nm in diameter, were intravenously administered into CD1 mice at a dose level of 200 µmol/kg and recovered from the blood 2 min post-injection by

* Present address: The University of Lausanne, Institute of Biochemistry, Lausanne, Switzerland

NATO ASI Series, Vol. H 82
Biological Membranes:
Structure, Biogenesis and Dynamics
Edited by Jos A. F. Op den Kamp

employing a "spin column" procedure as previously described in detail (Chonn et al., 1991). The proteins associated with the recovered liposomes were analyzed by SDS-polyacrylamide gel electrophoresis followed by silver staining to visualize the proteins or by immunoblot analysis. Further, the amount of protein associated with the liposomes was quantitated using the micro bicinchoninic acid protein assay.

Several molecular properties of blood proteins that are considered to have a major influence on their surface adsorption properties include surface charge (most charged residues reside on external surfaces of proteins), size (proteins and other macromolecules are thought to form multiple contact points when adsorbed to a surface), stability of the proteins in plasma (unfolding of proteins at the surface would increase the number of adsorption sites), and carbohydrate content (Horbett and Brash, 1987). As well, the relative concentration of the proteins in plasma should affect the distribution of proteins adsorbed on surfaces.

Our findings, however, suggest that electrostatic interactions do not play a dominant role in protein-membrane interactions in complex protein mixtures. As shown in Figures 1 and 2, liposomes composed of different anionic phospholipids, but having similar overall net surface charge, have very different abilities to interact with soluble proteins. This is dramatically demonstrated by the observation that LUVs composed of 20 mol% bovine liver phosphatidylinositol have a 6 fold greater capacity to bind proteins than those composed of 20 mol% plant phosphatidylinositol. This finding suggests that fatty acyl composition of the phospholipid can markedly influence blood protein/membrane interactions. Further, these membranes exhibit very different biological properties, as measured here by their clearance property from the circulation. It is apparent that membranes that are highly reactive with soluble blood proteins are cleared very rapidly.

As shown in Figure 2, size does not appear to be an overriding factor differentiating the surface activity of soluble proteins; a complex profile of proteins of varying molecular sizes is associated with LUVs. Further, there are blood proteins which have a specific affinity for membranes composed of certain anionic phospholipids, namely cardiolipin, phosphatidic acid or phosphatidylserine. A striking example here is a protein that migrates with a molecular weight corresponding to approximately 53 000. The levels of this protein binding to cardiolpin- or phosphatidic acid-containing LUVs corresponds to similar or even greater levels than those for albumin (66 kDa). By N-terminal region protein sequence analysis and

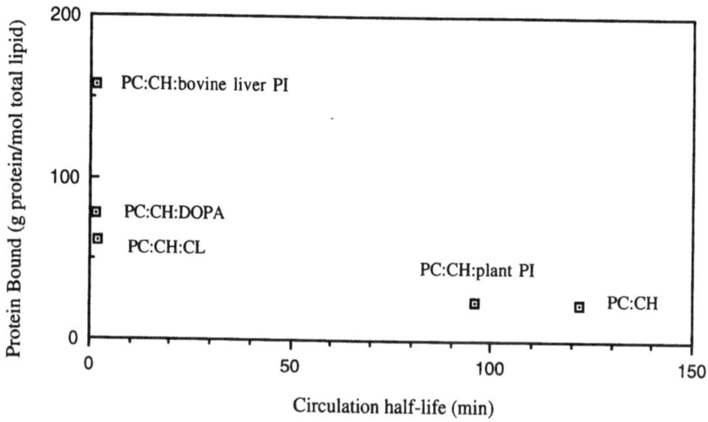

Figure 1. ***Amount of protein associated with liposomes composed of various anionic phospholipids and the relationship to circulation half-life.***

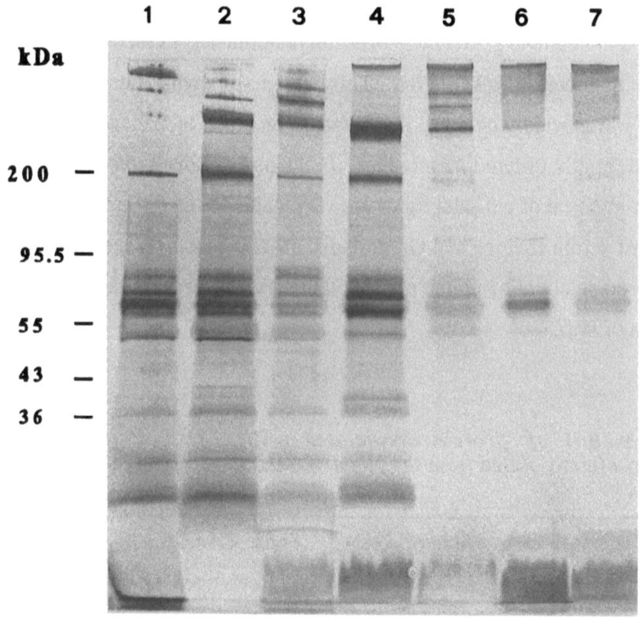

Figure 2. ***Silver-stained nonreducing SDS-polyacrylamide gels of proteins associated with liposomes recovered from the circulation of mice after 2 min post-injection.*** The lanes contain 25 nmol of total lipid of liposomes composed of the following: PC:CH:CL (35:45:10, lane 1), PC:CH:DOPA (35:45:20, lane 2), PC:CH:DOPS (35:45:20, lane 3), PC:CH:PI (bovine liver, 35:45:20, lane 4), PC:CH:PI (plant, 35:45:20, lane 5), PC:CH:PG (35:45:20, lane 6) and PC:CH (55:45, lane 7).

immunological analysis, this protein has been identified as the mouse equivalent to human β2-glycoprotein I. The reported values for the concentration of β2-glycoprotein I in rats and humans is approximately 0.2 mg/ml, 200 fold less than the plasma concentration of serum albumin. If one assumes that the association of albumin to these vesicles is non-specific, then this finding would indicate that β2-glycoprotein I is greatly concentrated on these anionic membranes. It is interesting to note that β2-glycoprotein I has recently been shown to be a cofactor for the binding of some antiphospholipid antibodies. Inasmuch as cardiolipin, phosphatidic acid and phosphatidylserine normally reside intracellularly and become expressed when cells undergo programmed cell death or apoptosis, such protein-membrane interactions may be involved in distinguishing "nonself" from "self" membranes.

In subsequent studies, we further investigated the effect of fatty acyl composition on the protein binding ability of LUVs composed solely of phosphatidylcholines. The total amount of protein bound to saturated phosphatidylcholine LUVs increased as the length of the fatty acyl chain increased (Table 1). This was a somewhat unexpected result because conceptually, it was widely believed that it would be more difficult for proteins to insert into tightly packed, highly ordered membranes. In support of this hypothesis is the observation that vesicles composed of gel state lipids are very stable to the release of entrapped solutes in the presence of serum (Senior and Gregoriadis, 1982). However, our findings indicate that LUVs composed solely of saturated phospholipids exhibit very rapid clearance kinetics from the circulation of mice.

Table 1. ***Amount of protein associated with LUVs composed of saturated phosphatidylcholines isolated from the blood of CD1 mice.***

Liposome composition	Fatty Acid Species	Phase Transition Temperature (°C)	Protein Binding Index (g/mol lipid)[a]
DMPC	14:0/14:0	23	23 ± 4
DPPC	16:0/16:0	41.5	48 ± 6
DSPC	18:0/18:0	54.5	96 ± 8
DAPC	20:0/20:0	66	101 ± 9

[a] Values represent average and standard deviation from 2 independent determinations, each using 8 mice.

Why blood proteins should strongly interact with solid phase saturated PC vesicles is not inherently obvious. The lack of net surface charge and the uniformity of the lipid head group in the liposomes studied suggests that membrane fluidity and fatty acyl chain composition are significant determinants of liposome-protein interactions. It has been demonstrated that homogeneous vesicles composed of gel state phosphatidylcholines in the absence of cholesterol develop packing defects upon cooling below their phase transition temperatures. These defects are thought to expose hydrophobic domains on the surface of the bilayer that increase the contact between water and the hydrophobic fatty acyl chains.

Table 2. ***Influence of cholesterol on the protein binding ability of distearoylphosphatidylcholine (DSPC) LUVs.***

Composition of LUVs	Protein Binding Index (g/mol lipid)
DSPC	96 ± 8
DSPC:CH (9:1)	86 ± 6
DSPC:CH (8:2)	54 ± 5
DSPC:CH (7:3)	23 ± 3
DSPC:CH (6:4)	27 ± 4
DSPC:CH (5:5)	28 ± 4

[a] Values represent average and standard deviation from 2 independent determinations, each using at least 4 mice.

The inclusion of increasing amounts of cholesterol in DSPC LUVs dramatically reduces the protein binding ability (Table 2). Cholesterol is known to eliminate the sharp gel-liquid crystalline phase transition in homogeneous saturated phosphtidylcholine vesicles resulting in a permanent liquid crystalline state at cholesterol concentrations greater than 30 mol%. Thus, the inclusion of cholesterol likely eliminates potential membrane defects and consequently reduces the likelihood of the interaction of blood proteins with these liposomes.

These findings suggest that protein-membrane interactions can be regulated by alterations in the lipid composition of membranes. Exposure of hydrophobic domains in the membrane as a result of fatty acyl packing defects or by inclusion of certain anionic phospholipids leads to membrane surfaces that are highly interactive with soluble proteins. This interaction leads to protein unfolding on the surface of the membranes resulting in stable protein-membrane interactions. The general properties of membrane-reactive proteins, determined from studies involving simple systems, do not for the most part apply to complex protein mixtures.

Acknowledgments

This work was funded through grants from the Medical Research Council of Canada and the National Cancer Institute of Canada.

References

Chonn, A., Semple, S. C. and Cullis, P. R. (1991) Separation of large unilamellar liposomes from blood components by a spin column procedure: towards identifying plasma proteins which mediate liposome clearance in vivo. Biochim. Biophy. Acta 1070, 215-222.

Hope, M. J., Bally, M. B., Webb, G. and Cullis, P. R. (1985) Production of large unilamellar vesicles by a rapid extrusion procedure: characterization of size, trapped volume and ability to maintain a membrane potential. Biochim. Biophys. Acta 812, 55-65.

Horbett, T. A and Brash, J. L. (1987) Proteins and interfaces: current issues and future prospects. In Proteins at Interfaces: Physicochemical and Biochemical Studies (Brash, J. L. and Horbett, T. A., eds) pp. 1-33, American Chemical Society, Washington, D. C.

Nayar,R., Hope, M. J. and Cullis, P. R. (1989) Generation of large unilamellar vesicles from long-chain saturated phosphatidylcholines by extrusion technique. Biochim. Biophys. Acta 986, 200-206.

Senior, J. and Gregoriadis, G. (1982) Stability of small unilamellar liposomes in serum and clearance from the circulation: the effect of phospholipid and cholesterol components. Life Sce. 30, 2123-2136.

Protein Sorting and Glycolipid-Enriched Detergent-Insoluble Complexes in Epithelial Cells *

K. Fiedler, P. Dupree and K. Simons
Cell Biology Programme
European Molecular Biology Laboratory
Heidelberg, Germany

Epithelial cells form a permeability barrier between the external and internal environment of the body and are involved in physiological processes such as secretion, absorption and ion transport (Simons and Fuller, 1985; Rodriguez-Boulan and Nelson, 1989; Wandinger-Ness and Simons, 1991). Simple epithelial cells are organized into a cell monolayer with an apical and basolateral cell surface separated by tight junctions. These two plasma membrane domains do not only contain distinct proteins but they also have different compositions of lipids (van Meer et al., 1987; van Meer and Burger, 1992). The exoplasmic leaflet of the apical cell surface is highly enriched in glycosphingolipids (GSL), presumably providing protection against the external environment. The mechanisms involved in the generation and maintenance of epithelial cell polarity have been extensively studied using MDCK (Madin-Darby canine kidney) cells as a model system (McRoberts et al., 1981; Rodriguez-Boulan, 1983). Apically and basolaterally destined proteins and lipids have been found to be sorted from each other in the *trans*-Golgi network (TGN) and packaged into separate vesicular carriers followed by selective targeting to the cell surface (Griffiths and Simons, 1986; Kobayashi et al., 1992).

Model: Glycosphinglipid-Enriched Rafts as a Sorting-Platform for Apical Proteins

GSLs have been postulated to form dynamic, liquid-crystalline-like clusters within the surrounding fluid phospholipids in the membrane bilayer. This is supposedly due to the unique property of the frequently hydroxyl-modified ceramide backbone and the sugar headgroup moieties to form intermolecular hydrogen bonds (Pascher, 1976; Thompson and Tillack, 1985). This led us to propose that the sorting of both proteins and lipids to the apical cell

* The work was supported by the Boehringer Ingelheim Fonds (K.F.), the Royal Society (P.D.) and SFB 352 of the DFG.

NATO ASI Series, Vol. H 82
Biological Membranes:
Structure, Biogenesis and Dynamics
Edited by Jos A. F. Op den Kamp

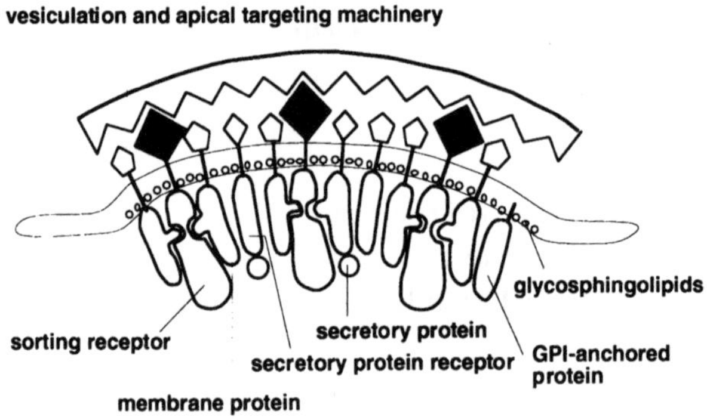

Figure 1: Hypothetical "apical" raft in the *trans*-Golgi network of MDCK cells. The patch will vesiculate into an apical transport vesicle. (Modified from Simons and Wandinger-Ness, 1990).

surface is intimately related (Simons and van Meer, 1988; Simons and Wandinger-Ness, 1990) and mediated by a co-clustering or raft formation of GSLs and apically sorted proteins in the TGN (Fig.1). The inclusion of apical proteins might involve a direct interaction with the GSLs or an association with a putative sorting-receptor bearing an inherent affinity for GSL-enriched rafts, in turn leading to an exclusion of basolaterally destined cargo from the sorting-platform. This process must be regulated and allow the release or mobilization of delivered proteins at the cell surface (Hannan et al., 1993). The machinery components involved might be specific for the apical pathway or effect a similar function in the transcytotic pathway from the basolateral to the apical membrane domain.

Detergents as a Tool to Study Raft Formation

In our attempts to identify components involved in protein sorting in MDCK cells we have been using the zwitterionic detergent CHAPS (3-[(3-cholamidopropyl)dimethylammonio]-2-hydroxypropanesulfonate) to extract TGN-derived-exocytic carrier vesicles isolated from perforated cells. The apical marker protein influenza hemagglutinin (HA) was found to be included into a CHAPS-insoluble complex, yet excluding the basolateral marker VSV (vesicular stomatitis virus) G protein (Kurzchalia et al., 1992). In agreement with our hypothesis, complex formation (analyzed by pulse-chase experiments and CHAPS extraction) did not occur in the earliest stage of the secretory pathway, the endoplasmatic reticulum (ER), but first in the late Golgi-apparatus (Fiedler et al., 1993). Similarly, the apically targeted glycosyl phosphatidylinositol (GPI) anchored protein placental alkaline phosphatase (PLAP) was shown to become Triton X-100 insoluble in the medial/trans Golgi (Fig.2) during its itinerary to the cell surface (Brown and Rose, 1992).

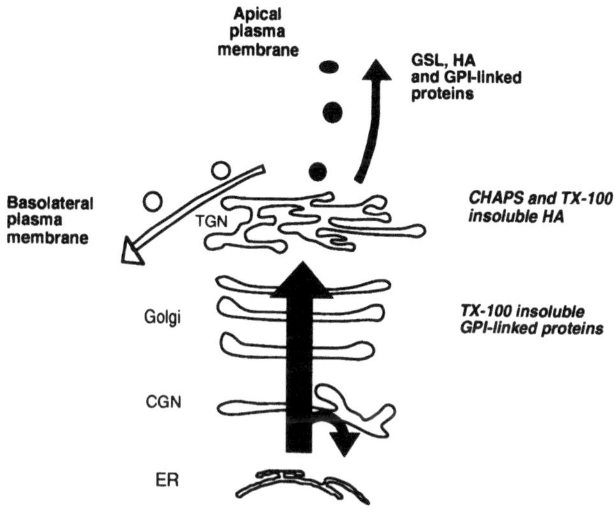

Figure 2: Schematic depiction of the development of detergent-insoluble complexes during passage through the secretory pathway in a polarized epithelial cell.

Protein Composition of the Detergent-Insoluble Complexes

In order to compare the membrane protein composition of the CHAPS-insoluble complex and the Triton X-100 insoluble residue described by Brown and Rose (1992) we used a cellular membrane fraction isolated from metabolically ^{35}S-labeled influenza virus infected MDCK cells (Fiedler et al., 1993). The membranes were either solubilized with 20 mM CHAPS on ice and centrifuged through a 0.9 M sucrose cushion or solubilized with 1% Triton X-114 on ice and floated to the interface of a 1.1/0.15 M sucrose gradient. The insoluble residues were then extracted with 1% Triton X-114 at 37°C and subjected to a temperature-induced phase separation. The detergent phases were analyzed by two-dimensional (2-D) gels (Fig.3).

The protein composition of the detergent-insoluble complexes was very simple compared to the starting material and showed overall similar compositions. Both complexes included influenza HA, but there were large quantitative differences for individual vesicular proteins. The proteins were identified by comparison with the 2-D gel patterns obtained after immunoisolation of carrier vesicles (Wandinger-Ness et al., 1990) and assigned as specifically apical or common to both vesicle types (designated in Figure 3 as A and C, respectively). The CHAPS complex was enriched relative to the Triton complex for VIP21 (*V*esicular *I*ntegral-membrane *P*rotein of *21* kDa) and C14 and contained small quantities of other vesicular proteins such as VIP36

(*V*esicular *I*ntegral-membrane *P*rotein of *36* kDa; previously named C9 in Wandinger-Ness et al., 1990), C11 and C13. In contrast, the latter were very abundant in the Triton-insoluble material. In addition, A26 and C10 were present as minor components in both complexes.

In further experiments we showed that, as expected for cargo proteins, GPI-anchored molecules were present in the Triton-insoluble complex, however, they were only minor components and relatively depleted from the CHAPS-insoluble complex.

Lipid Composition of the Detergent-Insoluble Complexes

The lipid content of the CHAPS-insoluble material was investigated (Fiedler et al., 1993) in order to determine whether glycolipids were enriched, similarly as reported for the Triton-insoluble material, containing a ratio of phospholipids:cholesterol:glycolipids of ~1:1:1 (Brown and Rose, 1992). The overall lipid content was strikingly lower compared to that of the Triton-insoluble membrane fraction, which was in good agreement with the buoyant density of 1.21 g/cm^3 for the CHAPS-insoluble pellet versus 1.08 g/cm^3 for the Triton-insoluble fraction. On the other hand, the lipid composition was very similar; glycerophospholipids were depleted and sphingomyelin was enriched (Table I). Cerebrosides and the ganglioside GM3 were similarly enriched, but in contrast, the Forssman antigen and lactosylceramide, which were enriched in the Triton-insoluble fraction, were depleted from the CHAPS-insoluble pellet. Interestingly, it has been reported that artificial liposomes of the same lipid composition as the Triton-insoluble fraction, containing no protein, are also insoluble in Triton-X 100 (Brown, 1992), suggesting that the lipids themselves are responsible for detergent insolubility. Therefore, Triton may leave

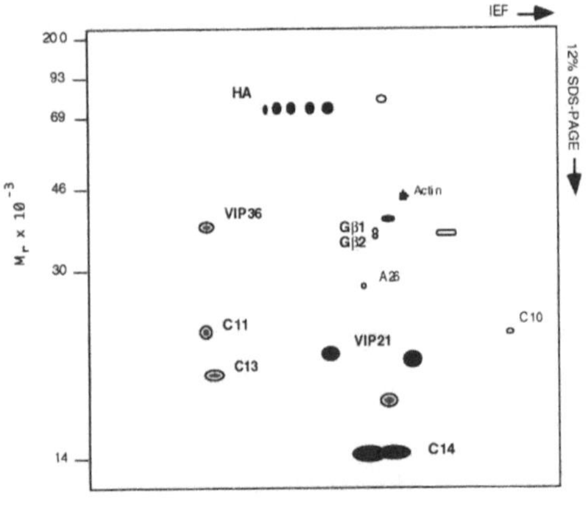

Figure 3: Schematic view of the protein composition of the CHAPS and Triton X-114 insoluble comlexes. Dark shaded forms indicate proteins relatively enriched in the CHAPS-insoluble complex, grey forms denote other vesicular proteins, and unfilled forms represent unknown proteins. An asterisk marks the position of actin. (Modified from Fiedler et al., 1993)

Table I Lipid composition of MDCK cells and the CHAPS-insoluble complex (from Fiedler et al., 1993)

Lipid	MDCK cells nmol/10^8 cells	%	CHAPS complex nmol/10^8 cells	%	Percent Insoluble
Phospholipids	2735		5.8		
Phosphatidylcholine		44.6		9.1	< 0.1
Phosphatidylethanolamine		25.8		7.7	< 0.1
Phosphatidylserine		8.5		7.8	< 0.2
Phsophatidylinositol		6.0		0.5	< 0.1
Cardiolipin		3.0		2.0	< 0.2
Sphingomyelin		12.1		72.9	1.3
Cholesterol	652		5.1		0.8
Triglyceride	63		0.47		0.8
Cerebrosides	22		0.58		2.6
Lactosylceramide	17		0.01		< 0.1
Forssman antigen	19		0.02		< 0.1
GM3	19		0.23		1.2

the putative glycolipid domains in cellular bilayers insoluble. In contrast, the CHAPS-insoluble pellet seems to be more dominated by protein-protein interactions and the small amount of lipid left may be bound to protein. Therefore, CHAPS seems to isolate preferentially proteins interacting with each other from the glycolipid rafts.

Protein Purification and Characterization

To identify proteins present in the detergent-insoluble complexes and other vesicular proteins we used preparative two-dimensional SDS-IEF/SDS-PAGE to resolve MDCK membrane fractions (unpublished results). Coomassie stained spots from up to 40 gels were excised and concentrated using a concentration-elution gel (Rasmussen et al., 1991), followed by in-gel trypsin digestion and amino acid microsequencing. VIP21/Caveolin, isolated from a CHAPS insoluble pellet, has been described earlier and was found to be localized to the Golgi complex, intracellular vesicles, and the cell surface (Kurzchalia et al., 1992). Interestingly, over 90% of the VIP21/Caveolin was concentrated in plasma membrane invaginations called caveolae (Dupree et al., 1993; Rothberg et al., 1992). Furthermore, we have identified the β1 and β2 subunit of heterotrimeric G-proteins as major components of the Triton-insoluble complex (unpublished results). Together with the finding that upon antibody addition the β-adrenergic receptor, a G-protein coupled receptor, colocalizes with VIP21/Caveolin to caveolae, this suggests that molecules involved in signal-transduction processes may be concentrated in glycolipid-enriched rafts. VIP36, a protein found in apical and basolateral exocytic carrier vesicles and the CHAPS and Triton complex has been analyzed in more detail. The encoding cDNA has been cloned from an MDCK cDNA library (unpublished results). The amino acid sequence suggests that VIP36 is a type I integral membrane protein with a N-terminal cleaved signal sequence and a C-terminal membrane anchor with a ten amino acid cytoplasmic tail. The 31 kDa luminal/exoplasmic domain was found to be potentially homologous to leguminous plant lectins (Sharon and Lis, 1990) with a 25% identity within 206 amino acids (unpublished results). Key residues involved in lectin metal and sugar binding are conserved. Also the secondary structure prediction of VIP36 is in good agreement with the known tertiary structure of leguminous plant lectins (Sharon, 1993). In addition, a 42 kDa protein, enriched in apical exocytic carrier vesicles, has been found to be homologous to a previously identified intestine specific annexin (ISA; Wice and Gordon, 1992), thus representing a member of a novel epithelial specific family of annexins, presumably involved in membrane fusion during exocytotic/endocytotic events.

Glycolipid Rafts as a Versatile Mechanism for Segregating Proteins in Membrane Bilayers

A key feature of our working model of protein sorting in epithelial cells is the cooperative sorting of apically destined proteins and glycolipids (Simons and van Meer, 1988; Simons and Wandinger-Ness, 1990). A prediction of this model is that the onset of glycolipid clustering or raft formation should occur in the Golgi, shortly following the biosynthesis of sphingolipids (Schwarzmann and Sandhoff, 1990; van Meer and Burger, 1992). In agreement with this, apical proteins in tranport, e.g. influenza HA (Skibbens et al., 1989; Fiedler et al., 1993) and GPI-linked proteins (Brown and Rose, 1992; Garcia et al., 1993), were found to be included into detergent-insoluble complexes in the Golgi and these were enriched in glycosphingolipids, establishing the relationship between detergent insolubility and glycolipid microdomains.
The total protein composition of the CHAPS- and Triton-insoluble complexes, analyzed by two-dimensional gel electrophoresis, was very simple, and several proteins previously identified in post TGN carrier vesicles from MDCK cells were present. VIP21/Caveolin, the first raft component isolated (Kurzchalia et al., 1992), was surprisingly found not only in the TGN and intracellular vesicles, but was highly concentrated in caveolae on the cell surface (Dupree et al., 1993). Caveolae are structures found in all cell types (Severs, 1988), known to contain GPI-anchored proteins (Anderson et al., 1992) and are possibly enriched in glycophingolipids (Montesano et al., 1982; Parton et al., 1988). Thus, caveolae have many intriguing similarities to glycolipid rafts of apical carrier vesicles (Rothberg et al., 1990; Lisanti and Rodriguez-Boulan; 1991, Brown, 1992). The presence of VIP21/Caveolin in both structures makes a functional connection between them and suggests that glycolipid raft-based sorting is a more general phenomenon that mediates the segregation of proteins into subdomains within the lipid bilayer. Moreover, it implies that, in addition to membranes of the secretory pathway, surface

caveolae are also contributing to the composition of the Triton- and CHAPS-insoluble complexes (Fiedler et al., 1993; Sargiacomo et al., 1993).

The identity of factors governing the specificity of association with apically targeted glycolipid rafts or caveolae remains an open question. Interestingly, the capability of epithelial cells to sort GPI-linked proteins correctly towards the apical cell surface not only correlates with the preferential apical sorting of glycolipids but also with the presence or absence of VIP21/Caveolin (Zurzolo et al., 1993; Sargiacomo et al., 1993). However, since VIP21/Caveolin has no external domain that could bind to GPI-anchored proteins (Dupree et al. 1993) other factors must be involved (cf. Sargiacomo et al., 1993 for an alternative view).

VIP36, a novel component of exocytic carrier vesicles and glycolipid-enriched rafts was found to be potentially homologous to leguminous plant lectins. We speculate that it might bind to sugar residues of glycosphingolipids and/or GPI-anchors. Only further work will show whether VIP36 might provide the missing link between the luminal proteins and/or glycolipids and the cytoplasm and exerts a function in protein sorting in epithelial cells.

Other activites that may involve glycolipid rafts are signal transduction processes (Brown, 1992; Cinek and Horejsi, 1992; Sargiacomo et al., 1993) or the formation of the myelin membrane (Pfeiffer et al., 1993). Although clustering of glycosphingolipids has been demonstrated in several systems (Thompson & Tillack, 1985; Curatolo, 1987; Boggs, 1987; Masserini et al., 1988), more work is necessary to define the specificity of the interactions leading to clustering among lipids and among lipids and proteins and to elucidate their manifold biological functions.

Anderson, R.G.W., Kamen, B.A., Rothberg, K.G. and Lacey, S.W. (1992) *Science*, **255**, 410-411.

Boggs, J.M. (1987) *Biochim. Biophys. Acta*, **906**, 353-404.

Brown, D. and Rose, J. (1992) *Cell*, **68**, 533-544.

Brown, D.A. (1992) *Trends Cell Biol.*, **2**, 338-343.

Cinek, T. and Horejsi, V. (1992) *J. Immunol.*, **149**, 2262-2270.

Curatolo, W. (1987) *Biochim. Biophys. Acta*, **906**, 111-136.

Dupree, P., Parton, R.G., Raposo, G., Kurzchalia, T.V. and Simons, K. (1993) *EMBO J.*, **12**, 1597-1605.

Fiedler, K., Kobayashi, T., Kurzchalia, T.V. and Simons, K. (1993) *Biochemistry*, **32**, 6365-6373.

Garcia, M., Mirre, C., Quaroni, A., Reggio, H. and Le Bivic, A. (1993) *J. Cell Science*, **104**, 1281-1290.

Griffiths, G. and Simons, K. (1986) *Science*, **234**, 438-443.

Hannan, L.A., Lisanti, M.P., Rodriguez-Boulan, E. and Edidin, M. (1993) *J. Cell Biol.*, **120**, 353-358.

Kobayashi, T., Pimplikar, S., Parton, R., Bhakdi, S. and Simons, K. (1992) *FEBS Lett.*, **300**, 227-231.

Kurzchalia, T., Dupree, P., Parton, R., Kellner, R., Virta, H., Lehnert, M. and Simons, K. (1992) *J. Cell Biol.*, **118**, 1003-1014.

Lisanti, M.P. and Rodriguez-Boulan, E. (1990) *TIBS*, **15**, 113-118.
Masserini, M., Palestini, P., Venerando, B., Fiorilli, D., Acquotti, D. and Tettamanti, G. (1988) *Biochemistry*, **27**, 7973-7978.
McRoberts, J.A., Taub, M. and Saier, M.H.J. (1981) In Sato, G. (eds.), *The Madin-Darby canine kidney (MDCK) cell line*. Alan Liss, New York.
Montesano, R., Roth, J., Robert, A. and Orci, L. (1982) *Nature*, **296**, 651-653.
Parton, R.G., Ockleford, C.D. and .Critchley, D.R. (1988) *Brain Res.*, **475**, 118-1127.
Pascher, I. (1976) *Biochim. Biophys. Acta*, **455**, 433-451.
Pfeiffer, S.E., Warrington, A.E. and Bansal, R. (1993) *Trends Cell Biol.*, **3**, 191-197.
Rasmussen, H.H.et al. (1991) *Electrophoresis*, **12**, 873-882.
Rodriguez-Boulan, E. (1983) In *Membrane biogenesis, enveloped RNA viruses, and epithelial polarity*. Alan Liss, New York, Vol. 1.
Rodriguez-Boulan, E. and Nelson, J. (1989) *Science*, **245**, 718-725.
Rothberg, K., Heuser, J.E., Donzell, W.C., Ying, Y.-S., Glenney, J.R. and Anderson, R.G.W. (1992) *Cell*, **68**, 673-682.
Rothberg, K.G., Ying, Y.S., Kamen, B.A. and Anderson, R.G.W. (1990) *J. Cell Biol.*, **111**, 2931-2938.
Sargiacomo, M., Sudol, M., Tang, Z. and Lisanti, M.P. (1993) *J. Cell Biol.*, **122**, 789-807.
Schwarzmann, G. and Sandhoff, K. (1990) *Biochemistry*, **29**, 10865-10871.
Severs, N.J. (1988) *J. Cell Science*, **90**, 341-348.
Sharon, N. (1993) *TIBS*, **18**, 221-226.
Sharon, N. and Lis, H. (1990) *FASEB*, **4**, 3198-3208.
Simons, K. and Fuller, S.D. (1985) *Ann. Rev. Cell Biol.*, **1**, 243-288.
Simons, K. and van Meer, G. (1988) *Biochemistry*, **27**, 6197-6202.
Simons, K. and Wandinger-Ness, A. (1990) *Cell*, **62**, 207-210.
Skibbens, J.E., Roth, M.G. and Matlin, K.S. (1989) *J. Cell Biol*, **108**, 821-832.
Thompson, T.E. and Tillack, T.W. (1985) *Ann. Rev. Biophys. Biophys. Chem.*, **14**, 361-386.
van Meer, G. and Burger, K.N.J. (1992) *Trends Cell Biol.*, **2**, 332-337.
van Meer, G., Stelzer, E.H.K., Wijnaendts-van-Resandt, R.W. and Simons, K. (1987) *J. Cell Biol.*, **105**, 1623-1635.
Wandinger-Ness, A., Bennett, M.K., Antony, C. and Simons, K. (1990) *J. Cell Biol.*, **111**, 987-1000.
Wandinger-Ness, A. and Simons, K. (1991) In Hanover, J. and Steer, C. (eds.), *The Polarized Transport of Surface Proteins and Lipids in Epithelial Cells*. Cambridge University Press, Cambridge.
Wice, B.M. and Gordon, J.I. (1992) *J.Cell.Biol.*, **116**, 405-422.
Zurzolo, C., van't Hof, W., van Meer, G. and Rodriguez-Boulan, E. (1993) *EMBO J.*, in press.

Structural and Functional Consequences of Acylation of a Transmembrane Peptide

T.C.B. Vogt, J.A. Killian and B. de Kruijff

Center for Biomembranes and Lipid Enzymology
Department of Biochemistry of Membranes
Utrecht University
Padualaan 8
3584 CH Utrecht
The Netherlands.

CONTENTS

SUMMARY

In this minireview we summarize our studies on the consequences of acylation for the structure and function of the transmembrane helical peptide gramicidin A. It is shown that acylation of the C-terminal ethanolamine group does not largely affect the polypeptide structure and conformation. However, the lifetime of the cation specific channels formed in black lipid membranes is greatly increased upon acylation. The conformation of the covalently attached acyl chain in bilayers of dimyristoylphosphatidylcholine was analyzed by ^{2}H NMR in oriented systems using palmitoylgramicidin in which the acyl chain was either per- or specifically deuterated. From C8 towards the methyl end the order and dynamics of the covalently

NATO ASI Series, Vol. H 82
Biological Membranes:
Structure, Biogenesis and Dynamics
Edited by Jos A. F. Op den Kamp

attached acyl chains follow closely that of the phospholipids in a liquid crystalline bilayer. In contrast, the carboxyl part of the acyl chain is highly ordered and takes up a defined conformation. The results are discussed in the light of the structural and functional consequences of acylation of transmembrane proteins.

INTRODUCTION

Many proteins contain covalently attached lipids. Glypiation, isoprenylation and fatty acylation appear to be dominant ways by which proteins in their life span can become modified by lipids. Fatty acylation can occur at different locations and in different ways (Schmidt, 1989). Myristoylation via a peptide bond on an N-terminal glycine is often encountered in water-soluble proteins. Palmitoylation is more commonly encountered in membrane proteins and can occur via a thioester bond to a cysteine or via an oxygen-ester bond to a threonine residue. Virtually nothing is known about the structural and functional consequences of acylation of proteins but in view of the different types of acylation, most likely these will be diverse. Fatty acylation can be expected to influence protein folding and localization. This in turn will affect the function of such proteins in cellular processes as has been suggested in case of signal transduction and membrane fusion. Fig. 1 schematically represents the various types of acylation and putative localizations of the fatty acyl residue with respect to the protein and the lipid part of a biomembrane.

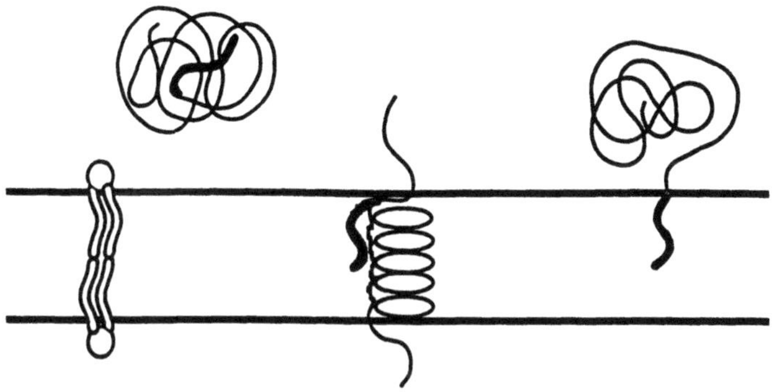

Fig. 1. Schematic representation of the various types of acyl proteins.

To get insight into the consequences of fatty acylation for protein structure and function, we carried out a series of studies using gramicidin A as a model peptide. We selected a transmembrane peptide because we were particularly intrigued in the consequences of palmitoylation of transmembrane proteins, where the fatty acid residue is attached to a part of the protein which is already embedded in the membrane. An example is the extremely hydrophobic surfactant protein Sp-C which is palmitoylated at two cysteines (Beers and Fisher, 1992).

HCO- L-Val- Gly- L-Ala- D-Leu- L-Ala- D-Val- L-Val- D-Val- L-(Trp-D-Leu)$_3$- L-Trp- $NHCH_2CH_2OH$

Fig. 2. Gramicidin in its channel conformation.

The choice of gramicidin is based on a number of considerations: (1) it is a very hydrophobic polypeptide, which spans in its channel conformation the membrane as an N-N $\beta^{6.3}$ helical dimer (Fig. 2); (2) the conformational and channel forming properties of the molecule are well studied and understood; (3) much insight has been obtained in the ways gramicidin can interact with surrounding lipids; (4) the polypeptide occurs in the natural mixture also in an acylated form (gramicidin K), in which the free hydroxyl of the C-terminus ethanolamine group is esterified to fatty acids, and (5) this acylation site is suitable for covalent attachment of acyl chains and its interfacial location mimics the location of the acylation site in palmitoylated membrane proteins. For a recent review on the conformational properties of gramicidin and lipid-peptide interactions of gramicidin the reader is referred to Killian (1992).

EFFECT OF ACYLATION ON THE STRUCTURE AND CHANNEL FUNCTION OF GRAMICIDIN

Gramicidin A isolated from the natural gramicidin mixture could be esterified with a range of fatty acids resulting in a series of acylgramicidins, which differed from the parent molecule only in the presence of an acyl chain esterified to the hydroxyl group of the C-terminal ethanolamine (Vogt et al., 1991). As expected, acylation further increased the hydrophobicity of the polypeptide as reflected in increased retention times in reversed phase chromatography.

Insight into the interfacial behavior and molecular dimensions of the acylgramicidin was obtained by monolayer techniques (Vogt et al., 1991). Compression of a monomolecular layer of gramicidin at the air-water interface results in the

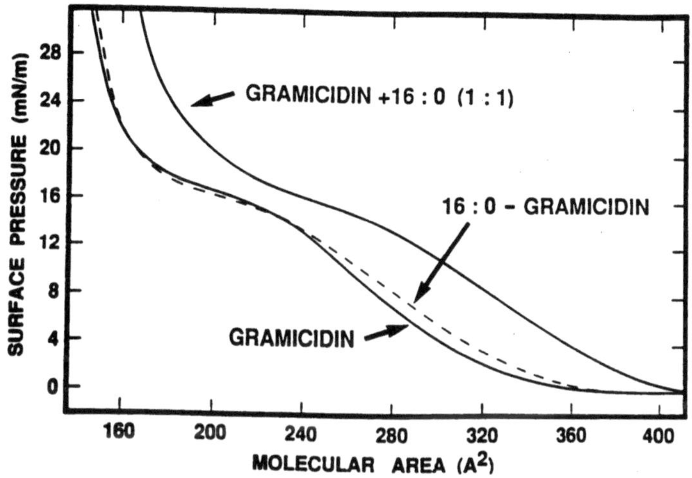

Fig. 3. Force-area curves of the (palmitoyl)gramicidin A at the air-water interface.

force-area curve depicted in Fig. 3. The compression curve has a deflection around 15 mN/m. Surprisingly, the pressure-area curve of palmitoylgramicidin is nearly identical to that of the free peptide. For monolayers of an equimolar mixture of gramicidin and palmitic acid, the force-area curve is shifted to larger areas in accordance with the behavior of the individual molecules.

These findings lead to two suggestions: (1) the interfacial properties of gramicidin are not influenced by palmitoylation and (2) the acyl chain does not contribute to the molecular area and will probably run closely along the peptide molecule where it fills up an existing cavity caused by the protruding tryptophans. The molecular areas at 19 mN/m correspond to the limiting molecular area of gramicidin and are listed for gramicidin and various acylgramicidins and control systems in Table I.

Table I. Molecular areas at the air-water interface and single channel lifetimes (at 100 mV) of (acyl)gramicidins. For details see Vogt et al. (1991, 1992).

System	molecular area (Å^2) at 19 mN/m	channel lifetime (sec)
Gramicidin A	179	0.8
Lauroylgramicidin A	179	3.9
Myristoylgramicidin A	177	4.0
Palmitoylgramicidin A	170	3.7
Stearoylgramicidin A	180	3.0
Oleoylgramicidin A	171	3.6
Gramicidin A + myristic acid[1)]	203	N.D.[2)]
Gramicidin A + palmitic acid[1)]	209	1.0
Gramicidin A + stearic acid[1)]	203	N.D.[2)]

[1)]Equimolar mixtures
[2)]N.D., not determined

The observation that the limiting molecular area is independent of the acyl chain length suggests an orientation of acylgramicidins with their C-terminus towards the aqueous phase, with the acyl chain running parallel to the molecule. The increase in molecular area of 20-30 Å^2 in the non-covalently bonded control situation reflects the area occupied by a free liquid-crystalline acyl chain and emphasizes the large differences in acyl chain organization between the covalently and non-covalently bonded situation. Circular dichroism analysis of (acyl)gramicidins incorporated in a lipid bilayer demonstrated that the $\beta^{6.3}$ helical conformation of gramicidin A (the channel conformation) is also the preferred conformation of the acylated species (Vogt et al., 1991).

The best known property of gramicidin A is its ability to form cation-selective channels in membranes. In black lipid membranes at low gramicidin concentrations single channels can be readily detected as defined rectangular fluctuations in the current upon applying a transmembrane potential (Fig. 4).

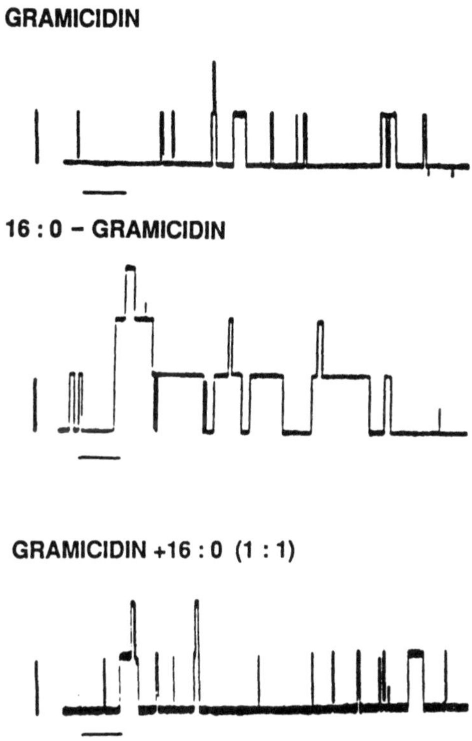

Fig. 4. Single channel recordings of (palmitoyl)gramicidin A.

At an applied voltage of 100 mV the single channels of gramicidin have an average duration of 0.8 seconds (Table I) and a conductance of ~15 pS for the system analyzed (Vogt et al., 1992). In the presence of an equimolar amount of free palmitic acid, the single channel characteristics are hardly influenced but covalent coupling of this and other long chain fatty acids causes a 4 to 6-fold increase in lifetime without strongly affecting the channel's conductance (Vogt et al., 1992). Very similar behavior has been observed for the acylated species isolated from the natural mixture (Williams et al., 1992). The results directly imply that the channels formed by the acyl gramicidins and by gramicidin A are structurally and conformationally equivalent. How then can we explain the increased lifetime of the channel? It is a general believe that channels arise by docking of the N-termini of two monomers in opposing monolayers of a membrane.

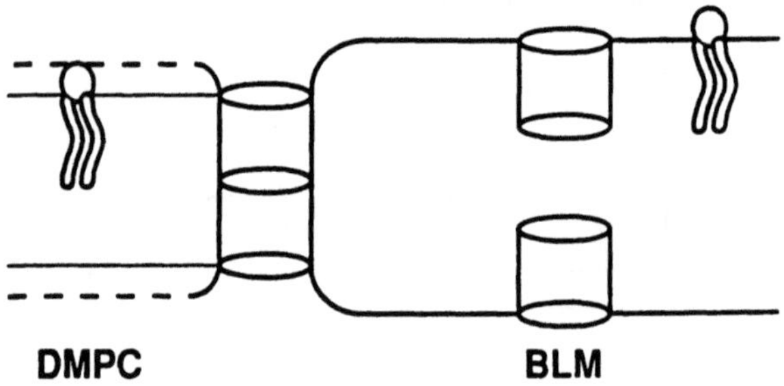

Fig. 5. The gramicidin channel in membranes of different thickness.

Fig. 5 schematically illustrates this in a cartoon which is to scale with respect to the dimensions of the gramicidin monomer and the thickness of the hydrophobic part of a bilayer made of dimyristoylphosphatidylcholine (DMPC) and the thicker black lipid membrane (BLM). In this latter case the favorable energetics of docking (formation of 6 hydrogen bonds in the hydrophobic center of the bilayer) is counterbalanced by the unfavorable energetics of deformation of the bilayer arising from dimple formation around the mouth of the channel. Within such a model the increased lifetime of the channel upon acylation of gramicidin can then be rationalized in terms of a stabilization by the acyl chains of the membrane deformation (Fig. 6).

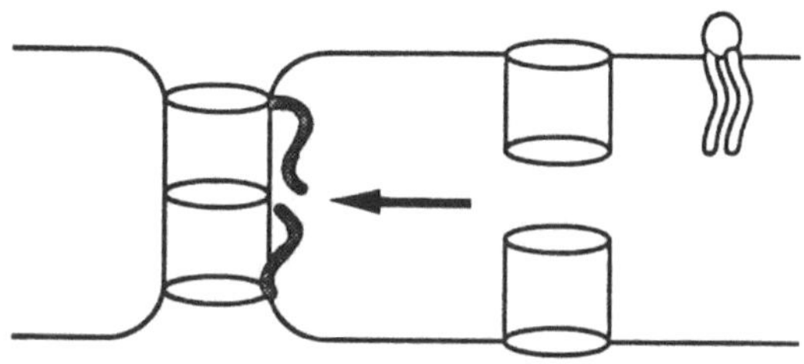

Fig. 6. Acylation of gramicidin favors channel formation by decreasing packing constrains around the channel.

THE STRUCTURE AND DYNAMICS OF THE COVALENTLY ATTACHED ACYL CHAIN

'Does acylation of transmembrane polypeptides alter the features of an acyl chain and thereby give rise to new properties of the molecule?' This is one of the most intriguing questions for this type of acyl proteins. We have addressed this question by synthesizing a series of acylgramicidins in which either at specific positions or at all positions, deuterium was incorporated in the acyl chain (Vogt et al., 1993). ^{2}H NMR of macroscopically oriented bilayers containing these molecules enables one to get quantitative information on the order and the dynamics of the various parts of the covalently attached acyl chain. This is due to the fact that the quadrupolar interaction within the nucleus of a deuteron is only partially averaged by the anisotropic motions which occur in a bilayer. A prerequisite for an accurate analysis of the ^{2}H NMR spectra is that the acyl gramicidin molecule within the bilayer is well oriented and undergoes fast (on the ^{2}H NMR time scale) long axis rotation.

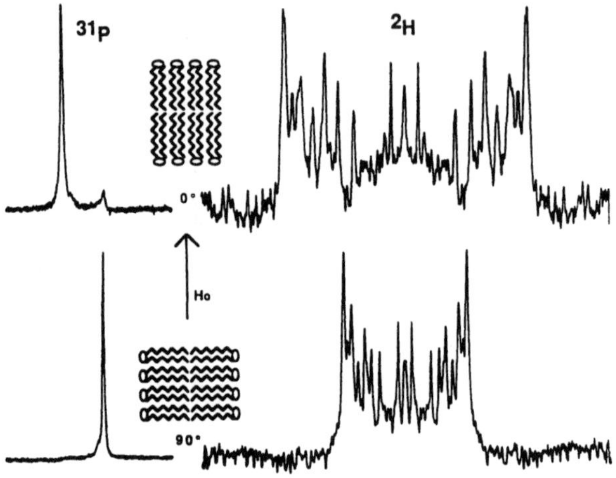

Fig. 7. ^{31}P and ^{2}H NMR spectra of [$^{2}H_{31}$-16:0]-gramicidin in DMPC bilayers oriented between glass plates at angles of 0° and 90° between the normal to the glass plates and the magnetic field H_0. [$^{2}H_{31}$-16:0]-gramicidin:DMPC 1:15 (molar), temperature 40°C.

Fig. 7 demonstrates that this is the case for the d_{31}-16:0 gramicidin/DMPC system oriented between glass plates. The sharp peaks at 29 and -14.5 ppm in the ^{31}P NMR spectrum of the sample oriented at an angle of 0° and 90°, respectively between the glass plates normal and the direction of the magnetic field, prove that the majority of the phospholipids are oriented parallel to this normal and undergo fast axial rotation, typical for a liquid crystalline bilayer. The sharp peaks in the ^{2}H NMR spectrum of the perdeuterated palmitoylgramicidin originate from the different deuterons present in the acyl chain and demonstrate a uniform orientation in the sample. The two-fold larger values of the residual quadrupolar splitting ($\Delta\nu_q$) of all resonances in the 0° orientation spectrum, as compared to the 90° orientation spectrum, demonstrate that the acylgramicidin molecules undergo fast axial rotation, presumably like gramicidin around the helix axis which is aligned parallel to the glass plates normal. Palmitoylgramicidin, deuterated at specific positions of the acyl chain, gives rise to more simple spectra which together with those obtained for the perdeuterated species allow for a quantitative analysis of the conformational and motional properties of the entire palmitoyl chain.

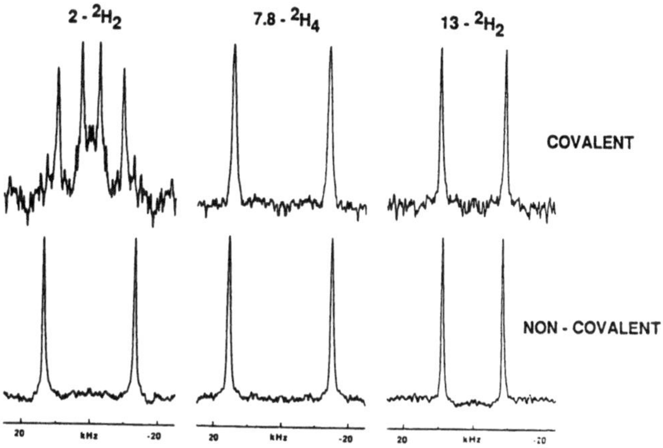

Fig. 8. ^2H NMR spectra of specifically labeled ^2H-16:0 gramicidin A in oriented DMPC bilayers (molar ratio 1:15). Spectra were recorded at 40°C at a 90° orientation of the bilayer normal with respect to the magnetic field. The bottom spectra were obtained for the non-covalent control samples.

Fig. 8 shows as example ^2H NMR spectra recorded at a 90° sample orientation of oriented DMPC bilayers containing $2d_2$, $7,8d_4$- and $13d_2$-16:0-gramicidin (top) and as a control, the equivalent system containing equimolar amounts of the free corresponding fatty acid and gramicidin (bottom). The spectra immediately reveal a number of interesting features. For $13d_2$-16:0 gramicidin a spectrum with a single quadrupolar splitting is observed with a value which is nearly identical to that of the control system. This demonstrates that the 2 deuterons are motionally equivalent and that the acyl chain has an identical order at this position for the two systems. A similar conclusion can be drawn for the 7,8-d4 species which give rise to larger quadrupolar splittings because the chain is more ordered in this region.

Remarkably, at the 2-position a doublet is observed for the covalently attached acyl chain in contrast to the expected singlet for the control situation. This together with the low signal-to-noise ratio and the T1 relaxation behavior (Vogt et al., 1993) demonstrates that this part of the covalently attached acyl chain has a rigid structure thereby making the two deuterons non-equivalent. The data allow to assign order parameters (S_{C-D}) to the individual deuterons using the measured values of $\Delta\nu_q$ (Seelig and Niederberger, 1974). The results are depicted in Fig. 9.

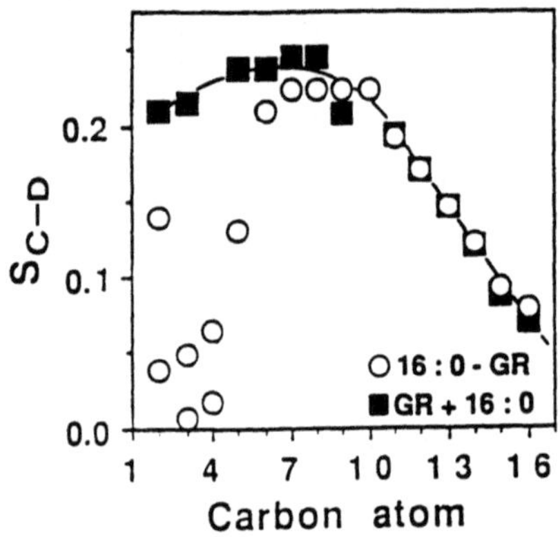

Fig. 9. Order parameter profile of the palmitoyl acyl chain either attached or not attached to gramicidin A present in DMPC oriented bilayers (1:15, molar ratio) at 40°C.

The order parameter profile of the non-covalently bonded control sample is typical for a liquid-crystalline bilayer with a plateau region from C2 to C8 and from there a gradually decreasing order towards the center of the bilayer. From C8 to C16 the order parameters of the covalently attached palmitoyl chain are very similar to the control situation. This then demonstrates that the order of this part of the acyl chain is determined by the surrounding phospholipids and not by the polypeptide which runs immediately next to the acyl chain. Strinkingly different order parameters (smaller values and different values for individual deuterons) are observed for the C2-C5 region of palmitoylgramicidin as compared to the control.

Analysis of the dynamics of the covalently attached palmitoyl chain by ^2H NMR T_1 relaxation measurements confirmed that the methyl part of the acyl chain has motional properties very similar to lipids in a bilayer whereas the carboxyl terminal part of the acyl chain is much more rigid. Because this part is more rigid, the $\Delta\nu_q$ values can be analyzed in terms of torsion angles. The analysis revealed six most probable solutions of the conformation of the part of the acyl chain which is attached to the gramicidin molecule (Vogt et al., 1993). All were characterized by an angle of 39°-49° between a line connecting C1 and C6 and the long axis of the gramicidin molecule. Around C6, the acyl chain will make a bend and will run parallel to the gramicidin molecule. Such conformations allow the acyl chain to run over the channel rim at the C-terminus of the gramicidin molecule. Molecular modelling and ^2H NMR studies of palmitoylgramicidin containing specifically ^2H labeled amino acid side chains (T.C.B. Vogt and R.E. Koeppe II, unpublished observations) suggest that the acyl chain runs in between the side chain of Trp-9 and Leu-10 and close to Trp-9. This is supported by 2D ^1H NMR studies on the naturally occuring species gramicidin K, which suggest close proximity of the acyl chain to Leu-10 (Taylor et al., 1991). The model presented in Fig. 10 is the best fit to all available data and shows to scale the palmitoylgramicidin molecule in half of a DMPC bilayer.

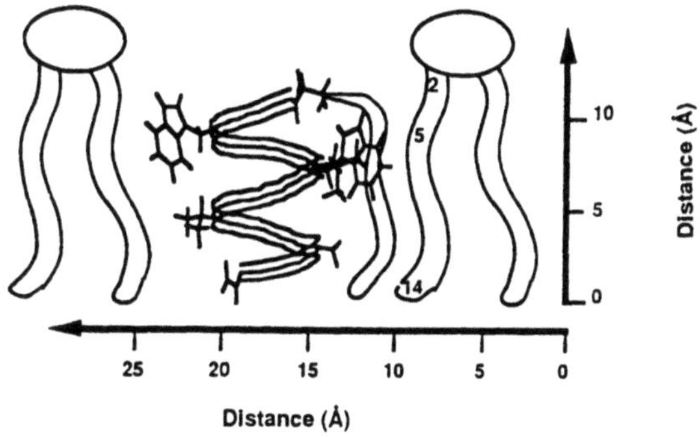

Fig. 10 A schematic representation of a 16:0-gramicidin molecule incorporated in a DMPC bilayer.

We speculate that the bend of the acyl chain is involved in the stabilization of the gramicidin channel discussed above.

The biological Significance

Extrapolation of the results obtained on acyl gramicidin to acylated transmembrane polypeptides leads to a number of suggestions on the function of acylation.

The covalent attachment of the acyl chain to an interfacial residue of the protein can be expected to give rise to a specific conformation of the acyl chain which is different from that of the surrounding lipids. This could enable specific molecular interactions to occur with other membrane components. Alternatively, the acyl chain could locally change the conformational and motional properties of the polypeptide side chains and thereby change polypeptide function or localization. Finally, the theory offered to explain the effect of acylation of gramicidin on channel lifetime suggests that the covalent combination of a rigid transmembrane component (the polypeptide) and a flexible lipid makes it energetically possible for a membrane protein to temporarily or more permanently reside in an environment of high surface curvature (Fig. 11).

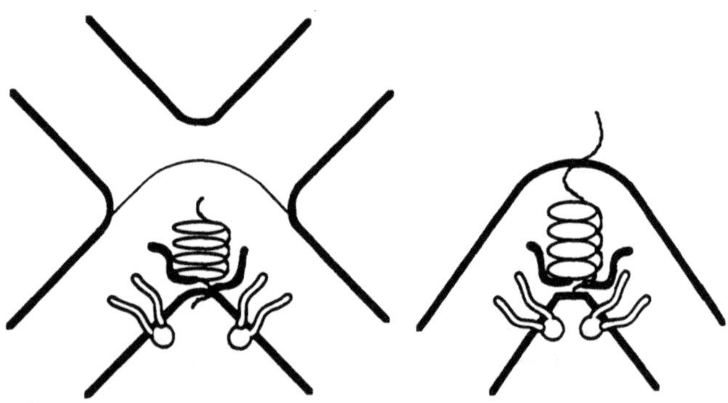

Fig.11 Hypothetical representation of acyl proteins in membranes with highly curved interfaces such as in tubular myelin (left) or in virus envelopes (right).

Such situations often occur in biological processes. Examples include the highly interwoven bilayer network formed by tubular myelin. At the nexus of the intersecting bilayers, a palmitoylated transmembrane protein (such as Sp C) could stabilize the unfavourable lipid packing conditions. Alternatively, in the assembly of viruses, acylation of transmembrane polypeptides could facilitate the formation of the envelope. The small size of the virion and the potentially irregularly shaped nucleocapsid could cause a high bending strain, in particular to the inner monolayer, which could in part be relieved by the presence of the covalently attached acyl chain. In this light it is of interest to note that in viral membrane proteins, the cysteines to which the palmitic acid chain are coupled, are localized at the inner membrane interface (Schultz et al., 1988).

ACKNOWLEDGEMENT

We would like to thank Drs. O.S. Andersen, R.A. Demel and R.E. Koeppe II for their contributions to the work described in this manuscript. Karin Brouwer is thanked for the preparation of the manuscript.

REFERENCES

Beers MF and Fisher AB (1992) Surfactant protein C - A review of its unique properties and metabolism. Am J Physiol 63: 1151-1160

Killian JA (1992) Gramicidin and gramicidin-lipid interactions. Biochim Biophys Acta 1113:391-425

Schmidt MFG (1989) Fatty acylation of proteins. Biochim Biophys Acta 988:411-426

Schultz AM, Henderson LE and Oroszlan S. (1988) Fatty acylation of proteins. Annu Rev Cell Biol 4:611-647

Seelig J and Niederberger W (1974) Deuterium labeled lipids as structural probes in liquid crystalline bilayers. J Am Chem Soc 96:2069-2072

Taylor MJ, Mattice GL, Hinton JF and Koeppe RE II (1991) NMR studies of acylated gramicidin in d6-DMSO solution and d25-SDS micelles. Biophys J 59:319a

Vogt TCB, Killian JA, Demel RA and De Kruiff B (1991) Synthesis of acylated gramicidins and the influence of acyation on the interfacial properties and conformational behavior of gramicidin A. Biochim Biophys Acta 1069:157-164

Vogt TCB, Killian JA, De Kruijff B and Andersen OS (1992) Influence of acylation on the channel characterstics of gramicidin A. Biochemistry 31:7320-7324

Vogt TCB, Killian JA and De Kruijff B (1993) The structure and dynamics of the acyl chain of a palmitoylated transmembrane polypeptide. Biochemistry, submitted

Williams LP, Narcessian EJ, Andersen OS, Waller GR, Taylor MJ, Lazenby JP, Hinton JF and Koeppe RE II (1992) Molecular and channel forming characteristics of gramicidin Ks: a family of naturally occuring gramicidins. Biochemistry 31:7311-7319

NON-CRYSTALLOGRAPHIC METHODS TO STUDY MEMBRANE PROTEINS

Anthony Watts,
Department of Biochemistry,
University of Oxford,
OXFORD, OX1 3QU,
UK

Conventional high resolution NMR methods cannot be applied readily to the study of macromolecular complexes. The complication arises because of the intrinsic slow motion of the whole complex with respect to the applied magnetic field causing extensive line-broadening due to the anisotropy of the magnetic interactions with the applied field. However, the anisotropy of certain nuclei (^2H and ^{31}P) can be exploited and put to good use when studying membrane proteins using solid state, static NMR methods on oriented membranes. In an alternative approach, NMR samples are set spinning at 54.7° (the "magic" angle) to the applied field. Here the anisotropic magnetic interactions are averaged to give high resolution-like NMR spectra, but from very large complexes. Using these approaches, specific molecular details can be obtained for large integral membrane proteins in bilayers, by studying small, labelled molecules or prosthetic groups bound in specific sites within the protein to enable deductions to be made about the protein binding or interaction site.

Solid state, static (^2H & ^{31}P)-NMR

Deuterium is a nucleus which has been exploited in a range of systems. It can substitute for protons without perturbation of the chemistry of the system and has reasonable NMR sensitivity. In a system where no residual motion occurs, the bond vector for a specifically placed CD_3-bond can be deduced from the NMR spectrum.

To exploit this potential, retinal has been deuterated at a number of defined sites (Fig. 1) and incorporated into bacteriorhodopsin (BR). In

NATO ASI Series, Vol. H 82
Biological Membranes:
Structure, Biogenesis and Dynamics
Edited by Jos A. F. Op den Kamp

oriented purple membranes, it has been possible to record ^{2}H-NMR spectra which vary very significantly in shape with orientation of the membrane patches with respect to the applied field. Computer simulation of the NMR spectra, assuming a conal distribution of the methyl groups around the membrane normal and incorporation of line-broadening parameters to account for the mosaic spread and instrumental broadening, permits determination of the bond vectors for the retinal within the protein. The mosaic spread can be deduced rather exactly and independently phosphorus-31 NMR experiments on the same samples, in which the spectra from the lipid phosphates give a chemical shift anisotropy which is very sensitive to the orientation of the sample with respect to the membrane normal in a well defined way. The line-broadening which occurs as a direct result of mosaic spread of the whole sample, can then be determined and put into the deuterium NMR spectral simulation. This then leaves only one line-broadening parameter to be included in the simulations. The orientation of the CD_3-bond can be now be deduced rather accurately. Indeed, a difference of 1kHz, out of a range of between 0 and 40kHz (ie: 1 part in 40), gives an angular accuracy of 1 -2° in the bond vector orientation. Such accuracy is not readily attainable by many other methods, except high resolution X-ray methods.

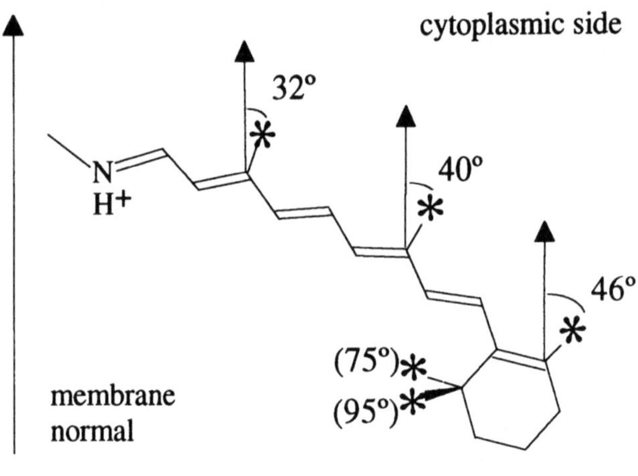

Figure 1. *Retinal, showing the sites of specific deuteration as used in deuterium NMR studies of bacteriorhodopsin containing the labelled chromophore. The bond vectors have been measured by an ab initio approach, to reveal a slightly curved polyene chain when the chromophore is in the protein, and a 6-(S)-trans cyclohexene ring.*

The structure of the retinal within the protein has now been resolved (Fig. 1) using these approaches to reveal a slightly curved polyene chain and a 6-(S)-trans cyclohexene ring lying approximately perpendicular to the membrane surface, which confirms information gained from other methods. The chromophore is tilted along its length, beginning at about 23° with respect to the membrane normal, and finishing at about 45° to the normal at the ring end of the retinal (Ulrich and Watts, 1993; Ulrich *et al.*, 1992). The methods now offers potential for studying small molecules within proteins which can also be oriented into patches. The number of cases where this is possible may be limited at the moment, but significant effort is being invested in forming 2D-arrays of membrane proteins for electron microscopic structural studies, and such sample are ideal for this kind of NMR.

In another approach to the study of protein structure, we have selectively deuterated cytochrome *c* at the following sites:

- all 19 lysines,

- all the exchangeable amides have been replaced with deuterons;

- the 3 histidines have been deuterated; and,

- met-65 and met-80 have been deuterated.

The deuterium NMR spectrum from amide exchanged lyophilized protein is broad and typical for a structure with amide deuterons in a helical structure. This protein and its associated non-exchanging deuterons is quite stable in solution and as the recrystallized or amorphous protein. However, as soon as the protein is brought into contact with anionic cardiolipin bilayers, the amide deuterons are released into the aqueous phase. This implies that the protein unfolds to some degree on interaction with the membrane (Spooner and Watts, 1991a).

The ^{2}H-NMR spectra from the labelled lysine residues of the protein when bound to the membrane is very narrow and not typical for a protein which is constrained by a bilayer membrane (Spooner and Watts, 1991a). This suggests a high degree of local motion around the labelled groups. This motion cannot be due to whole protein motion in solution or in the membrane since the labelled histidines give a spectrum typical of a restrained protein. It therefore appears that the protein can undergo large motional excursions at its periphery as well as in the secondary structural parts, but still retain a membrane bound state.

Further insight into the protein-lipid association come from ^{31}P-NMR studies of the relaxation properties of the cardiolipin phosphate groups Spooner and Watts, 1991b). Unusually for this nucleus, although observed in a limited number of other cases, the ^{31}P-NMR spin-lattice relaxation time as a function of temperature, shows a minimum at 25 °C for protein-free bilayers. On addition of protein, whether in the oxidized or reduced form, the relaxation of the phosphate nucleus is enhanced by up to 100-fold. This suggests that the heme group is enhancing the phosphate relaxation through a direct paramagnetic interaction.

In summary, the release of otherwise stable amide protons (deuterons) into solution from a folded protein on interaction suggests a highly dynamic reversible unfolding and refolding of the protein at the bilayer surface. Some surface groups become very mobile during interaction with the membrane surface and the heme cleft opens sufficiently to permit access to lipid phosphates; the paramagnetic enhancement of the phosphate relaxation is due to the induction of appreciable low-lying short-lived electronic excited states for both states of the protein, probably involving the ligand perturbation of met-80, as shown in other NMR experiments in which this group is ^{13}C labelled (see below).

Magic angle spinning (MAS), solid state ^{13}C-NMR

In an alternative approach to the static solid state NMR method, NMR samples are set spinning at 54.7° (the "magic" angle) to the applied field. Here the anisotropic magnetic interactions are averaged to give high resolution-like NMR spectra from very large complexes which would normally not be accessible by high resolution NMR (Smith and Griffin, 1988).

In a study on cytochrome *c*, the complex of protein bound to bilayers gives a broad, ill-resolved ^{13}C-NMR spectrum when observing natural abundance carbon-13 by conventional high resolution NMR. However, when the sample is subjected to MAS NMR methods, by spinning at a speed of > 2.5kHz, the resonances from the lipid chains (Spooner and Watts, 1992) and specifically labelled protein residues, met-65 and met-80, are distingushable

(Fig. 2). By chosing pulse conditions (using cross polarization) in the NMR experiments which can discriminate slow and fast molecular motions in the complex, it is possible to observe a differential interaction with the lipid acyl chains by cytochrome *c*, with a reduced mobility for the glycerol, carbonyl and methyl groups, in decreasing order of perturbation, and a significant increase in motion for the olefinic carbons of the acyl chains (Spooner and Watts, 1992). This observation therefore indicates that the protein may indeed penetrate the bilayer to some degree, and that this penetration does not last for an appreciable time on the NMR time-scale (ie: the protein is in the bilayer for less than ms).

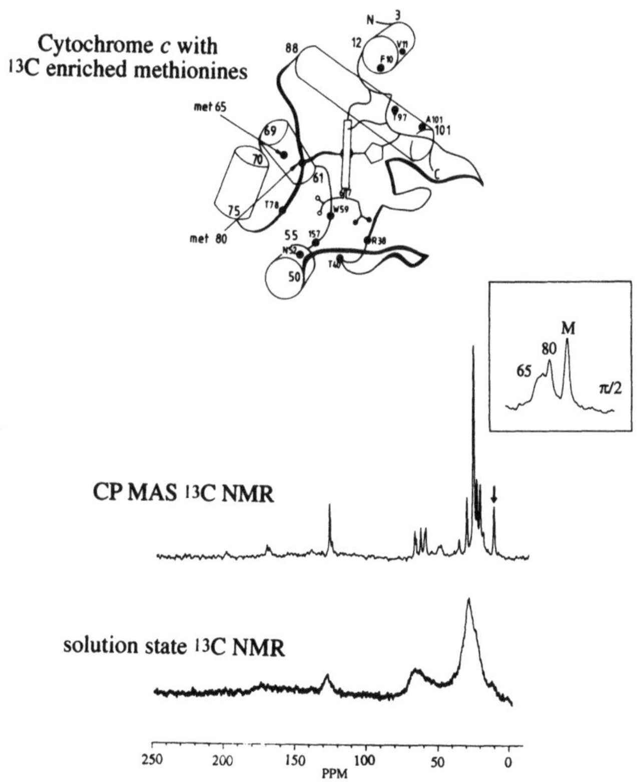

Figure 2. *Diagrammatic respresentation of the structure of cytochrome c from X-ray studies (above). The NMR spectra recorded conventionally by static, high resolution 13C-NMR (lower spectrum) showing the lack of resolution which can be obtained when then sample is studied by MAS NMR (upper spectrum) The resonances from isotopically labelled met-60 and met-80 are shown by the arrow and highlighted in the insert.*

The MAS NMR spectra for ^{13}C labelled met-80 and met-65, show that the met-80 is much more influenced by the interaction with the cardiolipin bilayers than the met-65, which is in one of the α-helices of the protein. Such a broadening is due to the increased motion in this residue when free from the heme ligand, whereas the met-65 is constrained on the helix.

Further studies using this approach have been carried out with a large integral protein, the sugar transporter from *E.coli*. GalP (54kDa; 12 - 14 transmembrane passes) is expressed in membranes to > 55% of the total membrane protein and will transport glucose, which has been labelled with 13C in the C1 position and can be observed by MAS NMR methods; conventional NMR would give no resolved resonances. In the solid and when in the protein binding site, the glucose can be observed at a level of just 250 nmoles. In addition, it is possible to resolve the predominance of the β-enantiomeric form, which implies that this is the form of the sugar in the binding site. One big advantage of the MAS NMR approach (Fig. 3), is that the free, isotropically moving sugar molecules, which are not constrained, are invisible in the NMR experiment, unlike in the high resolution approach in which the opposite is true and the free molecule often dominates the spectrum.

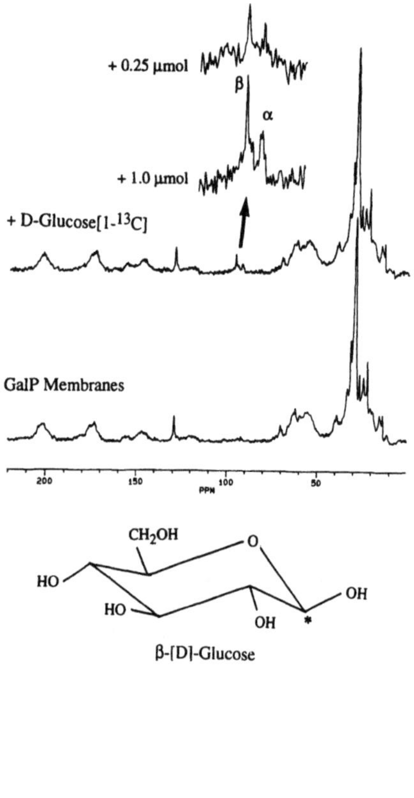

Figure 3. *Magic angle spinning, natural abundance ^{13}C NMR spectrum of E.coli plasma membranes in which the sugar transporter, GalP has been expressed at amplified levels of up to 55% of the total membrane protein (lower spectrum). When specifically labelled ^{13}C glucose is added to the sample, only that sugar which is bound in the protein is observed (upper spectrum) at ~90 - 100ppm, with the isotropically tumbling, non-bound sugar being invisible. The predominance of the β-enantiomer of the sugar in the binding site is seen from the higher peak height for this resonance.*

When the protein is loaded with ^{13}C-glucose, it is possible to add a non-competing sugar (L-galactose) and observe no competition with the labelled glucose, but then add either D-galactose, of inhibitors such as forskolin or cytocalaisin, and displace the glucose (Spooner *et al.*, 1993). This proves that the glucose really is bound to the active sugar binding site and is not in some other, solid environment in the NMR tube.

Finally, we have studied an analogue of the specific inhibitor of the H^+/K^+-ATPase, SCH 28080, which has been labelled in two different places with ^{13}C (Fig. 4). By now using a modification of the MAS NMR approach, called rotational resonance, which introduces the averaged dipolar couplings back into the NMR spectrum in a controlled way, the distance between the two labels can be determined rather precisely. For the solid compound, the ^{13}C-^{13}C distance has been determined to be 4.1 ± 0.3Å, which permits some constraints about the molecule to be introduced in model building.

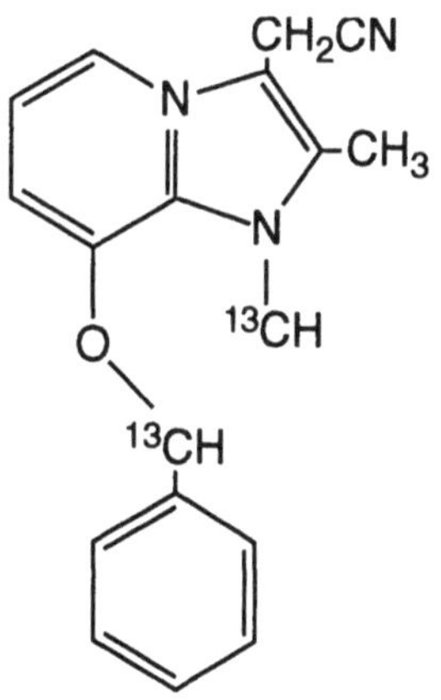

Figure 4. *SCH 28080, a specific inhibitor of the gastric H^+-K^+-ATPase showing the places where ^{13}C atoms have been introduced to enable rotational resonance MAS ^{13}C-NMR experiments to be performed to deduce the ^{13}C-^{13}C distance in the solid (D. Middleton, D.Phil thesis, 1993).*

Now the plan is to perform the experiment when the molecule is bound to the protein in the membrane, and investigate the conformation of the small molecule in its binding site. With one more labelled pair, it should be possible to give a fairly precise conformation of the binding site, and fit this to the parts of the protein chain which are thought to be involved in binding from photoaffinity labelling experiments.

CONCLUSIONS

In summary, it is possible now to investigate the conformation and geometry of small molecules bound into the binding site of large integral protein in membranes. This approach should permit some rational ideas about drug design, receptor-ligand interactions and protein binding sites to be gained using the power of NMR, where crystallography has so far not been successful.

ACKNOWLEDGEMENTS

The work described here has been supported by SERC (UK), MRC (UK), EC, and Wellcome Foundation. Valuable comments on the manuscript from Anne S. Ulrich and Paul J.R. Spooner are gratefully acknowledged.

REFERENCES

Smith SO and Griffin RG (1988) Ann. rev. Phys. Chem. 39 : 511-535.

Spooner PJR and Watts A (1991a) Reversible unfolding of cytochrome c upon interaction with cardiolipin bilayers. 1. Evidence from deuterium NMR measurements. Biochemistry 30 : 3871-3879.

Spooner PJR and Watts A (1991b) Reversible unfolding of cytochrome c upon interaction with cardiolipin bilayers. 2. Evidence from phosphorus-31 NMR measurements. Biochemistry 30 : 3880-3885.

Spooner PJR and Watts A (1992) Cytochrome c interactions with cardiolipin in bilayers: A multinuclear magic-angle spinning NMR study. Biochemistry 31 : 10129-10138.

Spooner PJR, Rutherford N, Watts A and Henderson PJF (1993) NMR observation of substrate in the binding site of an active sugar H^+ symport protein in native membranes. PNAS (submitted).

Ulrich AS and Watts A (1993) ^2H NMR lineshapes of immobilized uniaxially oriented membrane proteins. Solid state NMR 2 : 21-36.

Ulrich AS, Heyn MP and Watts A (1992) Structure determination of the cyclohexene-ring of retinal in bacteriorhodopsin by solid-state deuterium-NMR. Biochemistry 31 : 10390-10399.

Preferred Local Conformations of Peptides comprising the Entire Sequence of Bovine Pancreatic Trypsin Inhibitor (BPTI) and their Roles in Protein Folding.

Johan Kemmink and Thomas E. Creighton
European Molecular Biology Laboratory
Meyerhofstrasse 1
D-6900 Heidelberg
Germany

Unfolded single-domain globular proteins generally fold spontaneously into their native three-dimensional structures. As a consequence, the three-dimensional fold of a polypeptide chain must be determined by its primary structure, the amino-acid sequence. Considering the time scale of the folding process, which is of the order of 10^{-1}-10^{+2} s for small proteins, folding is unlikely to proceed by a random sampling of all possible conformations. The folding pathway of a protein is determined by (1) the nature of the initial unfolded protein under refolding conditions and (2) the kinetic intermediate states through which the protein molecules pass transiently when folding.

The true kinetic intermediates in protein folding are not usually populated to a substantial extent, for there is usually not discernable a lag period in appearance of the native protein, during which the steady-state concentration of a true kinetic intermediate would be generated (Creighton, 1988a, 1990). Such kinetic intermediates undoubtedly occur during protein folding, but they would be expected to be less stable than the unfolded protein, and therefore to unfold more rapidly than they complete folding, or to occur after the rate-determining step, but this is believed generally to occur late in folding (Creighton, 1988a, 1990). Such folding intermediates can be detected and characterized only if folding is coupled to a phenomenon like disulphide bond formation (Creighton, 1978, 1990).

It is difficult to characterize in detail an unfolded protein, especially under folding conditions, where it is only transient, but peptide fragments are now recognized as providing valuable information (Dyson *et al.*, 1988a,b, 1992a,b; Shortle, 1993). The absence of the remainder of the protein prevents acquisition of the native conformation, makes it possible to study the conformational properties of the peptide, both kinetically and at equilibrium, and simplifies the structural analysis. Any nonrandom conformation present in a peptide fragment that is adopted rapidly, relative to the rate of refolding of the entire protein, should be present in the unfolded protein also, unless it is disrupted by alternative, energetically more favoured

NATO ASI Series, Vol. H 82
Biological Membranes:
Structure, Biogenesis and Dynamics
Edited by Jos A. F. Op den Kamp

conformations involving the remainder of the polypeptide chain.

The best-characterized protein folding pathway is that involving disulphide bond formation in reduced bovine pancreatic trypsin inhibitor (BPTI†) (Creighton, 1978, 1990, 1992). Native BPTI contains three disulphide bonds (between Cys5 and Cys55, Cys14 and Cys38, and Cys30 and Cys51), which stabilize the native conformation to varying extents. The most productive pathway of folding of BPTI going from the unfolded fully reduced state (R) to the folded fully oxidized state (N) is:

R ---> (30-51) ---> (30-51,5-14) or (30-51,5-38) ---> $(30\text{-}51,5\text{-}55)_N$ ---> N

where the intermediates are designated by the residue numbers of the cysteine residues paired in disulphide bonds.

Detailed structural information is available for the fully folded protein (Deisenhofer & Steigemann, 1975) and for all the intermediates present in the above scheme (van Mierlo *et al.*, 1992, 1993a,b; States *et al.*, 1987). The native structure of BPTI can be considered to consist of four roughly parallel segments of polypeptide chain, residues 1 to 15, 16 to 28, 29 to 44, and 45 to 58, packed together to form a very stable core of hydrophobic sidechains (Figure 1; Deisenhofer & Steigemann, 1975). The first segment is irregular, but contains one turn of 3_{10}-like helix at residues 3 to 7. The second and third segments each comprise a single β-strand, residues 18 to 24 and 29 to 35, respectively, of the β-sheet of the protein, with a type I β-turn between them. The fourth segment includes three turns of α-helix, from residue 47 to 56. The other residues have much less regular conformations, and the residues at the N- and C-termini are more flexible and tend to adopt multiple conformations in solution (Wagner *et al.*, 1987; van Mierlo *et al.*, 1991a).

The intermediates (30-51), (30-51,5-14) and (30-51,5-38) all have similar partially folded conformations. The folded portions of these molecules consist of the β sheet and the C-terminal α helix packed together to form a hydrophobic core as in native BPTI. The other segments of the polypeptide chain are unfolded or much more flexible than in native BPTI (van Mierlo *et al.*, 1993a,b). The two-disulphide intermediate $(30\text{-}51,5\text{-}55)_N$ is characterized by a native-like conformation, designated by the symbol N.

---

† Abbreviations used: BPTI, bovine pancreatic trypsin inhibitor; CD, circular dichroism; COSY, 2D scalar correlated spectroscopy; DQ, double quantum spectroscopy; DQF, double quantum filter; NMR, nuclear magnetic resonance; NOE, nuclear Overhauser effect; NOESY, 2D NOE spectroscopy; ROESY, 2D rotating frame Overhauser effect spectroscopy; TOCSY, total correlation spectroscopy; $P_{i\text{-}j}$, peptide corresponding to residues i and j of BPTI; $P_{i\text{-}j}$ (XkY), peptide $P_{i\text{-}j}$ with amino acid X at position k replaced by Y; p.p.m., parts per million; δ, proton chemical shift; 1D, one-dimensional; 2D, two-dimensional.

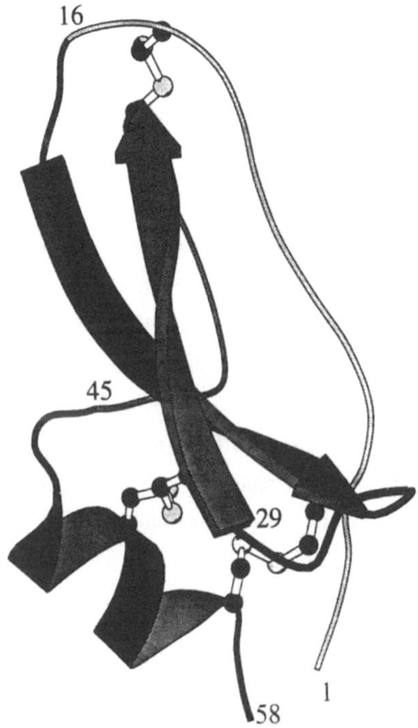

Figure 1. Schematic illustration of the folding topology of the native conformation of BPTI (Deisenhofer & Steigemann, 1975). The peptides P_{1-15}, P_{16-28}, P_{29-44} and P_{45-58}, representing the individual major structural segments, are indicated. The drawing was created using the program MOLSCRIPT (Kraulis, 1991).

Relative to the detailed structural knowledge of native BPTI and of all the major disulphide intermediates, very little is known about the nature of the initial fully reduced protein. It appears from many of its properties to be very unfolded and to approximate a random coil, even in otherwise folding conditions (Creighton, 1988b), although fluorescence energy transfer measurements indicate otherwise (Amir & Haas, 1988; Gottfried & Haas, 1992). To search for examples of local interactions that might be significant for the initial folding events in reduced BPTI, and to characterize its conformational properties further, the conformations of overlapping peptides P_{1-15}, P_{13-21}, P_{16-28}, P_{24-32}, P_{29-44}, P_{41-51}, and P_{45-58} of the entire BPTI sequence have been examined by two-dimensional NMR techniques and by circular dichroism. The various local interactions discovered were characterized by replacing the responsible residues in peptides designated as P_{i-j} (X*k*Y). The results are compared with the corresponding spectra of reduced BPTI and with the conformations of the native protein and of the disulphide intermediates in refolding. In this way, it is possible to elucidate the roles in folding of the various local interactions.

MATERIALS AND METHODS

Sample preparation

The peptides were synthesized using 9-fluorenyl-methoxycarbonyl chemistry and purified to > 95% homogeneity by reversed phase high performance liquid chromatography. The normally internal N and C termini were blocked with acetyl and amide groups, respectively. The normal terminal groups of BPTI were left unblocked in the peptides P_{1-15} and P_{45-58}. Due to an error during synthesis peptide P_{29-44} had a free amino terminus, but an amide group at the C terminus. Cysteine residues at positions 5, 14, 30, 38, 51 and 55 were replaced by serine to avoid problems from oxidation of cysteine residues.

Reduced BPTI was prepared by reduction of native BPTI (Trasylol®, Bayer AG) in 6 M guanidinium chloride, 10 mM dithiothreitol, 0.1 M Tris-HCl (pH 8.7), 1 mM EDTA, followed by isolation by gel filtration and lyophilization.

All NMR measurements were performed on samples containing 4 to 8 mg of peptide dissolved in 90% ^1H_2O/10% ^2H_2O (v/v). The pH was adjusted to 4.6 by adding small amounts of KOH or HCl solutions. Reduced BPTI with free cysteine thiols was at a concentration of 3 mM and at pH 3.5 for solubility reasons, and included 5 mM dithiothreitol to keep the protein reduced.

^1H NMR spectroscopy

^1H NMR spectra of all the peptide were recorded on Bruker AMX-500 and AMX-600 spectrometers using a spectral width of 10.0 p.p.m. The ^1H resonances were assigned using DQF-COSY (Rance *et al.*, 1983), TOCSY ($\tau_m \approx 60$ ms) (Braunschweiler & Ernst, 1983) and NOESY ($\tau_m = 300$ ms) (Jeener *et al.*, 1979) spectra recorded at 271 K. In the case of tetrapeptides, ROESY ($\tau_m = 300$ ms) spectra (Bothner-By *et al.*, 1984) were recorded rather than NOESY. Proton chemical shifts at 271 K were determined using 1D and DQF-COSY spectra. Amide proton temperature coefficients were calculated from a least-squares fit of amide proton chemical shift *vs* temperature obtained from a series of DQF-COSY spectra recorded at 10 degree intervals from 271 to 301 K. On reduced BPTI, DQ spectra (Wagner & Zuiderweg, 1983) and TOCSY spectra were also recorded at 283K.

Circular dichroism spectroscopy

Ultraviolet CD spectra were recorded using a Jobin-Yvon CD6 spectrometer calibrated with (+) 10-camphorsulphonic acid. Peptide concentrations were determined by the measurement of tyrosine absorbance ($\varepsilon_{280} = 1280\ M^{-1}\ cm^{-1}$; Gill & von Hippel, 1989) in the case of peptides P_{1-15}, P_{13-21}, P_{16-28} and P_{29-44} and by measurement of phenylalanine absorbance ($\varepsilon_{257} = 197\ M^{-1}\ cm^{-1}$; Gill & von Hippel, 1989) in the case of peptides P_{41-51} and P_{45-58}.

RESULTS AND DISCUSSION

General considerations

NMR spectra of folded proteins are characterized by substantial dispersion of the chemical shifts of the protons, reflecting their unique environments in the three-dimensional folded structure. The observed backbone NOE patterns in folded proteins are largely determined by the secondary structure: (1) strong $C^{\alpha}H(i)/NH(i+1)$ NOEs and weak or absent $NH(i)/NH(i+1)$ NOEs for extended conformations and (2) strong $NH(i)/NH(i+1)$ NOEs and weak or absent $C^{\alpha}H(i)/NH(i+1)$ NOEs in the case of turn or helical conformations. Turns and helices also give rise to medium-range NOEs of the type $C^{\alpha}H(i)/NH(i+x)$, where $2 \leq x \leq 4$. Hydrogen bonding interactions involving backbone amide protons protect them against exchange with the solvent, which can be measured directly or via a low temperature-dependence of the δ_{NH}. The overall fold of the protein is characterized by long-range NOEs between groups distant in the covalent structure, but close in space.

In contrast, unstructured peptides are generally characterized by the occurrence of relatively strong $C^{\alpha}H(i)/NH(i+1)$ NOEs exclusively; $NH(i)/NH(i+1)$ NOEs are very weak or absent. Normally, no medium range NOEs or NOEs between groups distant in the covalent structure are observed (Dyson & Wright, 1991). The ^{1}H chemical shifts of chemically identical groups are poorly dispersed and are close to their 'random coil' values (Wüthrich, 1986). The amide proton chemical shifts are very temperature dependent, indicating a low level of solvent protection. Deviations from this behaviour are indicative of nonrandom conformational preferences of a peptide. The observed NOE patterns and the measured NMR parameters are often not indicative of a single type of secondary structure, but are the result of multiple conformations in rapid equilibrium. Indicative of such behaviour is the simultaneous presence of NOE cross-peaks typical of extended and turn or helical conformations. Interpretation of such average NMR data in terms of structure is difficult, because *a priori* no detailed information is available about the composition of the ensemble giving rise to a observed NOE pattern. Consequently, it is generally possible to make only qualitative interpretations in terms of tendencies to adopt the various types of backbone conformations (Dyson & Wright, 1991). Nevertheless, preferences for a certain type of structure adopted by a peptide can be derived from the characteristics of the CD spectrum, the observed backbone NOE patterns, the values of the ^{1}H chemical shifts and the values of the NH proton temperature coefficients.

^{1}H *assignments*

^{1}H resonance assignments for the peptides were made using TOCSY spectra for the identification of scalar coupled spin systems and NOESY (or ROESY) spectra for the identification of the sequential $C^{\alpha}H(i)/NH(i+1)$ connectivities. This procedure was very straightforward in all cases, except for the proline containing peptides, where multiple spectra were observed due to the occurrence of *cis-trans* isomerism of the X-Pro peptide bond. The spectra of peptide $P_{1\text{-}15}$ containing four proline residues were particularly complicated by this phenomenon. ^{1}H chemical shifts and the amide proton temperature coefficients of all the major peptides spanning the entire BPTI sequence are given in Kemmink *et al.* (1993) and Kemmink & Creighton (1993).

Conformational preferences of the polypeptide backbone

All the peptides studied here gave similar CD spectra typical of random coils, with only a minimum at less than 200 nm, except for $P_{45\text{-}58}$, which had an additional weak minimum at about 225 nm (Figure 2). The NMR spectra of the peptides confirmed that they were largely unfolded: The observed NOE patterns (Figure 3) are characteristic of a predominance of random coil-like conformations, while the proton chemical shifts are in almost all cases close to their expected random coil values (Wüthrich, 1986). On the other hand, the peptides did not give the NMR spectra expected for totally random coils. Instances of local interactions involving side chains were observed (see below), and the different backbone NOE patterns observed under identical conditions (Figure 3) indicate somewhat different average conformational preferences of the polypeptide backbone.

No 3_{10} helical conformation was apparent in $P_{1\text{-}15}$ from the expected $C^{\alpha}H(i)/NH(i+2)$ NOE cross peaks (Kemmink *et al.*, 1993), not even at residues 3 to 7. On the other hand, NOE cross peaks were observed between the amides of Asp3 and Phe4 and between the side chains of Asp3 and Leu6. Similar NOEs are also observed with the native protein, suggesting a tendency for this peptide segment to adopt a native-like conformation locally.

In NOESY spectra of $P_{13\text{-}21}$, cross peaks were observed between the $C^{\alpha}H$ of Ser14 and the NH of Ala16, as well as a sequential NH/NH contact between Lys15 and Ala16. This indicates a substantial preference for nonrandom conformations of these residues, but does not define their nature. In native BPTI, this segment is largely extended and part of the trypsin binding site.

A preference for extended conformations was apparent in the peptide $P_{16\text{-}28}$: 1, strong $C^{\alpha}H(i)/NH(i+1)$ NOE cross peaks were observed for the entire segment; 2, only a few weak $NH(i)/NH(i+1)$ were observed; and 3, no significantly lowered δ_{NH} temperature coefficients were measured. Residues 18 to 24 comprise one β-strand in native BPTI.

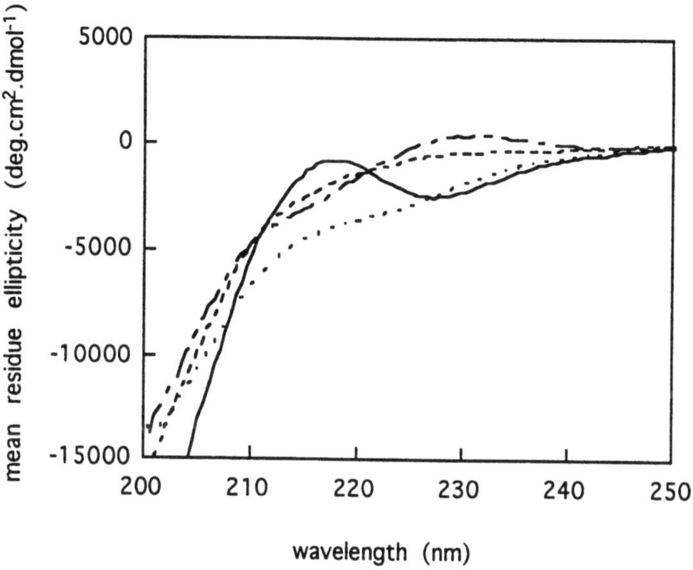

Figure 2. Representative CD spectra of peptides and of reduced BPTI (---) in 1 mM phosphate buffer (pH 7.0) at 0°C. The spectra of P_{16-28} (— - —) and P_{29-44} (– – –) are typical of the peptides studied here, except for P_{45-58} (——). None of the peptides had substantial CD at longer wavelengths.

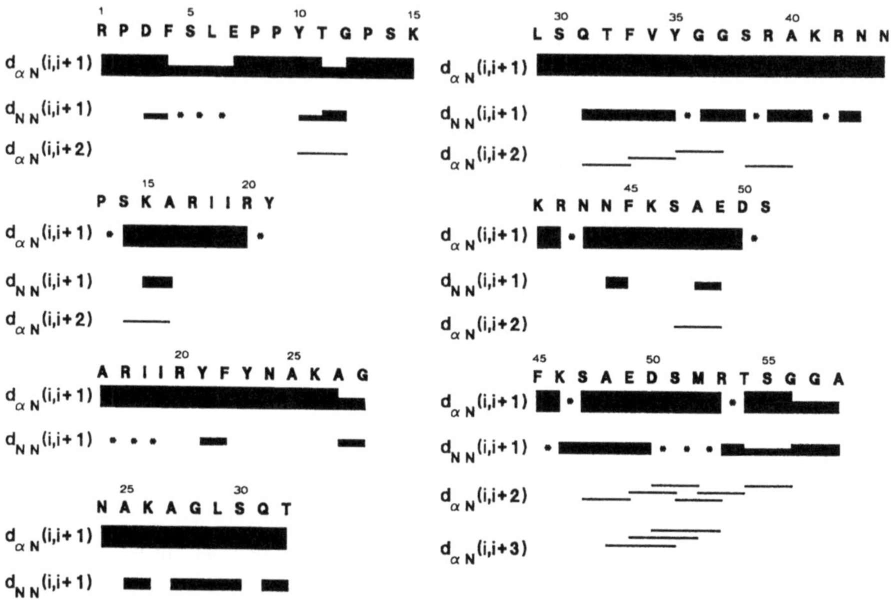

Figure 3. Schematic diagram summarizing the relative intensities of the NOE connectivities between the backbone NH (or $C^{\delta}H$ in the case of proline residues) and $C^{\alpha}H$ protons for all the peptides. An asterisk indicates cases of severe resonance overlap.

The other extended β-strand in BPTI, residues 29 to 35, is contained in peptide P_{29-44}. Strong $C^{\alpha}H(i)/NH(i+1)$ NOE cross peaks were observed, but also substantial $NH(i)/NH(i+1)$, and $C^{\alpha}H(i)/NH(i+2)$ NOE cross peaks (Figure 3). This indicates the presence of multiple conformations, including turns and 'nascent' helices, with no profound preference for one single type of conformation.

Residues Ala25 to Gly28 in native BPTI are involved in the type I β-turn connecting the two extended segments of the β-sheet. No medium-range NOE cross peaks of the expected type $C^{\alpha}H(i)/NH(i+2/3)$ were found, however, although they might have been masked by other peaks in the spectrum. In a 'classical' β-turn, the temperature coefficient of the fourth residue in the turn (Gly28 in this case) should be lowered due to hydrogen bond formation with the carbonyl of the first residue in the turn. The temperature coefficients measured in peptides P_{16-28} or P_{24-32} were not significantly decreased, however, so there was no unequivocal evidence for the presence of turn conformations in these peptides.

Only in peptide P_{45-58} were indications of a small preference for helical conformations apparent: 1, the presence of a number of $NH(i)/NH(i+1)$ contacts; 2, a number of almost contiguous medium-range NOE cross peaks of type $C^{\alpha}H(i)/NH(i+2)$ and $C^{\alpha}H(i)/NH(i+3)$; 3, lowered temperature coefficients of the amide protons of residues 49 to 57 (although those of Glu49 and Asp50 could be due to internal hydrogen bonding with their side chains). Residues 47 to 56 make up the major α-helix of folded BPTI. Strong $C^{\alpha}H(i)/NH(i+1)$ NOE cross peaks were also observed, however, which indicate that random coil conformations predominated. The CD spectrum of this peptide (Figure 2) was distinctive, although not typical of an α-helix. Similar CD spectra of peptides that appear slightly helical by NMR have been described and rationalized by Dyson *et al.* (1992a). Either the α-helices present are very short, frayed, or irregular, or the conformations present are more like an ensemble of individual reverse turns, described as 'nascent' helices (Dyson *et al.*, 1988b); single turns of α-helices are expected to be very unstable. In any case, such conformations are present in only a small fraction of the peptide molecules at any one time. A very similar CD spectrum was reported by Goodman & Kim (1989) for a peptide of residues 47 to 58 of BPTI, in which Cys51 and Cys55 were replaced by Ala, which is the residue with the greatest intrinsic helix propensity. This spectrum was also interpreted as indicating a small degree of helical conformation.

Local interactions involving aromatic side chains

(*i*) Amide of Gly(*i*+2). A very obvious local interaction between the aromatic side chain of Tyr10 and the NH of Gly12 was especially apparent in P_{1-15} as a result of the large consequent effect on δ_{NH} of Gly12 (6.70 instead of the normal 8.55 p.p.m. at 271K), its temperature dependence being opposite to normal, plus NOE cross peaks between appropriate protons (Kemmink *et al.*, 1993). The local nature of this interaction was confirmed by its demonstration in peptides $P_{8\text{-}13}$ and $P_{10\text{-}13}$, and by its disappearance when either the aromatic or glycine residue was replaced in peptides $P_{10\text{-}13}$ (Y10A) and $P_{10\text{-}13}$ (G12A). A similar interaction was observed in tetrapeptides $P_{10\text{-}13}$ (Y10F) and $P_{10\text{-}13}$ (Y10W) where the Gly12 δ_{NH} was 7.78 p.p.m. and 7.13 p.p.m., respectively. Therefore, an interaction is generally expected in unfolded proteins between an aromatic residue at position *i* and the NH of residue *i*+2 when it is Gly.

This interaction would also be expected between Tyr35 and Gly37 of BPTI. In peptide $P_{29\text{-}44}$, the Gly37 δ_{NH} was shifted upfield to 7.87 p.p.m., and NOE cross peaks were observed between Tyr35 and Gly37. A very similar value (7.60 p.p.m.) was found in the tetrapeptide $P_{10\text{-}13}$ (T11G). This interaction appears from the value of Gly δ_{NH} to be smaller in magnitude than that between Tyr10 and Gly12, presumably as a result of the extra flexibility provided by the intervening Gly residue.

(*ii*) Pro(*i*-2)-*cis*-Pro(*i*-1) side chains. One of the surprising observations with peptide $P_{1\text{-}15}$ was that a *cis* peptide bond preceding Pro8 or Pro9 caused the δ_{NH} of Gly12 to be much more normal (δ_{NH} = 8.04 p.p.m.) than when the peptide bond was *trans* (δ_{NH} = 6.70 p.p.m.). This implied that the interaction between Tyr10 and Gly12 had been disrupted. This seemed most plausible if an alternative interaction of the Tyr10 side chain was occurring in the *cis* isomer, and it was suggested that it might be interacting with the Pro8 and Pro9 side chains (Kemmink *et al.*, 1993). This has now been confirmed using peptide $P_{7\text{-}10}$, as well as the variants $P_{7\text{-}10}$ (E7A) and $P_{7\text{-}10}$ (Y10A). Multiple spectra were observed in each case, as expected from *cis-trans* isomerization of the X-Pro peptide bonds. The major isomer was always the isomer with both such bonds *trans*; less than 10% of the molecules had a *cis* peptide bond preceding Pro8. About 20% of the molecules of both peptides containing Tyr10 had a *cis* peptide bond between Pro8 and Pro9, and in each case the chemical shifts of the $C^{\alpha}H$ proton of Pro8 and of the $C^{\gamma}H_2$ and $C^{\delta}H_2$ protons of Pro9 were displaced upfield, relative to their values in the *trans* isomer. In the peptide with Ala replacing Tyr10, the amount of this *cis* isomer was considerably smaller (≈5%).

The magnitudes of the interactions between the aromatic groups of tyrosine residues and the NH group of Gly residues in the sequences of the type -Tyr-X-Gly- can be estimated from the δ_{NH} of the Gly residue. In the absence of this interaction, the random coil δ_{NH} value of Gly was found to be 8.55 p.p.m. In native BPTI, where the Tyr35 side chain is held in close proximity to the NH of Gly37, the latter's δ_{NH} is 4.3 p.p.m. (Tüchsen & Woodward, 1987). If it is assumed that the latter value is due primarily to the interaction with the Tyr35 aromatic ring being optimal and present all the time, the observed value of δ_{NH} can be used to estimate the magnitude of the equilibrium constant for the interaction. For example, in the case of the interaction between Tyr10 and Gly12 in $P_{1\text{-}15}$, Tyr10•Gly12, the observed δ_{NH} of 6.70 suggests that the equilibrium constant, $K_{10/12}$, for this interaction is:

$$K_{10/12} = \frac{[\text{Tyr10•Gly12}]}{[\text{no interaction}]} = \frac{8.55 - 6.70}{6.70 - 4.3} = 0.77 \qquad (1)$$

In other words, the interaction between the aromatic ring of Tyr10 and the Gly12 NH would be present about 44% of the time.

When there is a *cis* peptide bond between Pro8 and Pro9, the equilibrium constant for the interaction of these residues with Tyr10, $Pro_{8/9}$•Tyr10, can be quantified from its effect on the *cis-trans* equilibrium:

$$trans \overset{K_{t/c}}{<\text{---------}>} cis \overset{K_{8\text{-}10}}{<\text{---------}>} cis\ Pro_{8/9}\text{•}Tyr10 \qquad (2)$$

In the absence of Tyr10, the Pro8-Pro9 peptide bond was found to be *cis* roughly 5% of the time, so $K_{t/c} \cong 20$. In the presence of Tyr10, the *cis* form was present about 20% of the time, so

$$\frac{[cis] + [cis\ Pro_{8/9}\text{•}Tyr10]}{[trans]} = \frac{1 + K_{8\text{-}10}}{K_{t/c}} = \frac{1 + K_{8\text{-}10}}{20} = \frac{0.2}{0.8} \qquad (3)$$

Therefore, $K_{8\text{-}10} \cong 4$, and this interaction should be present about 80% of the time when the Pro8-Pro9 peptide bond is *cis*.

This analysis explains why the interaction of Tyr10 with Pro8 and Pro9 with a *cis* bond between them largely disrupts its alternative interaction with Gly12, for the former interaction is substantially stronger. The isomerization of this peptide bond would not be expected to affect directly the interaction between Tyr10 and Gly12. Consequently, the observed extent of the interaction between Tyr10 and Gly12, and the chemical shift of Gly12, when the Pro8-Pro9 peptide bond is *cis* can be predicted as in Equation 1:

$$cis\ Pro_{8/9}\text{•}Tyr10 \overset{1/K_{8\text{-}10}}{<\text{---------}>} cis \overset{K_{10/12}}{<\text{---------}>} cis\ Tyr10\text{•}Gly12 \qquad (4)$$

The interaction between Tyr10 and Gly12 would be predicted in this instance to be present only 13% of the time and the chemical shift of the Gly12 NH to be 7.98 p.p.m. This is remarkably close to the measured value of 8.04 p.p.m. (Kemmink *et al.*, 1993). The consistency of this analysis supports the validity of the measurement of $K_{8\text{-}10}$ from the chemical shift values (Equation 1), but does not prove it.

(*iii*) Ile(*i*-2) side chain. Indications of a local interaction between the sidechains of Ile19 and Tyr21 were found in both $P_{13\text{-}21}$ and $P_{16\text{-}28}$. The $C^{\gamma}H_3$ group of Ile19 resonated at 0.66±0.01 p.p.m., which is displaced upfield compared to the more normal value of 0.82±0.01 p.p.m. for the same group in Ile18. NOE cross peaks were also observed between the aromatic protons of Tyr21 and the sidechain protons of Ile19. Both phenomena indicate close proximity of the Tyr21 aromatic ring to the sidechain of Ile19. This was confirmed with the variant peptide P_{16-28} (Y21A); its $C^{\gamma}H_3$ group had a normal δ.

(*iv*) Ala(*i*+2) side chain. In peptide P_{16-28}, NOE cross peaks were observed between the Tyr23 aromatic protons and the $C^{\alpha}H$ and $C^{\beta}H_3$ protons of Ala25. The chemical shift of the Ala25 $C^{\alpha}H$ was shifted upfield slightly ($\delta = 4.12$, rather than 4.25 ± 0.02 in the other cases). That this was due to Tyr23 is shown by the chemical shift of the Ala25 $C^{\alpha}H$ proton being normal ($\delta = 4.27$ p.p.m.) in peptide P_{24-32}, where Tyr23 is absent.

Reduced BPTI

CD and ^1H NMR spectra of reduced BPTI, with six free cysteine thiol groups, were recorded under conditions very similar to those used with the peptides described above. Both spectra were characteristic of an unfolded polypeptide chain. There was very little dispersion of the NMR chemical shifts, and the spectrum was not assigned to individual protons.

Indications of the same local interactions observed in the peptides were also observed in reduced BPTI by NMR. For example, the fingerprint region of a TOCSY spectrum of reduced BPTI contained two $C^{\alpha}H$/NH cross peaks, characteristic of glycine spin-systems with degenerate $C^{\alpha}H$ protons, which was confirmed by the presence of remote connectivities in the DQ spectrum (Wagner & Zuiderweg, 1983), displaced upfield with respect to the NH chemical shift. Their positions in the spectrum at $\delta = 6.76$ and 7.88 p.p.m. were very similar to those found for the amide protons of Gly12 ($\delta_{NH} = 6.70$ p.p.m.) and Gly37 ($\delta_{NH} = 7.87$ p.p.m.) in the respective peptides P_{1-15} and P_{29-44}. In the aliphatic region of the reduced BPTI spectrum, one methyl group was prominent as the peak most shifted upfield ($\delta = 0.65$ p.p.m.). The TOCSY spectrum indicated that this resonance belongs to an isoleucine spin-system, and there are only two such residues in BPTI (Ile18 and Ile19). The $C^{\gamma}H_3$ group of Ile19 in both peptides in which it was present (P_{13-21} and P_{16-28}) was displaced upfield to the same position ($\delta = 0.67$ and 0.66 p.p.m., respectively). These close similarities strongly suggests that these resonances in reduced BPTI arose from the same protons as in the peptides and that the same responsible interactions were also present in the reduced protein.

The roles of the local interactions in the folding of BPTI

The conformations and NMR parameters of native BPTI and of all the major intermediates in the disulphide folding pathway of BPTI are known (van Mierlo *et al.*, 1991a,b, 1992, 1993a,b), so it is possible to follow these interactions during folding of BPTI and to infer their roles in folding.

An interaction in the unfolded protein that contributes to folding must be increased in magnitude in the relevant folding intermediates and in the fully folded state. Whatever contribution an interaction makes to the folded state of a species, that species must stabilize that interaction to the same extent. Conversely, interactions that become less stable in the intermediates or fully folded state hinder folding and decrease the stability of the folded state.

Table 1: Chemical shifts during folding of BPTI of groups involved in local interactions

Chemical shifts	**Gly 12** NH	**Gly 37** NH	**Ile 19** $C^{\gamma}H_3$	**Ala 25** $C^{\alpha}H$
Random-coil	8.55 ± 0.12	8.55 ± 0.12	0.83	4.25 ± 0.02
R (peptides)	6.70	7.87	0.66	4.12
(30-51)	6.78	7.21	0.70	3.63
(30-51, 5-14)	n.a.	7.23	0.71	3.63
(30-51, 5-38)	6.90	7.52	0.71	3.63
$(30\text{-}51, 14\text{-}38)_N$	7.32	4.37	0.73	3.71
$(5\text{-}55)_N$	7.18	n.a.	0.73	3.83
$(30\text{-}51, 5\text{-}55)_N$	7.17	n.a.	0.73	3.79
N	7.31	4.32	0.72	3.78

The chemical shifts of the various groups were measured in the peptides representing reduced BPTI, R, and in analogues of the disulphide intermediates in which free cysteine residues were replaced by serines. Values that were not determined are indicated by "n.a.". Intermediates with the native conformation are designated with subscript "N". The random coil shift values were measured in the same way in peptides in which no interaction appeared to be present. Measurements were made at -2°C, except for the native, N, and native-like species, which were at +10°C.

The interaction between Tyr35 and the NH of Gly 37 illustrates how an interaction assists folding. This interaction is greatly stabilized within the fully folded structure of BPTI, as the Tyr35 aromatic ring is held in close proximity to the Gly37 amide, producing a very anomalous δ_{NH} of 4.3 p.p.m. (Tüchsen & Woodward, 1987). Reciprocally, this interaction would be expected to stabilize the native conformation to the same extent. This is consistent with the 6.5 kcal/mol decrease in stability caused by replacing Tyr35 by otherwise similar Leu (Zhang & Goldenberg, 1993). This interaction is also present in the (30-51) intermediate, and it appears to become more stable as folding progresses along the most productive pathway, for it is preserved in the conformations of the intermediates and the value of δ_{NH} of Gly37 becomes progressively more anomalous (Table 1). Therefore, the interaction between Tyr35 and Gly37 appears to stabilize the intermediates progressively along the pathway.

Likewise, the interaction between Tyr23 and Ala25 is incorporated and strengthened within the native conformation of BPTI. The Tyr23 aromatic ring is held within the interior of the molecule, and the $C^{\alpha}H$ of Ala25 is pointing toward the face of the ring and only 0.26 nm from the nearest carbon atom of the ring. Consequently, the chemical shift of the Ala25 $C^{\alpha}H$ is displaced upfield, much more than in the peptides (Table 1). Similar interactions occur in all the productive intermediates (Darby *et al.*, 1992; van Mierlo *et al.*, 1992, 1993a,b), and the chemical shift of Ala25 $C^{\alpha}H$ is similar. Therefore, this interaction also appears to be important in the early stages of folding and in the final conformation.

In contrast, the interactions between Tyr10 / Gly12 (Kemmink *et al.*, 1993) and Ile19 / Tyr21 appear to be weakened or disrupted in the native conformation, for the chemical shifts of the Gly12 NH and Ile19 $C^{\gamma}H_3$ become less anomalous as folding progresses (Table 1). The Tyr21 side chain is fixed in the interior of the molecule and is separated from the Ile19 side chain by the side chain of Thr32. The Ile19 C^{γ} atom is no closer than 0.62 nm to any of the C atoms of the Tyr21 aromatic ring. The Gly12 NH and Ile19 $C^{\gamma}H_3$ groups are on the surface of the molecule, and neither is near other aromatic rings. Pardi *et al.*, (1983) have shown that the δ_{NH} of Gly12 is determined primarily by the proximity of Tyr10. Therefore, the interactions between these groups appear to be greatest in the reduced protein and are weakened or disrupted as folding proceeds. These interactions are only weak in the unfolded protein, so there should not be a great energetic cost to disrupting them.

General implications for protein folding

The nonrandom conformations that are usually observed in unfolded proteins, including those described here, are generally local and involve only a few groups close in the covalent structure. Therefore, the interactions are only weak and are not cooperative, in that they are independent of the other interactions present elsewhere in the molecule, and they are undoubtedly formed and broken rapidly. For example, even relatively complex and somewhat cooperative α-helices are formed, and unravelled, on a timescale of about 10^{-5} seconds (Creighton, 1993). Therefore, it is the stabilities of these interactions that are important, not their rates of formation.

Nonrandom conformations in the unfolded protein that are also present in the native conformation are undoubtedly important for the folding process, as their presence should decrease the conformational searching that must take place to find the native conformation. These nonrandom conformations include both the conformational tendencies of the polypeptide backbone and local noncooperative interactions of groups close in the covalent structure. In each case, the responsible interactions are only weak and can be either incorporated in the folded conformation or disrupted. Nevertheless, it is undoubtedly significant that the nonrandom conformations found in unfolded proteins are so often retained in the final folded conformation (e.g. Brown & Klee, 1971; Dyson *et al.*, 1992a,b).

ACKNOWLEDGMENTS

We thank Dominique Nalis and Richard Jacob for synthesizing and purifying the peptides, Nigel Darby, Luis Serrano and Ruud Scheek for assistance and advice, Monique van Straaten for technical assistance, and Bayer AG for a generous gift of BPTI (Trasylol®).

REFERENCES

Amir D and Haas E (1988) Reduced bovine pancreatic trypsin inhibitor has a compact state. Biochemistry 27: 8889-8893

Bothner-By AA, Stephens RL and Lee J (1984) Structure determination of a tetrasaccharide: transient nuclear Overhauser effects in the rotating frame. J. Amer. Chem. Soc. 106:811-813

Braunschweiler L and Ernst RR (1983). Coherence transfer by isotropic mixing: application to proton correlation spectroscopy. J. Magn. Reson. 53: 521-528

Brown JE and Klee WA (1971) Helix-coil transition of the isolated amino terminus of ribonuclease. Biochemistry 10: 470-476

Creighton TE (1978) Experimental studies of protein folding and unfolding. Progr. Biophys. Mol. Biol. 33: 231-297

Creighton TE (1988a) Toward a better understanding of protein folding pathways. Proc. Natl. Acad. Sci. USA 85: 5082-5086

Creighton TE (1988b). On the relevance of non-random polypeptide conformations for protein folding. Biophys. Chem. 31: 155-162

Creighton TE (1990) Protein folding. Biochem. J. 270: 1-16

Creighton TE (1992) Folding pathways elucidated using disulphide bonds. In Protein Folding (Creighton TE, ed.) pp 301-351, WH Freeman, New York

Creighton TE (1993) Proteins: Structures and Molecular Properties 2nd ed., WH Freeman, New York

Darby NJ, van Mierlo CPM, Scott GHE, Neuhaus D and Creighton TE (1992) Kinetic roles and conformational properties of the non-native two-disulphide intermediates in the refolding of bovine pancreatic trypsin inhibitor. J. Mol. Biol. 224: 905-911

Deisenhofer J and Steigemann W (1975) Crystallographic refinement of the structure of bovine pancreatic trypsin inhibitor at 1.5 Å resolution. Acta Crystallogr. sect. B 31: 238-250

Dyson HJ, Rance M, Houghten RA, Lerner RA and Wright PE (1988a) Folding of immunogenic peptide fragments of proteins in water solution. I. Sequence requirements for the formation of a reverse turn. J. Mol. Biol. 201: 161-200

Dyson HJ, Rance M, Houghten RA, Lerner RA and Wright PE (1988b) Folding of immunogenic peptide fragments of proteins in water solution. II. The nascent helix. J. Mol. Biol. 201: 201-218

Dyson HJ and Wright PE (1991) Defining solution conformations of small linear peptides. Ann. Rev. Biophys. Biophys. Chem. 20: 519-538.

Dyson HJ, Merutka G, Waltho JP, Lerner RA and Wright PE (1992a) Folding of peptide fragments comprising the complete sequence of proteins. Models for initiation of protein folding. I. Myohemerythrin. J. Mol. Biol. 226: 795-817

Dyson HJ, Sayre JR, Merutka G, Shin H-C, Lerner RA and Wright PE (1992b) Folding of peptide fragments comprising the complete sequence of proteins. Models for initiation of protein folding. II. Plastocyanin. J. Mol. Biol. 226: 819-835

Gill SC and von Hippel PH (1989) Calculation of protein extinction coefficients from amino acid sequence data. Anal. Biochem. 182: 319-326

Goodman EM and Kim PS (1989) Folding of a peptide corresponding to the α-helix in bovine pancreatic trypsin inhibitor. Biochemistry 28: 4343-4347

Gottfried DS and Haas E (1992) Nonlocal interactions stabilize compact folding intermediates in reduced unfolded bovine pancreatic trypsin inhibitor. Biochemistry 31: 12353-12362

Jeener J, Meier BH, Bachman, P and Ernst RR (1979) Investigation of exchange processes by two-dimensional NMR spectroscopy. J. Chem. Phys. 71: 4546-4553

Kemmink J, van Mierlo CPM, Scheek RM and Creighton TE (1993) Local structure due to an aromatic-amide interaction observed by ^{1}H-nuclear magnetic resonance spectroscopy in peptides related to the N terminus of bovine pancreatic trypsin inhibitor. J. Mol. Biol. 230: 312-322

Kemmink J and Creighton TE (1993) Local conformations of peptides representing the entire sequence of bovine pancreatic trypsin inhibitor (BPTI) and their roles in folding. Submitted

Kraulis PJ (1991) MOLSCRIPT: a program to produce both detailed and schematic plots of protein structures. J. Appl. Crystallogr. 24: 946-950

Pardi A, Wagner G and Wüthrich K (1983) Protein conformation and proton nuclear-magnetic-resonance chemical shifts. Eur. J. Biochem. 137: 445-454

Rance M, Sørensen OW, Bodenhausen G, Wagner G, Ernst RR and Wüthrich K (1983) Improved spectral resolution in COSY ^{1}H spectra of proteins via double quantum filtering. Biochem. Biophys. Res. Commun. 117: 479-485
Shortle D (1993) Denatured states of proteins and their roles in folding and stability. Curr. Opinion Struct. Biol. 3: 66-74
States DJ, Creighton TE, Dobson CM and Karplus M (1987) Conformations of intermediates in the folding of pancreatic trypsin inhibitor. J. Mol. Biol. 195: 731-739
Tüchsen E and Woodward C (1987) Assignment of asparagine 44 side chain primary amide ^{1}H-NMR resonances and the peptide amide N^{1}H resonance of glycine 37 in basic pancreatic trypsin inhibitor. Biochemistry 26: 1918-1925
van Mierlo CPM, Darby NJ, Neuhaus D and Creighton TE (1991a) The 14-38,30-51 double-disulphide intermediate in folding of bovine pancreatic trypsin inhibitor: a two-dimensional ^{1}H NMR study. J. Mol. Biol. 222: 353-371
van Mierlo CPM, Darby NJ, Neuhaus D and Creighton TE (1991b) A two-dimensional ^{1}H NMR study of the (5-55) single-disulphide folding intermediate of bovine pancreatic trypsin inhibitor. J. Mol. Biol. 222: 373-390.
van Mierlo CPM, Darby NJ & Creighton TE (1992) The partially-folded conformation of the 30-51 intermediate in the disulfide folding pathway of bovine pancreatic trypsin inhibitor. Proc. Natl. Acad. Sci. U.S.A. 89: 6775-6779
van Mierlo CPM, Darby NJ, Keeler J, Neuhaus D and Creighton TE (1993a) Partially-folded conformation of the (30-51) intermediate in the disulphide folding pathway of bovine pancreatic trypsin inhibitor: ^{1}H and ^{15}N resonance assignments and determination of backbone dynamics from ^{15}N relaxation measurements. J. Mol. Biol. 229: 1125-1146
van Mierlo CPM, Kemmink J, Neuhaus D, Darby NJ and Creighton TE (1993b) ^{1}H-NMR analysis of the conformational properties of the partially-folded non-native two-disulphide bonded intermediates (30-51,5-14) and (30-51,5-38) in the disulphide folding pathway of bovine pancreatic trypsin inhibitor. Submitted
Wagner G and Zuiderweg ERP (1983) Two-dimensional double quantum ^{1}H NMR spectroscopy of proteins. Biochem. Biophys. Res. Commun. 113: 854-860
Wagner G, Braun W, Havel TF, Schaumann T, Go N and Wüthrich K (1987) Protein structures in solution by nuclear magnetic resonance and distance geometry. The polypeptide fold of the basic pancreatic trypsin inhibitor determined using two different algorithms, DISGEO and DISMAN. J. Mol. Biol. 196: 611-639
Wüthrich K (1986) NMR of Proteins and Nucleic Acids, Wiley, New York
Zhang JX and Goldenberg DP (1993) An amino acid replacement that eliminates kinetic traps in the BPTI folding pathway. Submitted

Co-translational Modification, Stability and Turnover of Eukaryotic Proteins

Ralph A. Bradshaw, Jose Sy*, Albert E. Stewart, Richard L. Kendall@, Hubert Hondermarck# and Stuart M. Arfin
Department of Biological Chemistry
College of Medicine
University of California, Irvine
Irvine, California 92715-1700
U.S.A.

Eukaryotic function is tightly controlled through the complex mechanisms that regulate transcriptional and translational events. These processes are variously augmented by co- and post-translational modifications that affect function, location and ultimately turnover for each protein. Among the least well understood aspects are stability, including the process of folding, and degradation of both normal and damaged proteins. Since proteolytic destruction of proteins is as important as synthesis in determining the level of a biological activity, it represents a major cellular activity whose dissection is essential for the full appreciation of the regulation of eukaryotic cell function.

Proteolytic enzymes act on their substrates either at the termini (exo-) or internally (endo-) and can be readily divided into intracellular and extracellular categories. The endoproteases of the latter type, which function in a variety of physiological processes, such as digestion, fertilization and the turnover of extracellular messengers, are generally small, highly stable monomeric proteins containing a number of disulfide bridges. The intracellular endoproteases are varied in their location and are often less well defined in terms of cellular metabolism. Some are compartmentalized and are involved in the processing of precursors, such as the signal peptidases of the endoplasmic reticulum and the mitochondrion, or have other specialized roles important to the organelle in which they occur. Homologs of the extracellular aspartic acid and thiol proteases are found in lysosomes as members of the cathepsin family and these certainly function during cell lysis and in the turnover of proteins that are incorporated into the lysosomal compartment (Dice, 1987). An additional member of the thiol family, calpain, which occurs in two different isozymic forms, is widely distributed in the cytoplasmic compartment of eukaryotic cells and its activity is in part controlled through the binding of calcium ions to a regulatory subunit. Although many theories have been

---

[1]* Department of Chemistry, CSU Fresno, Fresno CA 92740
@ Present address: Department of Biochemistry, Merck Research Laboratories, Rahway NJ 07065
# Present address: Laboratoire de Biologie du Developpement, Universite de Lille 1, 59655 Villeneuve d'Ascq cedex France

NATO ASI Series, Vol. H 82
Biological Membranes:
Structure, Biogenesis and Dynamics
Edited by Jos A. F. Op den Kamp

advanced regarding possible cellular functions of the calpains, definitive roles are still lacking. The most important of the intracellular proteases with respect to turnover may be the proteasome, a high molecular weight complex, whose functions are at least in part connected to the ubiquitin system (Rivett, 1993). One form of this multi-subunit complex recognizes proteins which have been previously covalently tagged with a polyubiquitin chain and specifically degrades them. An independent protease, commonly referred to as ubiquitinase, releases the ubiquitin protein intact so that it may be recycled for further use (Jentsch et al., 1991). As with signal peptidase, this protease is highly specific and does not appear to function in any other capacity.

The attachment of ubiquitin, a 76 amino acid protein that, as the name implies, is found apparently universally in eukaryotes, requires a system of activating transferases. The first of these, commonly designated E1, activates the alpha-carboxyl group of the C-terminal glycine residue of ubiquitin via a mixed anhydride formed with ATP and transfers the activated ubiquitin to a sulfhydryl group of this enzyme to form a high energy thioester. Only a single copy of this enzyme exists in yeast and null mutations are lethal (McGrath et al., 1991). Subsequently, the activated ubiquitin is transferred to a sulfhydryl group of one or more members of a family of proteins, generally designated E2s, again forming a thioester. Some members of the E2 family of proteins can directly transfer the ubiquitin, via isopeptide linkage to the side chain of a lysine residue(s), to specific substrates for a myriad of purposes. Yeast contains multiple E2 moieties, and null mutations of any individual E2 has not proven to be lethal (Jentsch et al., 1991). At least one E2, and probably more, function in proteolytic degradation mechanisms.

One of the seemingly best defined intracellular degradation pathways was elucidated by Varshavsky and coworkers and has been called the N-end rule (Bachmair et al., 1986). As the name suggests, the degradation of substrates in this pathway depends on amino-terminal recognition and involves a third component of the ubiquitin-mediated pathway, E3 (Bartel et al., 1990), acting in concert with one or more E2s. Briefly, certain amino acids occurring in the amino-terminal position of a protein clearly destabilize it and lead to its rapid turnover, providing other conditions are also met. As shown in Fig. 1, the N-termini are recognized at two different sites, which are specific for basic (type I) and bulky hydrophobic residues (type II), respectively (Hershko and Ciechanover, 1992). Smaller, as well as acidic, side chains are not bound and, hence, do not contribute directly to this process. (However, a second E3 entity that apparently recognizes smaller and acetylated amino acids has been described (Gonda et al., 1989; Heller and Hershko, 1990)). Importantly, dipeptides containing amino-terminal basic or bulky residues will competitively inhibit this binding and block ubiquitin-mediated degradation of protein substrates in cell-free systems and, apparently, in whole cells (Reiss et al., 1988; Gonda et al., 1989; Baker and Varshavsky, 1991).

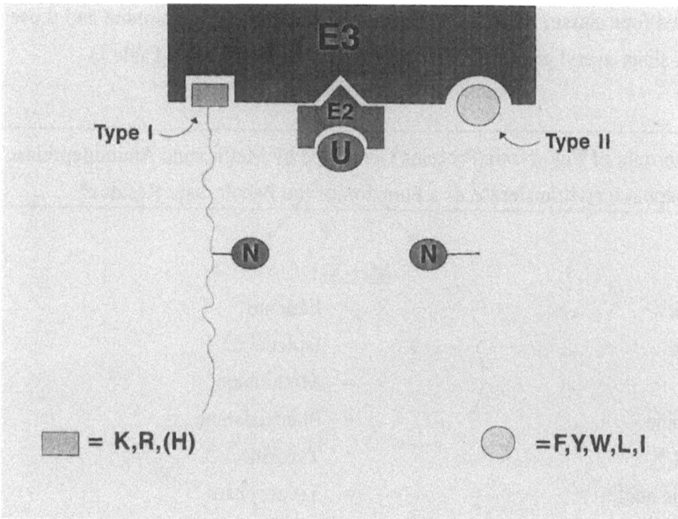

Fig. 1. Schematic representation of the auxiliary substrate recognition protein (E3) of the ubiquitin-dependent degradation pathway. The type I and II sites are entirely independent. E2 represents one or more ubiquitin-conjugating enzymes.

The amino-terminal residue of a protein is governed by the genetic coding sequence and co- and post-translational processing events. Since virtually all eukaryotic protein synthesis is initiated by methionine, the default situation (no processing) would result in proteins with this residue as amino terminus. To date, this has not been a common finding; however, the determination of the amino-terminal residue of a mature protein requires direct sequencing and the majority of the protein sequence database is now derived from nucleic acid sequences, so that exact knowledge of the true N-terminal residue of many proteins is lacking. The actual N-terminal residue of a protein is probably determined by the specificity of the methionine aminopeptidase (MAP) that is apparently ribosomally associated and acts co-translationally to tailor protein N-termini (Arfin and Bradshaw, 1988). Substrate specificity of MAP is determined almost exclusively by the size of the penultimate residue as determined either in situ (Huang et al., 1987; Boissel et al., 1988) or with homogeneous preparations

(Chang et al., 1990; Kendall and Bradshaw, 1992). Along with the companion enzyme(s), that transfers N-alpha-acetyl groups from acetyl-CoA, also as a co-translational process, there can be generated four classes of proteins: those with and without methionine and those with and without N-alpha-acetyl groups (Arfin and Bradshaw, 1988). (See Table I).

Table I. N-termini of Eukaryotic Proteins Generated by Methionine Aminopeptidase and N-alpha-acetyltransferase as a Function of the Penultimate Residue*

N-Acetyl-X-:	**Met-X-:**
X = Glycine	X = Leucine
= Alanine	= Isoleucine
= Serine	= Methionine
= Threonine	= Phenylalanine
N-Acetyl-Met-X-:	= Tyrosine
X = Aspartic acid	= Tryptophan
= Glutamic acid	= Histidine
= Asparagine	= Arginine
X-:	= Lysine
X = Cysteine	= Glutamine
= Proline	
= Valine	

*Taken from Huang et al. (1987) and Boissel et al. (1988) and corroborated by the protein sequence database.

Clearly, the sequences retaining methionine (or N-alpha-acetyl methionine) are those that were found to be destabilizing by the N-end rule (Bachmair et al., 1986), which provides an immediate rationale for the universal distribution of the ribosomal MAP, i.e. to protect proteins from premature degradation by the ubiquitin pathway. The addition of N-alpha-acetyl groups may provide further stability or, in the cases of methionine modification, may provide access to a special degradation pathway (see below).

Methionine Aminopeptidase.

Although eukaryotic cells contain a broad distribution of aminopeptidase activity, the removal of methionine is more restricted (Kendall, 1992). Importantly, the cleavage of Met-Pro bonds appears to be limited to the ribosomally associated enzyme involved in co-translational processing of proteins. Utilizing this observation, as well as the other specificities suggested in Table I, a purification protocol for a MAP with this substrate profile from porcine liver has been developed (Kendall and Bradshaw, 1992). The homogeneous enzyme has a molecular weight of 70,000, although a secondary form apparently arising from limited proteolysis during isolation with a molecular weight of 67,000 was also observed. Similar values were obtained under non-denaturing conditions suggesting the enzyme is monomeric, at least in solution. It is also activated by cobalt ion (over 2x relative to controls) and is inhibited by beta-mercaptoethanol and EDTA. No other metal ions increased activity (although some, such as zinc and manganese, were inhibitory at 5 mM).

The Km and kcat values for this enzyme were affected by the residue in the second position of the substrate. Using the hemoglobin series (N-terminal Val-X-His-), the measured kcat/Km for peptides with penultimate proline, valine and alanine were 130, 260 and 1523 $nM^{-1}s^{-1}$, respectively. However, the proline and valine peptides were bound better than the alanine peptides as judged by their Km values (0.28 and 0.22 mM, respectively, vs. 0.62 mM). This suggests that the larger residues provide better binding energies but are not oriented as well for turnover. It may be that the more bulky residues, that are not hydrolyzed at all, are still bound to MAP and, hence, could be effective competitive inhibitors (Kendall and Bradshaw, 1992).

MAP enzymes of similar specificity have also been isolated from bacteria and yeast (Ben-Basset et al., 1987; Miller et al., 1987; Chang et al., 1990). The prokaryotic enzymes are smaller (~30,000 Da) but also utilize Co^{++} as a cofactor. A crystal structure (Roderick and Matthews, 1993) of the MAP from E. coli has been determined which clearly indicates that this enzyme represents a new fold (not related to any previously determined enzyme structure). The yeast enzyme is intermediate in size, having a Mr of 43,000, as calculated from the predicted sequence (Chang et al., 1992). A catalytic domain that is related in sequence to the bacterial enzymes is preceded by an amino-terminal extension. The role of this segment is uncertain, but it may be important for ribosome association, a feature clearly absent from the bacterial enzymes. In this regard, the role of the prokaryotic MAPs is unclear as N-terminal processing, to the extent that it occurs, is a post-ribosomal event and there is no ubiquitin-mediated degradation in bacteria [although an N-end rule has been demonstrated in E. coli (Tobias et al., 1991)]. It is curious, therefore, that the specificity of this cobalt-dependent MAP has been maintained from the simplest organisms to man.

N-Acetyl Methionine Pathway.

When aspartic acid, glutamic acid or asparagine occupy the penultimate position, the initiator methionine is not only retained but it is also acetylated (Huang et al., 1987) (Table I). This situation provides a unique opportunity for degradation through N-terminal recognition and ubiquitinylation because the N-acetyl methionine moiety can be excised by the acylaminoacyl hydrolase that has been well characterized from several sources (Krishna and Wold, 1992). These enzymes have been reported to have a broad range of activities, removing N-acetyl amino acids from a variety of substrates, usually small peptides. Utilizing an octapeptide beginning with N-acetyl methionine-glutamic acid, we screened porcine liver extracts for an enzyme specific for release of the N-acetyl methionine. Activity was readily demonstrated, but, upon purification, it proved to be identical with the previously reported enzymes (A.E. Stewart and R.A. Bradshaw, unpublished observations), suggesting that it is capable of functioning to remove N-acetyl methionine residues from proteins in situ.

The exposure of penultimate acidic residues would in turn provide substrates for a well-defined enzyme, arginine-tRNA protein transferase (RTPT), that for many years did not have a defined physiological role (Ciechanover et al., 1988). This enzyme catalyzes the transfer of arginine from tRNA to the N-terminus of proteins with N-terminal aspartic or glutamic acid residues. Interestingly, this enzyme was identified with assays utilizing extracellular proteins, as these were the only ones known with acidic residues at the N-terminus. The reason for this is now clear; such residues are only exposed by the acylaminoacyl hydrolase and they are apparently immediately converted to N-terminal arginine by RTPT. Once the basic residue has been attached, the protein could be recognized by E3 and potentially be degraded by the proteasome system following ubiquitinylation. Regardless of its ultimate fate, it would not retain the acidic residue at the amino terminus.

An apparent anomaly of this proposed pathway are proteins with N-terminal N-alpha-acetyl Met-Asn sequences. Asparagine is not a substrate for RTPT and would necessarily require deamidation to aspartic acid before it could be arginylated. However, an enzyme catalyzing such a reaction has not been previously reported, although evidence has been presented for the deamidation of both asparagine and glutamine when present at the N-terminus of engineered proteins in yeast and reticulocyte lysates (Gonda et al., 1989; Varshavsky, 1992). Accordingly, we devised an assay utilizing the peptide Asn-His-Gly-Ala-Trp-Leu-Leu-Pro and screened pig liver extracts for activity that would convert it to Asp-His-Gly-Ala-Trp-Leu-Leu-Pro. These experiments have led to the identification of a 33 kDa enzyme, which we have designated protein N-terminal asparagine deamidase (PNAD), that specifically catalyzes the conversion of a broad spectrum of peptide substrates with N-terminal asparagine to the corresponding aspartic acid form. Internal asparagine residues are not modified (A.E. Stewart, S.M. Arfin and R.A. Bradshaw, manuscript in preparation). Clearly, the prediction of the existence of this enzyme and its subsequent identification provides additional support for the physiological relevance of the N-end pathway.

Interestingly, N-terminal glutamine residues are not deamidated by this enzyme and we have not yet detected another enzyme with this activity. In yeast, glutamine residues in the penultimate position direct the retention of methionine but without subsequent acetylation (Huang et al., 1987). Such proteins could potentially be destabilized by the action of another aminopeptidase and an as yet to be identified glutamine deamidase. However, they apparently are not part of the asparagine pathway and proteins with a Met-Gln sequence may not be degraded by the N-end rule pathway at all.

Protein Degradation in Cell-free Systems.

In order to evaluate the contribution of various components involved in N-terminal-mediated protein turnover, a cell-free protein synthesis system utilizing rabbit reticulocytes was established (Ciechanover et al., 1991). Using appropriate mRNA templates, proteins labeled with ^{35}S-methionine can be synthesized and their energy-dependent turnover determined. The reticulocyte lysate contains all of the necessary components for this process but can be made limiting in ubiquitin in order to clearly separate the synthesis and degradation phases.

Table II. Degradation of In-vitro Synthesized Proteins+

	% Degradation		N-terminal
	- ATP	+ ATP	
^{125}I-Lysozyme	5	21	Lys
^{35}S-Glutathione-S-transferase	0.3	3	Pro
^{35}S-Asparagine synthetase	5	37	Cys
^{35}S-Phosphoglycerate mutase	5	26	Ala

+ Cell-free system using rabbit reticulocyte lysates (Ciechanover et al., 1991). Degradation was measured at 37°C for 45 min.

Table II shows the ATP-dependent degradation of three proteins prepared in this manner and their N-termini. (The ^{125}I-lysozyme was a positive control and was not generated by cell-free synthesis). Little or no turnover was found for glutathione-S-transferase which has proline as the residue penultimate to the initiator methionine (and is found as the unblocked N-terminus in the mature protein). In contrast, degradation comparable to the control was found for asparagine synthetase (AS) and phosphoglycerate mutase (PGM), with N-termini of cysteine and alanine, respectively. In the native protein, the alanine of PGM is N-acetylated. The energy-dependent turnover of AS is consistent with the suggestion that N-terminal cysteine can also enter the E3 pathway after arginylation by PTRT (Gonda et al., 1989). To test this hypothesis, the energy-dependent degradation of AS was measured as a function of RNase treatment of the reticulocyte lysate. The proteolysis of proteins requiring arginylation (via charged tRNA), such as bovine serum albumin which has an N-terminal aspartic acid, is inhibited by this treatment as shown in Fig. 2.

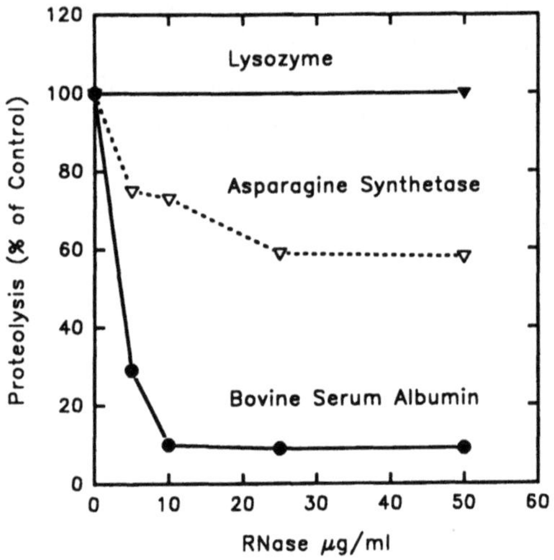

Fig. 2. Effect of RNAse treatment on proteolysis of selected substrates. Reticulocyte lysates were preincubated with the indicated amounts of micrococcal nuclease for 30 min at 30°C, followed by an additional 10 min of incubation with EGTA to inactivate the nuclease. Ubiquitin-dependent proteolysis of ^{125}I-lysozyme (N-terminal Lys), ^{125}I-bovine serum albumin (N-terminal Asp) and ^{35}S-asparagine synthetase (N-terminal Cys) was then measured. Proteolytic activity of 100% represents the amount of protein degraded in 45 min in lysates which were not treated with nuclease.

Lysozyme, with N-terminal lysine, does not require arginylation and is, therefore, degraded independently of the added RNase. AS was partially protected by the RNase treatment suggesting two species were subject to turnover, one dependent on tRNA and the other not. The tripeptide Glu-Val-Phe (EVF) acts to block arginylation and was also found to block a portion of AS (but not PGM) degradation (Sy, J., Arfin, S.M. and Bradshaw, R.A., unpublished observations). At present, the nature of the RNase-independent species is unknown. It may represent material already arginylated or it may be material degraded by an N-terminal independent pathway.

Support for this latter possibility is provided by the observed PGM turnover. When labeled PGM was prepared in the cell-free system and subjected to purification by chromatography on DE-52, two fractions were obtained. The material that was not retained was totally refractile to energy-dependent degradation. In contrast, the peak eluted with 0.5 M KCl was hydrolyzed. N-alpha-acetylation did not affect this distribution as determined with acetonyl-CoA and oxalacetate plus citrate synthase, inhibitors of N-alpha-acetylation. As expected, the tripeptide EVF did not inhibit the degradation of the bound material. As with the AS fraction that was not susceptible to RNase treatment, the nature of the DE-retained PGM is unknown. However, treatment of the degradation-resistant breakthrough peak with 6 M guanidine HCl followed by rapid dilution to initiate refolding generated 20% retained (and degradable) material (Sy, J., Arfin, S.M. and Bradshaw, R.A., unpublished observations). This suggests that the retained fraction is probably improperly folded protein that is susceptible to degradation, possibly via ubiquitin, but by an N-terminal independent pathway. The extent to which improper folding of any given protein may occur in situ and, thus, require degradation is unknown but may be greater than previously imagined.

Neurite Proliferation and the E3 Pathway.

Although the sophistication of the N-end rule pathway is compelling as a major cellular mechanism for specifically regulating protein turnover, little evidence has accumulated for its use in normal physiological functions. As a means of identifying proteins that are normally degraded in an N-terminal dependent fashion by the ubiquitin E3 pathway, we utilized various dipeptides that have been shown to be effective inhibitors of both the type I and II sites in E3 (Reiss et al., 1988; Gonda et al., 1989; Baker and Varshavsky, 1991) to "trap" proteins (and cause their accumulation) in yeast. However, 2D gel maps of extracts of these cells failed to reveal any such moieties (S.M. Arfin, unpublished observations). This may indicate that there is substantial redundancy in degradation pathways and that proteins that build up from the blockage of the E3 pathway can be turned over in other ways. It probably also is an indication that this pathway is of proscribed use for regulating normal cellular components.

Nonetheless, we have identified one process that does require N-end pathway participation, i.e. the proliferation of neurites from neuronal cells. PC12 cells, a cultured cell line from a rat pheochromocytoma, respond to various growth factors, such as nerve growth factor (NGF) and fibroblast growth factor (FGF), to produce a differentiated morphology highly similar to sympathetic neurons (Greene and Tischler, 1982). This induction is reversible and removal of the factor (or disruption of the signal transduction processes) results in the retraction of the neurites and reversion to the undifferentiated state. As shown in Fig. 3, treatment of PC12 cells with NGF and several dipeptides that are excellent inhibitors of protein degradation in a cell-free system (via the E3 pathway) results in the inhibition of neurite proliferation (Hondermarck et al., 1992).

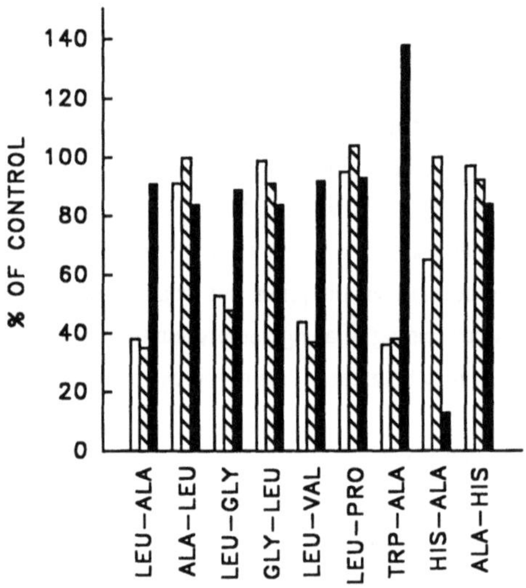

Fig. 3. Effect of dipeptides on NGF-induced neurite outgrowth in PC12 cells and on degradation of ^{125}I-labeled proteins in reticulocyte extracts. The open bars represent the percentage of cells bearing neurites after 72 h of treatment with the indicated dipeptide compared to control cells treated with NGF only. 62-84% of control cells expressed neurites after 72 h. The effect of the dipeptides on degradation of ^{125}I-beta-lactoglobulin (N-terminal Leu) and ^{125}I-BSA (N-terminal Asp) by reticulocyte extracts is represented by the hatched bars and solid bars, respectively. ATP-dependent degradation in the absence of dipeptides varied from 61-84%/h for ^{125}I-beta-lactoglobulin and 20-33%/h for ^{125}I-BSA. Taken from Hondermarck et al. (1992) with permission.

Thus, Leu-Ala, Leu-Gly, Leu-Val and Trp-Ala all show parallel effectiveness in the two assays. In contrast, the control peptides, Ala-Leu and Gly-Leu were unable to inhibit either activity. Interestingly, Leu-Pro was also ineffective in inhibiting both processes indicating that binding to the type II site in E3 is affected in part by the second residue of the bound sequence (see below). The type I inhibitor, His-Ala, was less effective but did cause some inhibition of neurite proliferation after 72 hr incubation (with NGF and dipeptide). The control peptide Ala-His was not inhibitory in either assay. Addition of Leu-Ala at 24 hr intervals after initiation of neurite outgrowth revealed that the inhibitors were able to prevent new neurite extension but were unable to reverse previously formed processes. Thus, the inhibitors are not blocking the binding or action (signal transduction) of the growth factor but rather the actual elaboration of the neurites themselves. Similar experiments demonstrated the effects to be reversible, i.e. removal of the dipeptide released the arrest of neurite outgrowth. Although the target proteins whose regulation (presumably degradation) is blocked by the E3 inhibitors is unknown as yet, it seems clear that the N-end rule is required in some fashion for the process of neurite growth.

As a result of our studies in PC12 cell responses, we have considerably extended the list of known inhibitors of E3 binding. These studies have demonstrated that second residues strongly influence binding to both sites, that D-amino acids are not tolerated in either position at the type II site but do not block binding at the type I (basic) site when in the penultimate position, and that penultimate prolines inhibit binding at both sites. By increasing our understanding of the binding to E3, it should be possible to better identify other physiological processes that use the N-end rule for regulation of one or more components (Sy, J., Arfin, S.M. and Bradshaw, R.A., manuscript in preparation).

<u>Summary</u>.

The importance and diversity of protein degradation in eukaryotic systems is well established but the specific details are just beginning to be defined. N-terminal recognition through E3 is clearly a highly specialized pathway that allows access to the ubiquitin-dependent proteasome. It appears to have very limited applications in the turnover of normal proteins and its chief role may be to eliminate foreign proteins that enter the cytoplasm by one or another mechanism (having escaped other surveillance systems such as lysosomes) or to degrade fragments that have arisen from intracellular proteolysis. However, we have provided evidence that at least one physiological process also utilizes this scheme and it is likely that others will be identified subsequently. At the same time, our studies suggest that degradation of improperly folded proteins (by N-terminal independent mechanisms) may be a more extensive process than previously thought. This would place even greater importance on understanding the mechanisms regulating protein turnover as failure to clear substantive amounts of improperly folded proteins would certainly be deleterious to cell physiology and activity.

ACKNOWLEDGMENT:

Work emanating from the authors' laboratory was supported by USPHS research grant DK32465. Ralph A. Bradshaw thanks the Alexander von Humboldt-Stiftung for support during the preparation of this manuscript and to attend the NATO ASI.

REFERENCES:

Arfin SM and Bradshaw RA (1988) Co-translational processing and protein turnover in eukaryotic cells. Biochemistry 27: 7979-7984.

Bachmair A, Finley D and Varshavsky A (1986) In vivo half-life of a protein is a function of its amino-terminal residue. Science 234: 179-186.

Baker RT and Varshavsky A (1991) Inhibition of the N-end rule pathway in living cells. Proc. Natl. Acad. Sci. USA 88: 1090-1094.

Bartel B, Wunning I. and Varshavsky A (1990) The recognition component of the N-end rule pathway. EMBO J. 9: 3179-3189.

Ben-Bassat A, Bauer K, Chang S-Y, Myambo K, Boosman A and Chang, S (1987) Processing of the initiator methionine from proteins: properties of the Escherichia coli methionine aminopeptidase and its gene structure. J. Bacteriol. 169: 751-757.

Boissel J-P, Kaspar TJ and Bunn HF (1988) Co-translational amino-terminal processing of cytosolic proteins. Cell-free expression of site-directed mutants of human hemoglobin. J. Biol. Chem. 263: 8443-8449.

Chang YH, Teichert U and Smith JA (1990) Purification and characterization of a methionine aminopeptidase from Saccharomyces cerevisiae. J. Biol. Chem. 265: 19892-19897.

Chang YH, Teichert U and Smith JA (1992) Molecular cloning, sequencing, deletion, and overexpression of a methionine aminopeptidase gene from Saccharomyces cerevisiae. J. Biol. Chem. 267: 8007-8011.

Ciechanover A, DiGiuseppe JA, Bercovich B, Orian A, Richter JD, Schwartz AL and Brodeur GM (1991) Degradation of nuclear oncoproteins by the ubiquitin system in vitro. Proc. Natl. Acad. Sci. USA 88: 139-143.

Ciechanover A, Ferber S, Ganoth D, Elias S, Hershko A and Arfin SM (1988) Purification and characterization of arginyl-tRNA protein transferase from rabbit reticulocytes. Its involvement in post-translational modification and degradation of acidic NH_2 termini substrates of the ubiquitin pathway. J. Biol. Chem. 263: 11155-11167.

Dice JF (1987) Molecular determinants of protein half-lives in eukaryotic cells. FASEB J. 1: 349-357.

Gonda DK, Bachmair A, Wunning I, Tobias JW, Lane WS and Varshavsky A (1989) Universality and structure of the N-end rule. J. Biol. Chem. 264: 16700-16712.

Greene LA and Tischler AS (1982) PC12 pheochromocytoma cultures in neurobiological research. Adv. Cell. Neurobiol 3: 373-414.

Heller H and Hershko A (1990) A ubiquitin-protein ligase specific for type III protein substrates. J. Biol. Chem. 765: 6532-6535.

Hershko A and Ciechanover A (1992) The ubiquitin system for protein degradation. Ann. Rev. Biochem. 61: 761-807.

Hondermarck, H, Sy J, Bradshaw RA, and Arfin SM (1992) Dipeptide inhibitors of ubiquitin-mediated protein turnover prevent growth factor-induced neurite outgrowth in rat pheochromocytoma PC12 cells. Biochem. Biophys. Res. Comm. 189: 280-288.

Huang S, Elliott RC, Lui PS, Koduri RK, Weickmann JL, Lee JH, Blair LC, Ghosh-Dastidar P, Bradshaw RA, Bryan KM, Einarson B, Kendall RL, Kolacz KH and Saito K (1987) The specificity of co-translational amino-terminal processing of proteins in yeast. Biochemistry 26: 8242-8246.

Jentsch S, Seufert W and Hauser HP (1991) Genetic analysis of the ubiquitin system. Biochim. Biophys. Acta 1089: 127-139.

Kendall RL (1992) The isolation and characterization of porcine liver methionine aminopeptidase. Ph.D. Thesis, University of California, Irvine.

Kendall RL and Bradshaw RA (1992) Isolation and characterization of the methionine aminopeptidase from porcine liver responsible for the co-translational processing of proteins. J. Biol. Chem. 267: 20667-20673.

Krishna RG and Wold F (1992) Specificity determinants of acylaminoacyl-peptide hydrolase. Protein Sci. 1: 582-589.

McGrath JP, Jentsch S and Varshavsky A (1991) UBA1: an essential yeast gene encoding ubiquitin-activating enzyme. EMBO J. 10: 227-236.

Miller CG, Strauch KL, Kukral AM, Miller JL, Wingfield PT, Mazzei GJ, Werlen RC, Graber P and Movva NR (1987) N-terminal methionine-specific peptidase in Salmonella typhimurium. Proc. Natl. Acad. Sci. USA 84: 2718-2722.

Reiss Y, Kaim D and Hershko A (1988) Specificity of binding of NH_2-terminal residue of proteins to ubiquitin-protein ligase. Use of amino acid derivatives to characterize specific binding sites. J. Biol. Chem. 263: 2693-2698.

Rivett AJ (1993) Proteasomes: Multicatalytic proteinase complexes. Biochem. J. 291: 1-10.

Roderick SL and Matthews BW (1993) Structure of the cobalt-dependent methionine aminopeptidase from Escherichia coli: A new type of proteolytic enzyme. Biochemistry 32: 3907-3912.

Tobias JW, Schrader TE, Rocap G and Varshavsky A (1991) The N-end rule in bacteria. Science 254: 1374-1377.

Varshavsky A (1992) The N-end rule. Cell 69: 725-735.

STRUCTURE AND MECHANISM OF PORINS

Georg E. Schulz
Institut für Organische Chemie und Biochemie
Albertstr. 21, D-79104 Freiburg im Breisgau
Germany

Porins form channels in the protective outer membrane of Gram-negative bacteria that are permeable for nutrients. They are usually homotrimers with subunit sizes ranging from 30 to 50 kDa and solute exclusion limits around 600 Da (Benz & Bauer, 1988; Jap & Walian, 1990). The channels are well permeable for polar solutes, but discriminate against nonpolar solutes of comparable sizes. Porins have been subdivided into specific and nonspecific ones (Nikaido, 1992). Specific porins show a comparatively large permeability at low, and saturation effects at high concentrations of the concerning solute. In contrast, non-specific porins function like inert holes showing diffusion rates that are proportional to the solute concentration at all levels.

The general architecture of porins has been first elucidated by electron microscopy (Engel et al., 1985, Jap et al., 1991). The first X-ray grade crystals were obtained from OmpF of E.coli (Garavito & Rosenbusch, 1980), the first atomic crystal structure was elucidated for a porin from a phototrophic bacterium (Weiss et al., 1990). An analysis of other crystals by molecular replacement methods using the Rb.capsulatus porin structure showed that the 16-stranded ß-barrel is present in quite a number of porins (Pauptit et al., 1991). Now, there are 4 porin structures known at atomic detail (Weiss & Schulz, 1992; Cowan et al., 1992; Kreusch et al., 1993). Among the integral membrane proteins, porins are particularly resistent against denaturants and proteases. Such a stability seems to be crucial for the formation of well-

NATO ASI Series, Vol. H 82
Biological Membranes:
Structure, Biogenesis and Dynamics
Edited by Jos A. F. Op den Kamp

ordered crystals. Particularly highly-ordered crystals diffracting to 1.8 Å resolution and better were obtained with the major porin of Rhodobacter capsulatus after running through a careful preparation procedure (Kreusch et al., 1991). The structure was solved by the usual method of multiple isomorphous replacement and amino acid sequence analysis (Schiltz et al., 1991). The quality of the established structure matches those of the best crystals from water-soluble proteins. The model of one subunit of the homotrimer contains all 301 amino acid residues of the polypeptide, 274 water molecules, 3 calcium ions, 3 octyltetraoxyethylene (C_8E_4) detergent molecules and 1 bound ligand. This ligand could not be identified yet. The following report focusses on this particular porin structure.

Each subunit of the trimer consists of a 16-stranded completely antiparallel ß-barrel and 3 short α-helices. All ß-strands are connected to their next neighbors forming the most simple ß-sheet topology (Fig.1). The loops a the bottom end of the barrel are short, whereas the loops at the top end are rather long. Accordingly, the ß-barrel has a smooth and a rough end. The longest loop at the top end contains 43 residues and runs into the interior of the barrel, where it constricts the pore to a small eyelet and stabilizes the barrel.

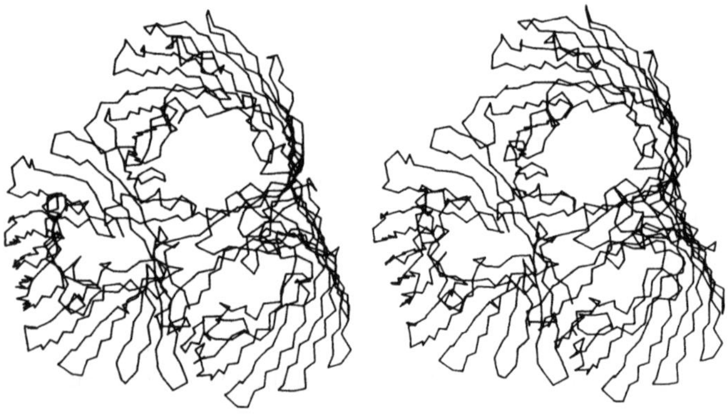

Figure 1: Stereo view of the C_α-backbone of trimeric porin from Rb.capsulatus taken from Weiss et al.(1991a). The view is from the external medium.

The porin architecture is shown in Figure 1. Around the exact threefold axis the subunits form a three-pronged star resembling hub and spokes of a wheel. The mobilities of the atoms in this central star as determined by the crystallographic temperature factors (Weiss & Schulz, 1992) are the lowest of the whole molecule. The star interior is nonpolar, six phenylalanines per subunit interdigitate tightly. The surface is polar. The star contains all six chain termini paired in salt-bridges between subunits. Consequently, a porin molecule is composed of a rigid central core constructed like a water-soluble protein, that is surrounded by three ß-barrel walls fencing off the membrane.

Figure 1 shows further that the central star is rather short along the threefold axis, which is perpendicular to the membrane plane. As a consequence there exists a common channel formed by all three subunits over a distance of about 15 Å. This common channel proceeds to the three diffusion-limiting eyelets formed between the central star and the long 43-residue loops that run into the ß-barrel interiors.

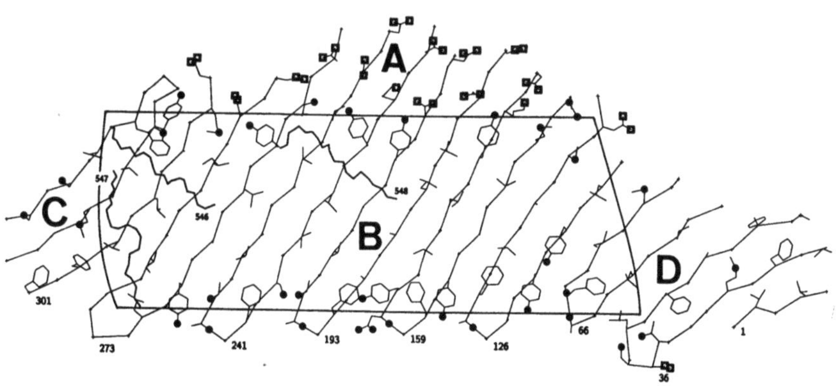

Figure 2: Outer surface of the ß-barrel projected onto a cylinder as taken from Weiss & Schulz(1992). The external medium is at the top and the periplasm is at the bottom. Ionogenic atoms are labeled by squares and polar atoms by dots. The surface areas marked by A, B, C, and D contain the ionogenic side chains cross-linking to the LPS molecules, the nonpolar side chains facing the nonpolar interior of the membrane, the lefthand-, and the righthand parts of the interface, respectively.

The longitudinal position of porin in the membrane can be derived from the outer surface shown in Figure 2. The nonpolar ring B with a height of 24 Å surrounding the trimer, certainly faces the nonpolar moiety of the membrane. The rough top end of the ß-barrel (see Fig.1) contains the larger loops with numerous charged side chains, while the smooth bottom end has only small loops with few polar residues. The top barrel end faces the external medium as derived from binding studies with antibodies and phages involving the large loops (Tommassen, 1988). This conclusion is in best agreement with the multitude of charged side chains at the top end which most likely interconnect to the acidic groups of the lipopolysaccharides (LPS) in the external half of the outer bacterial membrane. The salt bridge network integrates porin efficiently into the tough bacterial protection layer of noncovalently crosslinked LPS molecules, avoiding weak points at the protein : membrane interfaces.

Figure 2 shows furthermore two girdles of aromatic residues along the upper and lower border-lines between polar and nonpolar residues. The upper girdle contains mostly tyrosines pointing with their hydroxyl groups to the upper polar moiety of the membrane, while the lower girdle has phenylalanines pointing to the nonpolar membrane interior as well as tyrosines pointing to the polar part of the periplasmic half of the outer membrane. These patterns are significant, they occur also in the three recently determined other porin structures (Cowan et al., 1992; Kreusch et al., 1993). — Furthermore, there are four rare type-II′ ß-turns at the bottom end of the barrel. The peptide amides in these turns point toward the membrane where they can hydrogen bond to the phosphodiesters of the periplasmic half of the membrane.

The packing scheme in the exceptionally well-ordered crystals of the Rb.capsulatus porin (Kreusch et al., 1991) resembles the crystal packing of the photoreaction center (Deisenhofer & Michel, 1989) insofar as there are also polar crystal contacts. In the given space group R3 there exists only one contact type, a head to tail association, forming the three-dimensional array as indicated in Figure 3. This contact is obviously strong, it contains five hydrogen bonds and one salt bridge. The closest lateral distance between trimers occurs between nonpolar surfaces and is 15 Å. Any conceivable contact through detergent molecules should therefore be very weak providing only minor contributions to the crystals packing energy. The arrangement in the crystal corresponds to a stack of lipid bilayers containing porins. Accordingly, such bilayers could be present in the crystal. If this were the case, the crystals should show anisotropic electric conductivity.

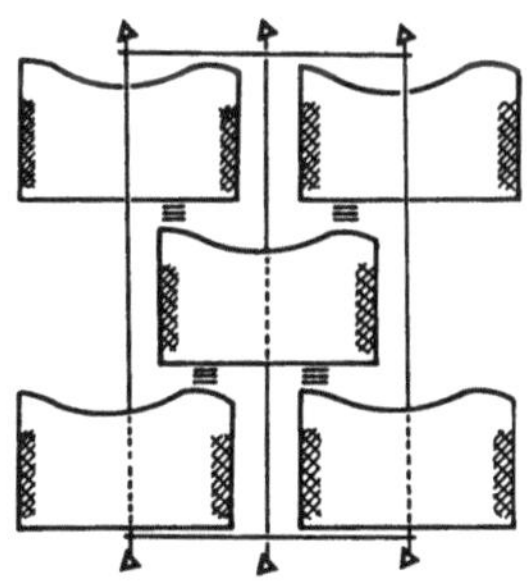

Figure 3: Molecular packing of porin in the crystals, the molecular threefold axes (▲) are crystallographic. The crystal contacts (≡) are all head to tail and polar. There is no lateral contact between the marked nonpolar surfaces.

Since the Tyr-Phe-pattern of Figure 2 has been observed for all four known porin structures, it most likely reflects a function. I suggest that these aromatics shield porin against fluctuations of the membrane and/or collisions with external effectors. Any transversal wave in the membrane or any push from the outside would rock the immersed porin as shown in Figure 4, exposing nonpolar protein surface to a polar membrane layer and polar protein surface to the nonpolar membrane interior. Both types of contact cause a large surface tension which is likely to scramble the polypeptide conformation. Since the aromatic side chains can rotate around their C_α-C_β-bonds much faster than the more massive trimer can rock, they are able to shield the respective surfaces against contact with the wrong counterpart (Fig.4). The strong

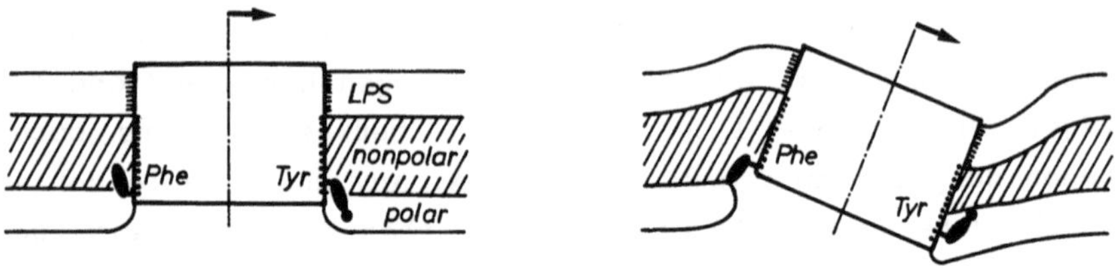

Figure 4: The suggested shielding rôle of phenylalanines and tyrosines during relative motions of porin and membrane.

association between the upper polar protein surface and the LPS layer explains the absence of Phe from the upper girdle: The cross-links to LPS anyway prevent the nonpolar membrane from reaching the upper polar protein surface.

The chain fold of trimeric porin containing the central rigid star with prongs that are connected by three ß-barrel walls suggests the following folding pathway: The three-pronged star folds initially like a water-soluble protein in the periplasm. Since this star contains all chain termini, the remaining chain segments form three large loops of about 200 residues each, which are suspended between the prongs. On contact with the nonpolar membrane interior, the three loops arrange themselves to the actually observed most simple ß-barrel topology with all strands antiparallel and connected to their next neighbors. This fold has an optimum neighborhood-correlation value (Schulz & Schirmer, 1979) allowing for a straight-forward folding process. On folding, the nonpolar side chains of the ß-barrel residues rotate to the outside facing the nonpolar moiety of the membrane. Subsequently, the large 43-residue loop inserts into the barrel, supports its center against membrane pressure and defines the pore size.

As shown in Figure 5 the eyelet of a channel is almost exclusively lined by ionogenic groups that also segregate into negatively and positively charged rims. The positive rim is at the surface of the central star and consists of half a dozen Arg, Lys and His side chains. The negative rim is mostly formed by Asp and Glu residues on the 43-residue loop running

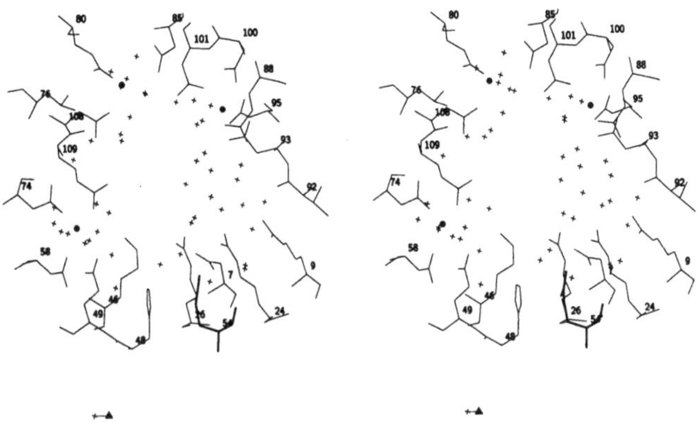

Figure 5: Stereo view from the external medium through the pore eyelet as taken from Weiss & Schulz(1992). The eyelet is lined by negatively charged side chains at its upper rim and positively charged ones at the lower rim. The molecular threefold axis (▲), water molecules (✗) and Ca^{2+} ions (●) are given.

into the ß-barrel. In the Rb.capsulatus porin the negative charges are partially compensated by two bound Ca^{2+} ions. The removal of these Ca^{2+} ions should change the permeability, as it has been actually observed for ATP (Carmeli & Lifshitz, 1989).

The positive and negative rims juxtaposed across the eyelet cause a transversal electric field (Weiss et al., 1991b), the strength of which is best estimated from two arginines participating in the positive rim. These arginines are at van-der-Waals distance showing very low mobilities at pH 7.2 in spite of their repelling positive charges. Such an arrangement is only possible if the charges are fixed by a strong electric field.

The high resolution X-ray structure analysis allows the assignment of reasonably well fixed water molecules (Weiss & Schulz, 1992). As shown in Figure 5, the eyelet contains quite a number of them. Water molecules orient themselves along the electric field as sketched in Figure 6. In the center of the eyelet there are mobile water molecules confined to a rather small cross section of about 4 Å by 5 Å. They form a hole

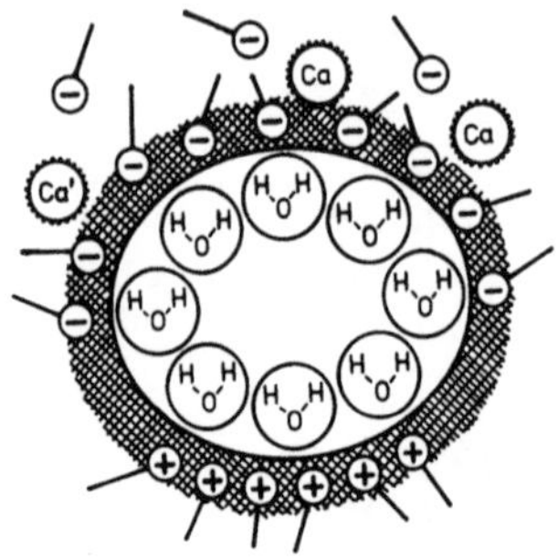

Figure 6: Sketch of the pore eyelet with a ring of bound water molecules that are oriented in the transversal electric field. The rim of negatively charged side chains is strengthened by two strongly (Ca) and one weakly (Ca') bound Ca^{2+} ion. The depth of the pore eyelet is about 6 Å.

through which a molecule with the cross section of an alkyl chain can permeate in principle. This is actually prohibited, however, by the strong transversal field as explained in Figure 7a. Field and eyelet behave like a charged capacitor that stores an electric energy proportional to the high dielectric constant of 80 of oriented water molecules multiplied by the volume. Any incoming solute with a lower dielectric constant, like about 2 for an alkyl chain, causes a decrease of the capacitor energy and thus generates a force expelling the intruder.

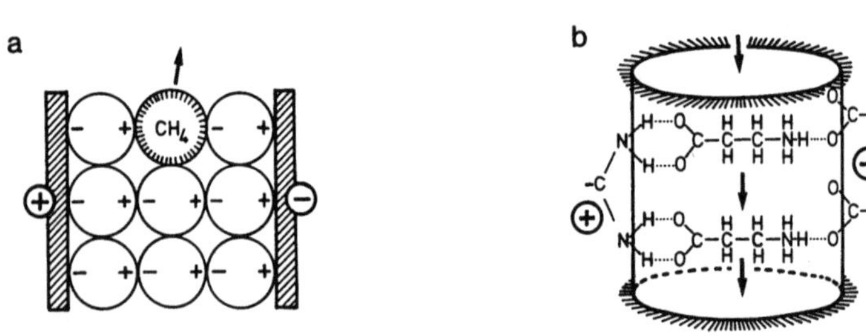

Figure 7: Diffusion through the pore eyelet. (a), Sketch of the pore eyelet as a capacitor explaining the energetical exclusion of nonpolar solutes. (b), Sketch of a permeating zwitter ion which remains oriented in the electric field.

Quite different is the diffusion of a polar solute as depicted in Figure 7b. Such a solute is oriented by the transversal electric field and will remain so over the whole diffusion distance. Consequently, the solute is oriented in the pore eyelet like a substrate bound to an enzyme. This reduces the entropy barrier of the diffusion process in the same way as this barrier is reduced for a chemical reaction on an enzyme surface. The diffusion is accelerated because there is no tumbling. In conclusion the transversal field acts as an electric separator, facilitating the permeation of polar solutes while fencing off nonpolar ones.

Porins incorporated into black lipid films may reduce their electric conductivity on application of a voltage around 100 mV across this film (Jap & Walian, 1990). Such behavior can be understood when considering the shape of the pore as can be visualized in Figures 1 & 5. The membrane channel has a rather large cross section over its whole length except for the short distance of about 6 Å at the diffusion-limiting eyelet. This implies that any voltage across the membrane (say 100 mV) lies essentially across the eyelet with its small cross section (certainly more than 90 mV). Since the energy gained by moving a single charge over 90 mV is about 10 kJ/mol and thus equivalent to the energy of a hydrogen bond, the strong longitudinal electric field along the eyelet can tear off any charged group that can be moved over the 6 Å length of the eyelet. This applies, for instance, to the ϵ-amino group of a lysine. Therefore, the voltage closure is most likely caused by a disruption of the eyelet structure, which does not correspond to a natural functional state of the porin.

Porins have been subdivided into specific and nonspecific ones (Nikaido, 1992). The structurally known Rb.capsulatus porin has previously been classified as nonspecific. In the crystal structure, however, this porin is ligated by a small molecule that has not yet been identified. Efficient binding of tetrapyrrols to this porin had been reported by Bollivar & Bauer (1992). The binding site observed in the crystal is rather nonpolar and located between the 43-residue loop running into the barrel and a little domain protruding to the external medium. The site is at the external end of the eyelet such that a large ligand can block diffusion. This situation suffices to cause a nonlinear diffusion characteristic, as they are observed for specific porins. Accordingly, the Rb.capsulatus porin is actually both, specific and nonspecific, depending on the examined solute. Generalizing this observation, I suggest that many if not all porins belong to both classes, but that the solutes picked up specifically have been identified only in very few cases.

ACKNOWLEDGEMENT

I thank A. Kreusch, E. Schiltz, J. Weckesser, M.S. Weiss and W. Welte for their essential contributions to the structure analysis of the Rb.capsulatus porin.

REFERENCES

Benz R and Bauer K (1988) Permeation of hydrophilic molecules through the outer membrane of Gram-negative bacteria. Eur.J.Biochem. 176: 1-19

Bollivar DW and Bauer K (1992) Association of tetrapyrrol intermediates in the bacteriochlorophyll *a* biosynthetic pathway with the major outer-membrane porin protein of Rhodobacter capsulatus. Biochem.J. 282: 471-476

Carmeli C and Lifshitz Y (1989) Nucleotide transport in Rhodobacter capsulatus. J.Bacteriol. 171: 6521-6525

Cowan SW, Schirmer T, Rummel G, Steiert M, Ghosh R, Pauptit RA, Jansonius JN and Rosenbusch JP (1992) Crystal structures explain functional properties of two E.coli porins. Nature(London) 358: 727-733

Deisenhofer J and Michel H (1989) The photosynthetic reaction center from the purple bacterium Rhodopseudomonas viridis. Science 245: 1463-1473

Engel A, Massalski A, Schindler H, Dorset DL and Rosenbusch JP (1985) Porin channel triplets merge into single outlets in the E.coli outer membranes. Nature(London) 317: 643-645

Garavito RM and Rosenbusch JP (1980) Threedimensional crystals of an integral membrane protein: an initial X-ray analysis. J. Cell Biol. 86: 327-329

Jap BK and Walian PJ (1990) Biophysics of the structure and function of porins. Quart.Rev.Biophys. 23: 367-403

Jap BK, Walian PJ and Gehring K (1991) Structural architecture of an outer membrane channel as determined by electron crystallography. Nature (London) 350: 167-170

Kreusch A, Weiss MS, Welte W, Weckesser J and Schulz GE (1991) Crystals of an integral membrane protein diffracting to 1.8 Å resolution. J.Mol.Biol. 217: 9-10

Kreusch A, Neubüser A, Schiltz E, Weckesser J and Schulz GE (1993) The structure of porin from Rhodopseudomonas blastica at 2.0 Å resolution. Submitted

Nikaido H (1992) Porins and specific channels of bacterial outer membranes. Mol.Microbiol. 6: 435-442

Pauptit RA, Schirmer T, Jansonius JN, Rosenbusch JP, Parker MW, Tucker AC, Tsernoglou D, Weiss MS and Schulz GE (1991) A common channel-forming motif in evolutionarily distant porins. J.Struct.Biol. 107: 136-145

Schiltz E, Kreusch A, Nestel U and Schulz GE (1991) Primary structure of porin from Rhodobacter capsulatus. Eur.J.Biochem. 199: 587-594

Schulz GE and Schirmer RH (1979) Principles of protein structure. Springer Verlag, New York

Tommassen J (1988) Biogenesis and membrane topology of outer membrane proteins in E.coli. In: Membrane biogenesis (Op den Kamp JAF, ed.). NATO ASI series H 16: 351-373, Springer Verlag, Berlin

Weiss MS, Wacker T, Weckesser J, Welte W and Schulz GE (1990) The three-dimensional structure of porin from Rhodobacter capsulatus at 3 Å resolution. FEBS Letters 267: 268-272

Weiss MS, Kreusch A, Schiltz E, Nestel U, Welte W, Weckesser J and Schulz GE (1991a) The structure of porin from Rhodobacter capsulatus at 1.8 Å resolution. FEBS Letters 280: 379-382

Weiss MS, Abele U, Weckesser J, Welte W, Schiltz E and Schulz GE (1991b) Molecular architecture and electrostatic properties of a bacterial porin. Science 254: 1627-1630

Weiss MS and Schulz GE (1992) Structure of porin refined at 1.8 Å resolution. J.Mol.Biol. 227: 493-509

THE PORE-FORMING TOXIN AEROLYSIN: FROM THE SOLUBLE TO A TRANSMEMBRANE FORM.

F. Gisou van der Goot[1], J. Thomas Buckley[2] and Franc Pattus[1]
[1]European Molecular Biology Laboratory
Meyerhofstrasse 1
6900 Heidelberg
Germany

INTRODUCTION

Bacterial pore forming toxins are an interesting model to study protein-membrane interactions including post translational insertion of proteins into membranes. These toxins have the remarkable feature of existing in two states either a water soluble form, which is secreted by the producing cell and can diffuse towards the target cell, and a membrane-bound form which forms the final transmembrane pore. This change in environment most certainly implies a major conformational change in the protein. Different types of protein seem to cope with this problem in different ways.

In a first approximation one can distinguish two types of pore-forming bacterial toxins. The first class is composed of α-helical proteins. They have been widely studied and include pore forming colicins, exotoxin A, the translocation domain of diptheria toxin and δ-endotoxins from *Bacillus thuringiensis*.These toxins constitute what has been refered to "inside-out" membrane proteins. In their soluble state, one or two hydrophobic helices are sheltered in the middle of the structure. Upon interaction with the membrane, either with a specific receptor or directly with the lipids, a large conformational change occurs which results in the exposure of the hydrophobic helices to the membrane core. A transmembrane potential is then necessary to induce channel formation (for review see Li, 1992, Lakey et al., 1992). Less is known about the second class of pore forming bacterial toxins includes aerolysin, alpha toxin of *Staphylococcus aureus* and streptolysin O. These toxins have a high degree of β-sheet structure which is not unusual for a soluble protein but very uncommon for membrane protein. The only

[2] Department of Biochemistry and Microbiology
University of Victoria
Box 3055
Victoria, B.C., V8W 3P6
Canada

NATO ASI Series, Vol. H 82
Biological Membranes:
Structure, Biogenesis and Dynamics
Edited by Jos A. F. Op den Kamp

well studied examples of β-sheet membrane proteins are the porins from the outer membrane of Gram-negative bacteria or from mitochondria. Toxins from this class have to oligomerize for channel formation to occur. As an example of this second class of toxin, we will describe the mechanism by which aerolysin induces cell lysis.

AEROLYSIN

Aerolysin is a pore-forming toxin produced by various strain of *Aeromonas* and is the major cause of the pathogenicity of these bacteria in mammals (Janda et al., 1984, Daily et al., 1981). The protein is synthesized as a preprotoxin containing a typical signal sequence at the amino-terminus that is cleaved off during transfer across the inner membrane of the producing bacteria (Howard and Buckley, 1985). Translocation of proaerolysin across the outer membrane from the periplasm requires a separate step, involving a group of more than 10 genes that comprise the general secretion pathway now known to occur in many Gram-negative bacteria (Jiang and Howard, 1991). Once outside the parent cell, the protoxin, which is completely inactive, diffuses towards the target cell where it binds with high affinity to a membrane receptor. The transmembrane protein glycophorin is the receptor on the surface of mammalian erythrocytes (Howard and Buckley, 1982). Subsequent maturation of the protein occurs by proteolytic cleavage at the C-terminus. Activation is followed by the formation of an oligomeric transmembrane channel that allows passive flux of ions and small molecules across the bilayer, leading to cytolysis of erythrocytes and other cells that are unable to repair the damage to their membranes.

The structure

Proaerolysin is secreted as a 470 amino acid protoxin, and we have shown by mass spectrometry that the protein is not post-translationaly modified (van der Goot et al., 1992). Proaerolysin has been crystallized (Tucker et al., 1990) and its structure has been recently determined by X-ray crystallography at 2.8 Å resolution (Parker et al., submitted). In close agreement with predictions based on circular dichroism measurements (CD), 40 % of the proaerolysin amino acid chain is in β-sheet and 17 % is in a-helices (van der Goot et al., 1992).

The primary sequence is very hydrophilic, and there are no hydrophobic stretches such as those normally encountered in transmembrane proteins.

The molecular weight of proaerolysin in solution measured by analytical ultracentrifugation was found to be 102 587 Da, corresponding to almost exactly twice the molecular weight of the monomer determined by mass spectrometry (van der Goot et al., 1993a). The fact that proaerolysin is a dimer in solution was further confirmed by cross-linking experiments with dimethyl suberimidate (figure 1). This results is less surprizing a posteriori since the asymmetric unit of the proaerolysin crystals also contains a dimer. The dimer could be dissociated by 0.05% SDS as shown both by ultracentrifugation and cross-linking experiments.

Dimerization of proaerolysin seems to confer stability to the protein. In fact, folding and dimerization seem to be tightly coupled. When the dimer is dissociated by SDS, all the rigid tertiary interactions are lost as shown by near ultraviolet circular dichroism (UV CD) (Figure 2) and the protein becomes sensitive to proteases (van der Goot et al, 1993a).

We have further characterized the monomeric state of aerolysin after dissociation of the dimer with SDS. Although the rigid tertiary interactions are lost, the secondary structure did not seem affected as shown by far UV CD measurements. The maximum emission wavelength of the tryptophan residues was not significantly affected and the protein exposed hydrophobic patches as witnessed by increased binding of the hydrophobic dye 1-anilino-8-naphtalen sulfonate (ANS). The proaerolysin monomer in SDS has the characteristics of a compact denatured state or molten globule.

The activation step

Activation is absolutely required for the toxin to be cytolytic, indeed proaerolysin is completely unable to lyse erythrocytes. To become active, the protein must be nicked within a narrow region of the sequence near the C-terminus and a variety of different proteases can be involved in this process. We have shown by mass spectrometry that trypsin cuts after K427 and chymotrypsin after R429 (van der Goot et al., 1992). The two forms of the mature toxin then produced are both cytolytic, but as we will see below they differ in activity. Since proaerolysin is a water-soluble inactive protein, whereas aerolysin can form transmembrane channels, one would expect activation to induce a major conformational shift in the molecule. Surprisingly activation has very little effect on the structure of the protein. First as proaerolysin is dimeric in solution, we had expected that activation would lead to the dissociation of the dimer, as this would have provided a clear route to the oligomeric form of the protein that

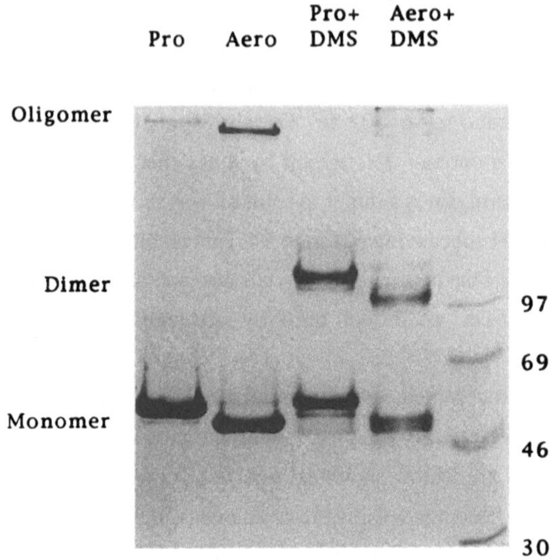

Figure 1: Cross-linking of proaerolysin and aerolysin with dimethyl suberimidate (DMS).
Proaerolysin and aerolysin (0.5 mg/ml) were crosslinked using 10 mM DMS in 0.1 M triethanolamine pH 8.5 (reproduced from van der Goot *et al.*, 1993b).

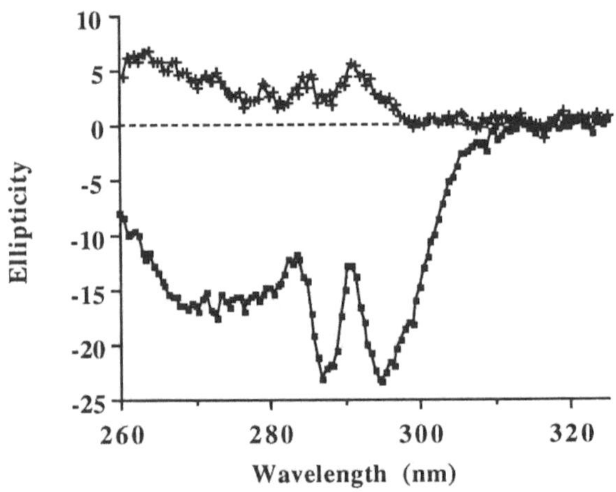

Figure 2: Near ultraviolet circular dichroic spectrum of wild type proaerolysin in the presence (crosses) and in the absence (filled squares) of 0.05 % SDS. Proaerolysin was diluted in 150 mM NaCl, 50 mM N-[2-hydroxyethy]piperazine-N'-[3-propanesulfonic acid] (EPPS), pH 7.4 to a concentration of 1.15 10^{-5} M. Ellipticities are expressed per mole of protein (10^3 deg.cm^2.$dmol^{-1}$).

generates the membrane pore. However ultracentrifugation and cross-linking experiments showed that the soluble form of aerolysin is also a dimer (figure 1). Far and near UV CD measurements suggested that both the secondary and the tertiary structure of the protein are not affected by the removal of the C-terminal peptide (van der Goot et al., 1992). In agreement with these observations, the maximum emission wavelength of the tryptophan residues was not significantly affected. The X-ray structure shows that the activation peptide is tightly linked to the rest of the protein by extensive hydrophobic contacts as well as by 17 hydrogen bonds (Parker et al., submitted). It is therefore unlikely that proteolytic cleavage would be sufficient by itself to dislodge the activation region. Moreover, displacement of the activation peptide would reveal a large hydrophobic patch, and this has not been observed experimentally. In fact, we have found that the aerolysin dimer is no more able to bind the hydrophobic dye 1-anilino-8-napthalene sulfonate (ANS) than proaerolysin (figure 3).

The exact role of the activation step is still unknown. It does not seem to destabilize the dimer as one might have expected. Thus the same concentration of SDS is needed to dissociate both the proaerolysin and the aerolysin dimer and they have the same stability when measuring denaturation by urea (van der Goot et al., 1993a).

Oligomerization

Although we do not know precisely what the effect of activation on the structure of aerolysin is, it does confer to the protein the ability to oligomerize. Oligomerization is absolutely required for the lytic activity of aerolysin (Howard and Buckley, 1985, Wilmsen, et al., 1990). *In vivo*, this step most probably takes place at the surface of the target cell, where the protein is concentrated by virtue of its affinity for the receptor. However oligomerization also can occur spontaneously in solution.

We could follow oligomerization *in vitro* by measuring the increase of light scattering (van der Goot et al., 1992) and thus identify factors that influence the rate of oligomerization. First the protease that activates the toxin is important. Thus aerolysin oligomerizes at a faster rate when it is produced by the action of chymotrypsin rather than trypsin. Low ionic strength, high protein concentration and increasing temperature also favor oligomerization. On the other hand, oligomerization can be completely prevented by the addition of by zinc (Wilmsen et al., 1990), or by raising the pH above 8 (unpublished data). Several single amino acid changes in the primary sequence of aerolysin have been made that also affect oligomerization. Substituting His 132 with Asp completely inhibits the formation of oligomers (Wilmsen et al., 1991), whereas changing Trp371 to Leu or Trp373 to Leu increases the ability of the protein to oligomerize 10 to 20 fold (van der Goot et al., 1993b).

Like other oligomeric proteins that have a high β-sheet content, such as poly C9 (Podack and Tschopp, 1982), P22 tail spike endorhamnosidase (Chen and King, 1991), the paired helical filaments involved in Alzheimer disease (Selkoe et al., 1982), and staphylococcal alpha-toxin (Bhakdi and Tranum-Jensen, 1991), the aerolysin oligomer is unusually stable. It is insensitive to proteases and cannot be disrupted by guanidine hydrochloride, guanidine thiocyanate, urea, high or low pH, SDS, reducing agents, or by heating (unpublished data). After polyacrylamide gel electrophoresis in sodium dodecyl sulphate the mobility of the oligomer indicates that it is pentameric or hexameric.

The far UV CD spectra of the oligomer is very similar to that of proaerolysin, an indication that there is little change in secondary structure upon oligomerization, but the near UV CD spectra are significantly different, demonstrating that there is some reorganization of domains or a modification of the position of the aromatic side chains (unpublished). This conformational change leads to the exposure of hydrophobic patches. This was observed by measuring the increase of the hydrophobic dye ANS after activation of the protoxin with trypsin (Figure 3). These patches are shielded in both proaerolysin and dimeric aerolysin. Upon oligomerization, hydrophilic surfaces of the protein that are otherwise exposed in the water become shielded and, conversely hydrophobic surfaces that were sheltered become exposed. This may provide the driving force for the initial penetration of the oligomer into the membrane (see below). Rearrangement of the tertiary structure of a protein without significant alteration of the secondary structure elements seems energetically economical, and it has been previously observed during membrane insertion of the pore-forming toxin colicin A (Lakey et al., 1991, van der Goot et al., 1991) and other pore-forming proteins (van der Goot et al., 1992).

We believe on the basis of the following observations that the transition from the dimer to the oligomer involves a passage through the monomeric state. Image analysis of two dimensional crystals of aerolysin indicates that monomers are arranged in parallel fashion to form the oligomer (Wilmsen et al., 1992), whereas in the dimer they are antiparallel. More evidence in favor of a dissociation of the dimer prior to oligomerization comes from the study of two point mutants W371L and W373L. As illustrated Figure 4, these two mutants are much more sensitive to SDS than wild type proaerolysin. As previously mentioned, these two mutants are able to oligomerize at far lower concentrations than wild type (van der Goot et al., 1993b). In contrast H132N, a mutant that is unable to oligomerize, has a similar stability in SDS as the wild type (figure 4). It thus appears that facilitating the dissociation of the dimer increases the ability to oligomerize. Finally it was important to trap a partially folded state of the monomer (see above). This contrasts with what is generally observed with dimeric enzymes for which a two state model, either folded dimer or completely unfolded monomer, is proposed

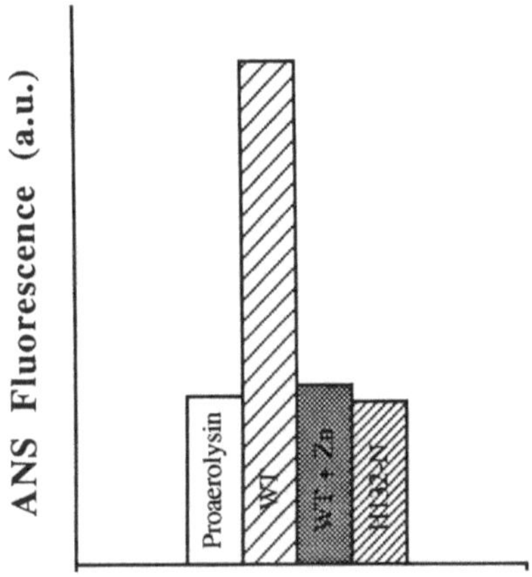

Figure 3: Binding of the hydrophbic dye 1-anilino-8-naphtalenesulfonate (ANS) The increase of fluorescense of ANS after activation of proaerolysin with trypsin was measured for wild type proaerolysin with and without 2.5 mM zinc acetate as well as with the H132-N mutant (in the absence of zinc). The protein was dilute in 20 mM Tris-acetate, pH 6.8, 37°C to a final concentration of 0.5 mg/ml. The final ANS concentration was 50 μM. The excitation wavelength was 380 nm and the emission wavelength 480 nm.

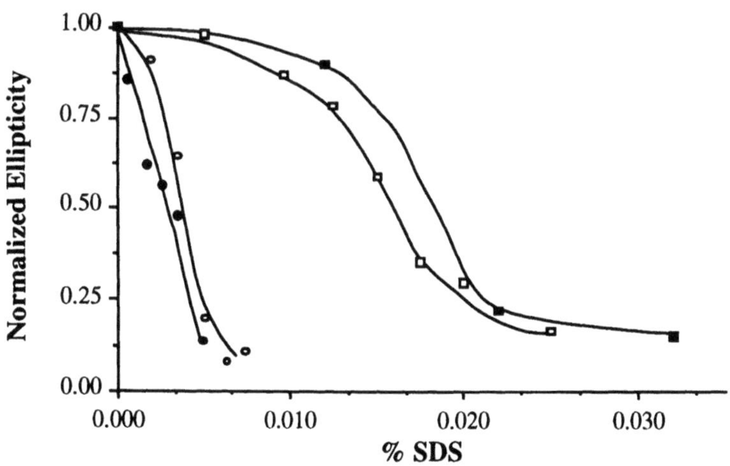

Figure 4: Effect of SDS on the tertiary structure of mutant and wild type proaerolysins. The ellipticity of proaerolysin at 294 nm was measured for the wild type (filled squares) toxin as well as H132N (open squares), W371L (filled circles) and W373L (open circles) point mutants. The normalized ellipticity is the ellipticity at 294 nm in the present of SDS over the ellipticity in the native state (0% SDS) (reproduced from van der Goot *et al.*, 1993b).

(Reece et al., 1991, Gittelman and Matthews, 1990, Dirr and Reinemer, 1991). This observation is important because considering the size of aerolysin, the complexity of its structure and the time scale at which oligomerization can occur, it seems extremely unlikely that the oligomer could form from fully unfolded monomers.

To reach a firm conclusion as to whether the dimer dissociates before oligomerization, one would have to trap the monomer as it undergoes the oligomerization process. This might however be very difficult as dimer dissociation and oligomerization are probably concerted events in order to protect the protein from destruction by proteases.

Aerolysin-membrane interactions

Little is known about the interactions between aerolysin and the lipid bilayer. Curiously we have never been able to measure any surface activity of aerolysin using lipid monolayers of various compositions. However experiments with artificial liposomes containing a chloride sensitive dye have shown that aerolysin induces chloride efflux (van der Goot et al., 1993b). A puzzling question about oligomeric pore-forming toxins is whether oligomerization occurs in solution outside the membrane prior to insertion or by lateral diffusion of the monomers in the membrane.

The study of aerolysin/membrane interactions is complicated by the fact that the sample is not homogeneous, it contains different populations: dimeric aerolysin, oligomeric aerolysin which tend stick together to form either soluble micelle-like structures (seen by electron microscopy) or aggregates. We have tried to identify which population is responsible for channel formation *in vitro*. Our current view, based on several indirect lines of evidence, is that only the oligomer is able to form channels in artificial liposomes (van der Goot et al., 1993b). We do not know whether the monomers that undergo oligomerization are free in solution of bound to the membrane. However he active form of aerolysin appears to be very minor sub-population suggesting that only the monomeric form of the oligomer, and not the micelle-like structures or the aggregates, are able to make a channel. This agrees well with fact that only the oligomeric form of aerolysin is hydrophobic.

These studies have been performed with artificial liposomes. The *in vivo* situation is however quite different. Indeed before lysing erythrocytes, aerolysin first binds to glycophorin. We do not know whether glycophorin is involved in this process. Glycophorin most probably enables to concentrate the toxin at the surface of the membrane but might also have an influence on the orientation of the protein or on dimer dissociation. This aspects have not yet been thoroughly investigated.

CONCLUSIONS

We now know many of the details in the steps required for the transformation of water-soluble proaerolysin into the transmembrane oligomeric aerolysin channel. Clearly the combination of structural, biochemical and molecular approaches that have now become available will help to explain some of the more puzzling aspects of this process and give us a better understand of the mechanism of protein-membrane interactions.

REFERENCES

Bhakdi, S. and Tranum-Jensen, J. (1991) Alpha-toxin of *Staphylococcus aureus*. Microbiol. Rev. 733-751.

Chen, B.-L. and King, J. (1991) Thermal unfolding pathway for the thermostable P22 tailspike endorhamnosidase. Biochemistry 30: 6260-6269.

Daily, O. P., Joseph, S. W., Coolbaugh, J. C. et al. (1981) Association of Aeromonas sobria with human infection. J. Clin. Microbiol. 13: 769-777.

Dirr, H. W. and Reinemer, P. (1991) Equilibrium unfolding of class pi glutathione S-transferase. Biochem-Biophys-Res-Commun 180: 294-300.

Gittelman, M. S. and Matthews, C. R. (1990) Folding and stability of trp aporepressor from *Escherichia coli*. Biochemistry 29: 7011-7020.

Howard, S. P. and Buckley, J. T. (1982) Membrane glycoprotein receptor and hole forming properties of a cytolytic protein toxin. Biochemistry 21: 1662-1667.

Howard, S. P. and Buckley, J. T. (1985) Activation of the hole forming toxin aerolysin by extracellular processing. J. Bacteriol. 163: 336-340.

Janda, J. M., Bottone, D. J., Sinner, C. V. and Calcaterra, D. (1984) Phenotype markers associated withgastrointestinal *Aeromonas hydrophila* isolates from symptomatic children. J. Clin. Microbiol. 17: 588-591.

Jiang, B. and Howard, S. P. (1991) Mutagenesis and Isolation of *Aeromonas-Hydrophila* Genes Which Are Required for Extracellular Secretion. J. Bacteriol. 173: 1241-1249.

Lakey, J. H., Gonzalez, M. J., van der Goot F.G.. and Pattus, F. (1992) The membrane insertion of colicins. Febs Lett. 307: 26-9.

Lakey, J. H., Massotte, D., Heitz, F., Dasseux, J. L., Faucon, J. F., Parker, M. W. and Pattus, F. (1991) Membrane Insertion of the Pore-Forming Domain of Colicin-A - A Spectroscopic Study. Eur. J. Biochem. 196: 599-607.

Li, J. (1992) Bacterial toxins. Current Opinion in Structural Biology 2: 545-556.

Podack, E. R. and Tschopp, J. (1982) Circular polymerization of the ninth component of complement. Ring closure of the tubular complex confers resistance to detergent dissociation and to proteolytic degradation. J. Biol. Chem. 257: 15204-15212.

Reece, L. J., Nichols, R., Ogden, R. C. and Howell, E. E. (1991) Construction of a synthetic gene for an R-plasmid -encoded dihydrofolate reductase and studies on the role of the N-terminus in the protein. Biochemistry 30: 10895-10904.

Selkoe, D. J., Ihara, Y. and Salazar, F. J. (1982) Alzheimer disease: insolubility of partially purified paired helical filaments in sodium dodecyl sulfate and urea. Science 215: 1243-1245.

Tucker, A. D., Parker,M.W., Tsernoglou, D. and Buckley, J.T. (1990) Crystallization of a proform of aerolysin, a hole-forming toxin from *Aeromonas hydrophila.* J Mol Biol 212: 561-562.

van der Goot, F. G., Ausio, J., Wong, K. R., Pattus, F. and Buckley, J. T. (1993a) Dimerization stabilizes the pore-forming toxin aerolysin in solution. J. Biol. Chem. (in press)

van der Goot, F. G., González-Mañas, J. M., Lakey, J. H. and Pattus, F. (1991) A 'molten-globule' membrane-insertion intermediate of the pore-forming domain of colicin A. Nature 354: 408-410.

van der Goot, F. G., Lakey, J. H. and Pattus, F. (1992) The molten globule intermediate for protein insertion or translocation through membranes. Trends in Cell Biology 2: 343-348.

van der Goot, F. G., Lakey, J. H., Pattus, F., Kay, C. M., Sorokine, O., Van Dorsselaer, A. and Buckley, T. (1992) Spectroscopic study of the activation and oligomerization of the channel-forming toxin aerolysin: Identification of the site of proteolytic activation. Biochemistry 31: 8566-8570.

van der Goot, F. G., Wong, K. R., Pattus, F. and Buckley, J. T. (1993b) Oligomerization of the channel-forming toxin Aerolysin precedes its insertion into lipid bilayer. Biochemistry 32: 2636-2642.

Wilmsen, H. U., Buckley, J. T. and Pattus, F. (1991) Site-directed mutagenesis at histidines of aerolysin from Aeromonas hydrophila: a lipid planar bilayer study. Mol. Microbiol. 5: 2745-2751.

Wilmsen, H. U., Leonard, K. R., Tichelaar, W., Buckley, J. T. and Pattus, F. (1992) The Aerolysin Membrane Channel Is Formed by Heptamerization of the Monomer. EMBO J 11: 2457-2463.

Wilmsen, H. U., Pattus, F. and Buckley, J. T. (1990) Aerolysin, a hemolysin from Aeromonas hydrophila, forms voltage-gated channels in planar bilayers. J Membrane Biol 115: 71-81.

Signal Sequences: Roles and Interactions by Biophysical Methods

Lila M. Gierasch
Department of Pharmacology
University of Texas Southwestern Medical School
5323 Harry Hines Boulevard
Dallas, Texas 75235-9041
U.S.A.

Signal sequences are essential for the efficient and selective targeting of nascent protein chains either to the endoplasmic reticulum, in eukaryotes, or to the cytoplasmic membrane, in prokaryotes (for a review, see Gierasch, 1989). Furthermore, signal sequences play a central, although poorly understood, role in the translocation of polypeptide chains across membranes. Despite their ability to perform multiple, common functions, signal sequences lack primary structural homology. Instead, they share several general properties (von Heijne, 1985): (1) an amino-terminal region with a net positive charge; (2) a hydrophobic core of about ten residues; (3) a locus six to eight residues preceding the cleavage site that often contains a prolyl or glycyl residue and has a relatively high predicted turn tendency; and (4) a motif at the cleavage site--AXA in prokaryotes or small-large-small in eukaryotes. These shared features and the finding that signal sequences are often transportable from one secreted protein to another with retention of function invite the study of isolated signal peptides using physical methods that may reveal the conformational and interactive propensities of these intriguing sequences. This information then can help to elucidate the likely roles of signal sequences in the export pathway. Also, the isolated signal peptides can serve as probes for the interactions of the signal sequence with proteins of the export pathway. By comparing properties of export-competent signal peptides with those of peptides corresponding to export-defective mutants, we have found that all functional signal peptides have both a capacity to form an α helix in membrane

NATO ASI Series, Vol. H 82
Biological Membranes:
Structure, Biogenesis and Dynamics
Edited by Jos A. F. Op den Kamp

environments and an ability to spontaneously insert into the acyl chain region of a membrane (Briggs & Gierasch, 1984; Briggs *et al.*, 1985; McKnight *et al.*, 1989; Hoyt & Gierasch, 1991a, 1991b). As principal methods, we have used circular dichroism (CD) and nuclear magnetic resonance (NMR) to determine the conformational behavior and fluorescence spectroscopy to analyze membrane insertion. Recently, we have developed a signal peptide of enhanced water solubility in order to apply the method of transferred nuclear Overhauser enhancements (trNOEs) to determine directly the conformation of a signal peptide in a lipid bilayer. We find from these studies and our fluorescence work that the hydrophobic core of the signal peptide resides well-embedded in the acyl chain region of the bilayer in a stable α helix, while the N-terminal region associates with the surface. The C-terminal segment adopts a less stable α helix and appears to spend some time at the interface. Fluorescence quenching results on the LamB signal peptide do not support a stable transmembrane arrangement; work is in progress to establish whether this behavior is general and what the influence is of additional residues past the cleavage site.

In this short review, our results on the conformational properties of signal peptides will be summarized, followed by a description of our findings on signal peptide/membrane interactions. Then, recent work on the conformations of signal peptides in bilayers using trNOEs will be presented. Lastly, the potential implications of these biophysical experiments for the functions of signal sequences *in vivo* will be suggested.

Conformational Properties of Signal Peptides

We have carried out conformational studies of two series of signal peptides: the wild-type and several mutants of the *E. coli* outer membrane proteins LamB and OmpA. Both circular dichroism (CD) and nuclear magnetic resonance (NMR) spectroscopies have yielded a consistent picture in which the signal peptides sample an ensemble of conformational states in aqueous solution, but favor an α helical conformation in interfacial or helix-promoting environments (Briggs & Gierasch, 1984; McKnight *et al.*, 1989; Hoyt & Gierasch, 1991a; Rizo *et al.*, 1993). Some defects in the abilities of the corresponding signal sequences to function *in vivo* can be correlated with a reduced tendency of the signal peptide to take up a helical conformation. To illustrate these approaches, representative data for the OmpA series are presented; these results were recently reported (Rizo *et al.*, 1993).

First, in Table 1 are shown the sequences of the OmpA signal peptides studied along with their export ability as determined in Masayori Inouye's laboratory. All of these display characteristic CD spectra for random conformations in aqueous buffer (Hoyt & Gierasch, 1991a). Addition of trifluoroethanol or sodium dodecyl sulfate leads to high contents of α helix for the

Table 1: Peptides studied

	peptide sequence[a]	export activity[b]
WT	M K K T A I A I A V A L A G F A T V A Q A / A P K D	++++
L6L8	L L	++++
Δ9	—	++++
Δ8	—	++
Δ6-9	——	0
I8N	N	0

[a] The first four residues of the mature OmpA protein (APKD) have been included in all peptides to improve solubility; in addition, these four residues can avoid truncation effects in the behavior of the C-terminus of the signal sequence. The slash indicates the cleavage site of the signal peptidase. [b] *In vivo* activity as judged by the abilities of the wild type and mutant signal sequences to mediate export of two heterologous proteins (Lehnhardt *et al.*, 1987; Goldstein *et al.*, 1990, 1991). Translocation of OmpA itself by several of these mutants is reported in Tanji *et al.* (1991). The relative activities do not change in the different systems.

wild-type, L6L8, Δ9, and Δ8 peptides, while the Δ6-9 and I8N variants, which show much reduced export ability *in vivo*, have lower helix contents, particularly in the SDS environment (Table 2). While CD gives an overall estimate of helix content, NMR enables the helical structure to be mapped onto the individual residues. We find in the OmpA signal peptides, as we did in the LamB signal by

Table 2: α-Helix content of LamB signal sequences

	% α-helix			
	TFE/Water		SDS	
peptide	CD	NMR	CD	NMR
WT	55-60	58	65-70	--
L6L8	60-65	52	60-70	71
Δ9	50-55	51	65-70	72
Δ8	50-55	51	60-65	66
Δ6-9	40-50	40	50-55	50
I8N	45-65	54	55-60	--

[a] All data were collected at 25 □C and apparent pH 2.5. CD data were acquired in TFE/water 1:1 (v/v) and 10-20 μM peptide concentrations, or in 40 mM SDS and 5 μM peptide concentration. NMR data were acquired in TFE-d_3/water 1:1 (v/v) or in 300 mM SDS-d_{25}, with 2 mM peptide concentration. CD content estimates from CD data were obtained by curve fitting to reference polylysine spectra (Greenfield & Fasman, 1969). NMR helix contents were calculated from Hα proton chemical shifts. The upfield shifts of Hα protons of residues in regions assumed to be helical were added, and an average conformational shift was calculated by dividing by the total number of peptide bonds. The average conformational shifts were divided by 0.35 ppm to obtain the α-helix percentages indicated (0.35 ppm is assumed to correspond to 100% α-helix).

NMR, that the highest content of α helix is in the hydrophobic core.

We conclude that the ability to adopt an α helical conformation in an interfacial environment like SDS micelles is a property of export-competent signal peptides. On the other hand, it appears from the results on both families of signal peptide that other properties are required for function, since tendency to take up a helical conformation does not adequately discriminate peptides with different levels of export activity.

Membrane Interactions of Signal Peptides

Several methods of assessing the abilities of the wild-type and mutant signal peptides to interact with lipid membranes have demonstrated that those peptides corresponding to functional signal sequences *in vivo* can insert spontaneously and deeply into a phospholipid membrane (for a review, see Jones *et al.*, 1990). Some differences exist between the LamB and the OmpA families of signal peptides, but the correlation of insertion capacity and function is retained in both families. The types of measurement carried out and the findings are briefly summarized here.

Originally, we used phospholipid monolayers to compare insertion capabilities of LamB signal peptides (Briggs *et al.*, 1985; McKnight *et al.*, 1989). The maximal surface pressure increase and the concentration required to reach this level were greatest for those signal peptides with high export activities, indicating that these peptides had high affinity for the monolayer and could insert well into the acyl chain region. Similar conclusions were reached for the OmpA signal peptides using the perturbation of steady-state fluorescence polarization anisotropy of diphenylhexatriene (DPH) as an indicator of bilayer insertion (Hoyt & Gierasch, 1991a). Interestingly, the LamB signal peptides do not cause a significant increase in DPH fluorescence polarization anisotropy. More recently, we have used Trp residues substituted at various positions within the various LamB signal peptides to analyze their membrane affinity and mode of interaction in greater detail (Jones *et al.*, in preparation). Both the blue-shift of the Trp fluorescence and the effectiveness of quenching by spin-labels introduced at different positions within the lipid acyl chains were used to assess the interaction. Comparing the affinities of mutant LamB signal peptides of differing hydrophobicity and charge revealed that much of the strength of the membrane interaction relies on electrostatic association; hydrophobicity contributes much less, but is still important. Furthermore, the parallax method of London and Chattopadhyay (1987) shows that the peptides clearly insert well within one leaflet of the bilayer (up to 10 A depth). However, an experiment using an aqueous quencher (Tempo-choline) and fluorescently labeled wild-type LamB signal peptide (labeled with NBD on a Cys residue at the penultimate position in the sequence) showed that the peptide does not traverse the membrane as a transmembrane helix, but instead both termini are on the same side (the side from which it inserts). By contrast, a model transmembrane peptide (of sequence KKKKKALALALALALALALALALAW-NH_2) stably adopts the expected transmembrane arrangement by the same measurement. Experiments are in progress to test the mode of insertion of the OmpA peptides by the same approach.

Several questions clearly remain: What is the molecular conformation of the inserted signal peptide? Which residues are taking up helix when the peptide inserts? How is the helical conformation accomodated in a kinked peptide that comes back out of the bilayer on the same side? These questions demand a method that can yield an image of the conformation of the inserted signal peptide. High resolution NMR would appear to be the method of choice, but tight association of a peptide with a lipid vesicle leads to extensive line-broadening of the peptide signals. In order to make use of the power of high resolution NMR techniques to describe the molecular conformation of the signal peptide, it was necessary to create experimental conditions that reduced the membrane affinity of the peptide. Then, we used the strategy of analyzing the signal of the free peptide (present in excess) in rapid exchange with the membrane-bound peptide. The time spent in the slowly-tumbling bound state enables the efficient development of peptide interproton nuclear Overhauser effects or NOEs (as contrasted with the inefficient NOE development in the rapidly tumbling free peptide). These so-called transferred NOEs (Clore & Gronenborn, 1982,1983) are related to interproton distances in the *bound* peptide, and thus can yield the desired picture of the molecular conformation of the membrane-inserted peptide.

In addition to finding conditions favoring fast exchange of the signal peptide, we also had to modify the peptide to enhance its water solubility. The addition of excess positive charge at the sites of the native positively-charged residues (native sequence: MMITLRKLPLAVAVAAGVMSAQAMA; modified peptide sequence, changes italicized: MMITLR*KRR*KLPLAVAVAAGMSAQAMA, Trp added as a fluorescent probe) raised the solubility adequately without perturbing the membrane interaction. Using membranes composed of only 1% POPG and 99% POPC, as large unilamellar vesicles (LUVs), resulted in the appropriate affinity for trNOE measurements. Indeed, we have now observed new signal peptide NOEs in the presence of LUVs and assigned them to residues within the sequence. The trNOE analysis confirmed that the peptide is largely α helical, with the helix extending from residue 8 through the hydrophobic core, and continuing to the C-terminus. The data indicate that the helix may be somewhat disrupted at the end of the hydrophobic core (Gly17). The N-terminal region of the peptide is tethered to the membrane surface, as manifested by its large trNOEs, but is not part of the helix.

In addition to the information we obtained from trNOEs, we were able to use NMR to learn about the depth of insertion of the peptide in a complementary fashion to our fluorescence experiments. We incorporated spin labeled lipids into the LUVs used in the NMR experiments and monitored line-broadening of specific crosspeaks in 2D spectra of the signal peptide. Increased line-broadening will occur in those parts of the molecular that insert into a region close to the spin label. By comparing the effectiveness of line-broadening by spin labels at different position along the acyl chains, we could deduce the approximate depth of penetration analogously to the approach used for quenching (London & Chattopadhyay, 1987). The data from the two methods are entirely consistent and yield a picture of the helical signal peptide inserted well into the acyl chain region, but not traversing the membrane.

Implications

The importance of the tendency of signal peptides to insert into a bilayer must be reconciled with current knowledge of the targeting and translocation apparatus. Clearly, protein interactions are critical at the point of targeting and during translocation, and we are now actively studying the recognition of the signal sequence by SecA, SRP54, and Ffh. Nonetheless, we believe that the spontaneous insertion of the signal sequence plays a role after release of the nascent chain from SecA or SRP, and before engagement of the chain with the translocation components. Further work is required to test this model and to develop a clear picture of the events in translocation.

ACKNOWLEDGEMENTS

The work described in this review is due to the efforts of several members of the laboratory, past and present. I mention in particular: Jeffrey Jones, Zhulun Wang, Josep Rizo, David Hoyt, and Jamie McKnight. This work was supported by grants from the NIH (GM34962), the Texas Advanced Research Fund, and the Robert A. Welch Foundation.

REFERENCES

Briggs, M. S., & Gierasch, L. M. (1984) *Biochemistry 23*, 3111-3114.
Briggs, M. S., Gierasch, L. M., Zlotnick, A., Lear, J. D., & DeGrado, W. F. (1985) *Science 228*, 1096-1099.
Chattopadhyay, A., & London, E. (1987) *Biochemistry 26*, 39-45.
Clore, G. M., & Gronenborn, A. M. (1982) *J. Mag. Res. 48*, 402-417.
Clore, G. M., & Gronenborn, A. M. (1983) *J. Mag. Res. 53*, 423-442.
Gennity, J., Goldstein, J., & Inouye, M. (1990) *J. Bioenerg. Biomembr. 22*, 233-270.
Gierasch, L. M. (1989) *Biochemistry 28*, 923-930.
Goldstein, J., Lehnhardt, S., Inouye, M. (1990) *J. Bacteriol. 172*, 1225-1231.
Goldstein, J., Lehnhardt, S., Inouye, M. (1991) *J. Biol. Chem. 266*, 14413-14417.
Greenfield, N., & Fasman, G. D. (1969) *Biochemistry 8*, 4108-4116.
Hoyt, D. W., & Gierasch, L. M. (1991a) *Biochemistry 30*, 10155-10163.
Hoyt, D. W., & Gierasch, L. M. (1991b) *J. Biol. Chem. 266*, 14406-14412.
Jones, J. D., McKnight, C. J., & Gierasch, L. M. (1990) *J. Bioenerg. Biomembr. 22*, 213-232.
Lehnhardt, S., Pollitt, S., & Inouye, M. (1987) *J. Biol. Chem. 262*, 1716-1719.
McKnight, C. J., Briggs, M. S., & Gierasch, L. M. (1989) *J. Biol. Chem. 264*, 17293-17297.
Rizo, J., Blanco, F. J., Kobe, B., Bruch, M. D., & Gierasch, L. M. (1993) *Biochemistry 32*, 4881-4894.
Tanji, Y., Gennity, J., Pollitt, S., & Inouye, M. (1991) *J. Bacteriol. 173*, 1997-2005.
von Heijne, G. (1985) *J. Mol. Biol. 184*, 99-105.

ASSEMBLY OF INTEGRAL MEMBRANE PROTEINS

Gunnar von Heijne
Department of Molecular Biology
Karolinska Institute Center for Structural Biochemistry
NOVUM
S-141 52 Huddinge
Sweden

How do proteins insert into membranes, and what characteristics of the nascent chain control their folding into the final 3D-structure? In the past few years, molecular genetics, biochemistry, and structural biochemistry have all contributed to our understanding of these processes, and we can now exert a certain control over membrane protein biogenesis. In this chapter, I will review the main ideas that are currently guiding the field, with particular emphasis on the role that positively charged amino acids play during membrane insertion and on some novel techniques that have made it possible to obtain information on helix-helix interactions in a membrane environment without having to resort to structure analysis by X-ray crystallography or NMR. I will only discuss integral membrane proteins of the helix-bundle class, *i.e.*, proteins whose membrane spanning segments are formed from long stretch of hydrophobic amino acids that fold into transmembrane α-helices. The so-called ß-barrel membrane proteins are dealt with in the chapter by Schulz in this volume.

Insertion of a transmembrane protein: The 'positive inside' rule

Conceptually, the biogenesis of an integral membrane protein may be divided into at least three distinct steps. First, the protein has to find the correct membrane; second, its transmembrane segments have to insert into that membrane in their correct orientations; and third, the transmembrane segments have to pack together to form a helix bundle in the membrane (and, concomitantly, the extra-membranous parts of the protein have to fold). The first step is a problem of protein targeting (Pugsley, 1989), and will not be discussed further here.

The most conspicuous feature of integral membrane proteins of the helix bundle type is the presence of long (~20 residues or more) stretches of hydrophobic amino acids that form the transmembrane α-helices. However, the amino acid sequence must also

NATO ASI Series, Vol. H 82
Biological Membranes:
Structure, Biogenesis and Dynamics
Edited by Jos A. F. Op den Kamp

encode signals that control the orientation of the transmembrane segments in the bilayer. Statistical studies of integral membrane proteins with experimentally determined topologies have demonstrated a universal correlation between the distribution of positively charged amino acids and the overall topology of the molecule: lysines and arginines abound in non-translocated parts of the protein, but are comparatively scarce in translocated parts. This 'positive inside' rule has been found to hold for bacterial inner membrane proteins (von Heijne, 1986), eukaryotic plasma membrane proteins (von Heijne and Gavel, 1988; Sipos and von Heijne, 1993), thylakoid membrane proteins (Gavel, et al., 1991), and organelle-encoded mitochondrial inner membrane proteins (Gavel and von Heijne, 1992), and thus clearly points to positively charged residues as being somehow involved in determining the orientation of the transmembrane segments. It should also be noted that there is no corresponding bias in the distribution of negatively charged amino acids (Asp and Glu), suggesting that it is the number or frequency of positively charged residues, rather than, *e.g.*, the net charge of a domain, that plays the decisive role.

An important aspect of the 'positive inside' rule is that it only applies to rather short domains (60-70 residues or less), whereas longer domains have a content of Arg+Lys indistinguishable from that of globular secretory proteins (von Heijne, 1986; von Heijne and Gavel, 1988), suggesting that the mechanism of membrane translocation may be different for short and long domains.

There now exist ample experimental evidence that positively charged amino acids indeed are major determinants of membrane protein topology. In our own work, we have used the *E. coli* inner membrane protein leader peptidase (Lep) as a model. This protein has two transmembrane segments, and is oriented with both its N-terminal and C-terminal end in the periplasm, Fig.1. This orientation can be inverted by removing most of the positively charged amino acids from the cytoplasmic P1-domain (mutant Lep') and then adding four extra lysines to the N-terminus (mutant Lep'-inv) (von Heijne, 1989). Further studies showed that as little as a single lysine can act as a 'topological switch' in certain contexts (Nilsson and von Heijne, 1990), that arginines and lysines are approximately equally efficient in this respect, that histidines can affect the topology when the cytoplasm is made more acidic (Andersson, et al., 1992), and that acidic residues normally have very little influence on the topology (Nilsson and von Heijne, 1990; Andersson, et al., 1992).

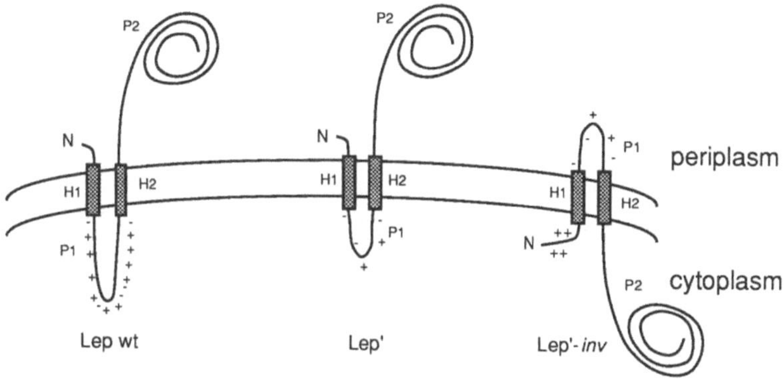

Fig.1 Orientation of Lep in the inner membrane. Also shown are two constructs that differ only by four N-terminal lysines and yet have opposite orientations.

More recently, we have constructed molecules with four transmembrane segments by duplicating the H1-H2 region of Lep. These '2xLep' constructs generally insert into the inner membrane as predicted by the distribution of positively charged residues, except when there is conflicting topological information in different parts of the molecule. One example is shown in Fig.2; in this case, our preliminary data indicate that the molecule solves the problem of conforming to the 'positive inside' rule by only inserting three of the four hydrophobic segments across the membrane (our unpublished data).

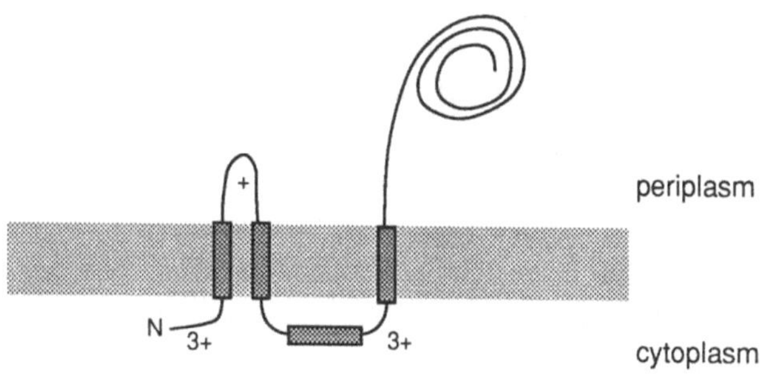

Fig.2 Topology of a 2xLep construct with a 'conflicting' charge distribution. Note that the 'positive inside' rule still applies, but that one hydrophobic segment has not been inserted across the membrane.

As note above, the 'positive inside' rule applies only to domains that are shorter than 60-70 residues. We (von Heijne, 1986) and others (Kuhn, 1988) suggested already some years ago that this might be related to the observation that the short periplasmic loop of the phage M13 procoat protein (as well as the short periplasmic loop in the Lep'-

inv mutant (von Heijne, 1989)) can insert across the inner membrane even under conditions when the *sec*-machinery is blocked (see Wickner's chapter for a review of the *sec*-machinery). We recently tested this notion by systematically lengthening the loop in the Lep'-inv mutant, and indeed found that the degree of *sec*-dependence increases with the length of the translocated loop up to a limiting length of ~60 residues (Andersson and von Heijne, 1993). It thus appears that one factor influencing the ability of the nascent chain to interact productively with the *sec*-machinery is the length of the translocated domain, and that large domains, by virtue of their reliance on the *sec*-machinery, can have a much higher frequency of charged amino acids than short loops.

Prediction of membrane protein topology

Hydrophobicity analysis methods allow the identification of candidate membrane-spanning segments in an amino acid sequence. However, one is often left with a situation where some of the transmembrane segments can be predicted with high confidence, whereas others have a marginal average hydrophobicity and cannot be predicted as transmembrane or extra-membranous with any certainty. On the other hand, we also know that the topology of any membrane protein should respect the 'positive inside' rule, and this observation can be used to improve the prediction methods quite substantially. We have thus proposed a scheme (von Heijne, 1992) where one first identifies 'certain' and 'putative' transmembrane segments from a hydrophobicity plot, and then constructs all possible topologies that include all the 'certain' transmembrane segments and either include or exclude all 'putative' ones. Finally, one ranks the topologies according to the overall bias in the distribution of Arg+Lys residues between the two sides of the structure. This scheme works surprisingly well for bacterial inner membrane proteins, and produces the correct topology for about nine proteins out of ten.

Topologies of eukaryotic membrane proteins are more difficult to predict, in part because the charge bias is less extreme than for bacterial proteins. However, reasonably reliable predictions can still be made if, in addition to the charge bias, one also takes into account that large cytoplasmic and extra-cytoplasmic domains tend to have somewhat different overall amino acid compositions, and that the net charge difference (also counting negatively charged residues) across the most N-terminal transmembrane segment correlates strongly with its orientation (Sipos and von Heijne, 1993).

Packing of transmembrane α-helices

Studies on bacteriorhodopsin have suggested that helix-bundle membrane proteins attain their native structure in two steps: first, they insert into the membrane and the transmembrane helices form; second, these pre-formed helices assemble into a helix bundle (Popot and de Vitry, 1990; Popot and Engelman, 1990). Since the transmembrane segments can be readily identified directly fro the amino acid sequence, at least for bacterial inner membrane proteins (see above), the remaining difficulties in the 'membrane protein folding problem' are largely problems of helix-helix interactions in a membrane environment. To date, there exist no reliable algorithms that allow the prediction of optimal helix-helix interfaces, and, indeed, very few such interfaces have even been experimentally mapped so far. The only two high-resolution structures available are the bacterial photosynthetic reaction center (Deisenhofer, et al., 1985) and bacteriorhodopsin (Henderson, et al., 1990), and, although they have allowed preliminary studies on helix-helix packings (Rees, et al., 1989a; Rees, et al., 1989b), more experimental data are clearly needed.

Since standard biophysical methods of structure determination have proved difficult to apply to membrane proteins, a number of workers are trying to use molecular genetics as an alternative approach. In a recent study (Lemmon, et al., 1992), the helix-helix interface in the glycophorin A dimer was mapped by saturation mutagenesis, using an assay based on the observation that the wildtype molecule forms a dimer that is stable in SDS and that hence can be resolved by SDS-PAGE. Surprisingly, the interface was found to be largely formed by hydrophobic amino acids.

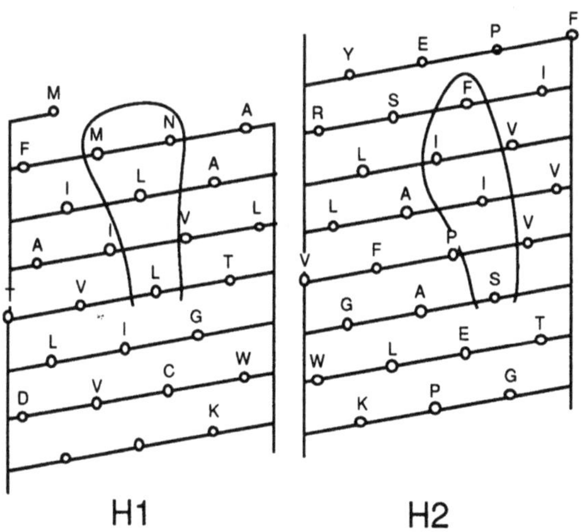

Fig.3 The H1-H2 helix-helix interface in Lep. The encircled residues have more than 50% of their surface area buried in the interface.

Helix-helix interfaces have also been mapped by placing cysteines in the transmembrane segments and looking for the formation of inter-helical disulfides. This technique has been used successfully on the Tar receptor (a dimer that forms a four-helix bundle in the membrane) (Pakula and Simon, 1992), and on Lep (Whitley, et al., 1993). In both cases, the interfaces were again found to involve mostly hydrophobic amino acids, Fig.3. On the other hand, inter-helix charge-pairs have been implicated in the assembly of the T-cell receptor complex (Bonifacino, et al., 1991; Cosson, et al., 1991; Lankford, et al., 1993).

Conclusion

As has been shown above, we now understand many of the features of a nascent membrane protein that guide its membrane insertion, although the cellular mechanisms that decode that information are still not fully characterized. The development of new methods to study helix-helix interfaces may lead to a better understanding of the 'packing rules', and thus to new possibilities for predicting not only the topology but in fact the entire 3D structure of integral membrane proteins.

References

Andersson, H., Bakker, E. , and von Heijne, G. (1992). Different Positively Charged Amino Acids Have Similar Effects on the Topology of a Polytopic Transmembrane Protein in *Escherichia coli*. J Biol Chem *267,* 1491-1495.

Andersson, H. , and von Heijne, G. (1993). *Sec*-Dependent and *sec*-Independent Assembly of *E. coli* Inner Membrane Proteins - The Topological Rules Depend on Chain Length. EMBO J *12,* 683-691.

Bonifacino, J. S., Cosson, P., Shah, N. , and Klausner, R. D. (1991). Role of Potentially Charged Transmembrane Residues in Targeting Proteins for Retention and Degradation Within the Endoplasmic Reticulum. EMBO J *10,* 2783-2793.

Cosson, P., Lankford, S. P., Bonifacino, J. S. , and Klausner, R. D. (1991). Membrane Protein Association by Potential Intramembrane Charge Pairs. Nature *351,* 414-416.

Deisenhofer, J., Epp, O., Miki, K., Huber, R. , and Michel, H. (1985). Structure of the protein subunits in the photosynthetic reaction centre of *Rhodopseudomonas viridis* at 3Å resolution. Nature *318,* 618-624.

Gavel, Y., Steppuhn, J., Herrmann, R. , and von Heijne, G. (1991). The Positive-Inside Rule Applies to Thylakoid Membrane Proteins. FEBS Lett *282,* 41-46.

Gavel, Y. , and von Heijne, G. (1992). The Distribution of Charged Amino Acids in Mitochondrial Inner Membrane Proteins Suggests Different Modes of Membrane Integration for Nuclearly and Mitochondrially Encoded Proteins. Eur J Biochem *205,* 1207-1215.

Henderson, R., Baldwin, J. M., Ceska, T. A., Zemlin, F., Beckmann, E. , and Downing, K. H. (1990). A Model for the Structure of Bacteriorhodopsin Based on High Resolution Electron Cryo-Microscopy. J Mol Biol *213,* 899-929.

Kuhn, A. (1988). Alterations in the extracellular domain of M13 procoat protein make its membrane insertion dependent on *secA* and *secY*. Eur J Biochem *177*, 267-271.

Lankford, S. P., Cosson, P., Bonifacino, J. S. , and Klausner, R. D. (1993). Transmembrane Domain Length Affects Charge-Mediated Retention and Degradation of Proteins Within the Endoplasmic Reticulum. J Biol Chem *268*, 4814-4820.

Lemmon, M. A., Flanagan, J. M., Treutlein, H. R., Zhang, J. , and Engelman, D. M. (1992). Sequence specificity in the dimerization of transmembrane α-helices. Biochemistry *31*, 12719-12725.

Nilsson, I. M. , and von Heijne, G. (1990). Fine-tuning the Topology of a Polytopic Membrane Protein. Role of Positively and Negatively Charged Residues. Cell *62*, 1135-1141.

Pakula, A. A. , and Simon, M. I. (1992). Determination of Transmembrane Protein Structure by Disulfide Cross-Linking - The Escherichia-Coli Tar Receptor. Proc Natl Acad Sci USA *89*, 4144-4148.

Popot, J.-L. , and de Vitry, C. (1990). On the microassembly of integral membrane proteins. Annu.Rev.Biophys.Biophys.Chem. *19*, 369-403.

Popot, J. L. , and Engelman, D. M. (1990). Membrane Protein Folding and Oligomerization - The 2-Stage Model. Biochemistry *29*, 4031-4037.

Pugsley, A. P. (1989). Protein targeting. San Diego: Academic Press.

Rees, D. C., DeAntonio, L. , and Eisenberg, D. (1989a). Hydrophobic Organization of Membrane Proteins. Science *245*, 510-13.

Rees, D. C., Komiya, H., Yeates, T. O., Allen, J. P. , and Feher, G. (1989b). The bacterial photosynthetic reaction center as a model for membrane proteins. Annu.Rev.Biochem. *58*, 607-33.

Sipos, L. , and von Heijne, G. (1993). Predicting the topology of eukaryotic membrane proteins. Eur J Biochem *213*, 1333-1340.

von Heijne, G. (1986). The distribution of positively charged residues in bacterial inner membrane proteins correlates with the trans-membrane topology. EMBO J *5*, 3021-3027.

von Heijne, G. (1989). Control of Topology and Mode of Assembly of a Polytopic Membrane Protein by Positively Charged Residues. Nature *341*, 456-458.

von Heijne, G. (1992). Membrane Protein Structure Prediction - Hydrophobicity Analysis and the Positive-Inside Rule. J Mol Biol *225*, 487-494.

von Heijne, G. , and Gavel, Y. (1988). Topogenic signals in integral membrane proteins. Eur J Biochem *174*, 671-8.

Whitley, P., Nilsson, L. , and von Heijne, G. (1993). A 3D model for the membrane domain of Escherichia coli leader peptidase based on disulfide mapping. Biochemistry *in press*.

HOW DO PROTEINS CROSS A MEMBRANE?

Bill Wickner and Marilyn Rice Leonard*
Molecular Biology Institute
University of California
Los Angeles, California 90024-1570
U.S.A.

The questions. Classical studies by Palade, deDuve, and colleagues established that membranes divide cells into distinct compartments, each with a unique set of resident proteins catalyzing distinct functions. Each compartment is either a membrane, with its own set of embedded proteins, or a soluble space surrounded by a membrane. A typical eukaryotic cell may have over 20 compartments, while a bacterium such as E. coli has four: the cytoplasm, inner [plasma] membrane, periplasm, and outer membrane. In contrast, almost all protein synthesis begins in the cytosol in all cells, in a basically spatially undifferentiated manner. The first question then is how proteins are targeted, either to remain in the cytosol or to the appropriate membrane for translocation. Having arrived there, the second question is one of translocation mechanism. Is it by radically changing its structure to pass from the aqueous cytosol to the hydrocarbon-like interior of a membrane, or by a proteinaceous transport system ("translocase")? In either case, what is the energy source for this transfer? Is it the energy of protein synthesis pushing the chain out of the ribosome, a pulling force on the other side, electrophoresis, or is metabolic energy coupled to protein translocation by translocase? In Fig. 1, these questions are illustrated, with a fig leaf both conveying the attractive quality of the hidden solution and covering our ignorance about the ultimate answers.

These problems have been tackled over the last 2 decades by studies of dog pancreas endoplasmic reticulum, yeast endoplasmic reticulum, mitochondria, chloroplasts, and E. coli. The latter offers an unrivaled ease of combining biochemistry, cell fractionation, physiology, and genetics. In this essay, we present genetic (Bieker et al., 1990; Schatz & Beckwith, 1990) and biochemical (Wickner et al., 1991) approaches which illuminated the structure and function of E. coli preprotein translocase (Brundage et al.,

---

* After July, 1993; Department of Biochemistry, Dartmouth Medical School, Hanover, NH 03755. Work in the authors' laboratory is supported by the Insititute of General Medical Sciences, National Institutes of Health, USA

NATO ASI Series, Vol. H 82
Biological Membranes:
Structure, Biogenesis and Dynamics
Edited by Jos A. F. Op den Kamp

1990). Recent studies which show the homology between one subunit of translocase, the SecY protein, and integral membrane proteins of the yeast and mammalian endoplasmic reticulum (Gorlich et al., 1992) illustrate the generality of E. coli translocase function and heighten the interest in understanding its mechanism.

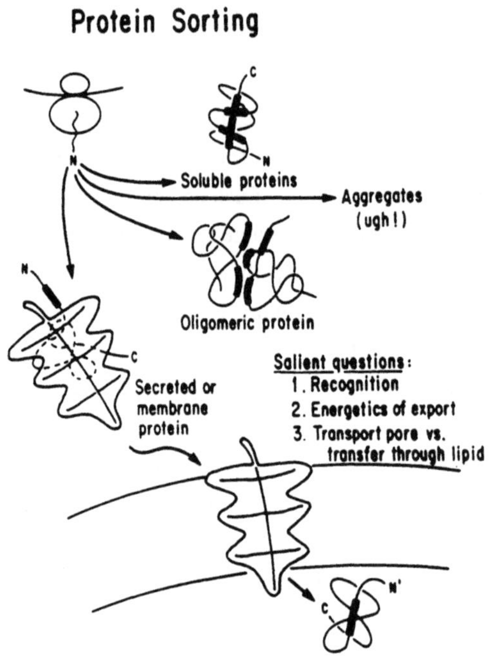

Figure 1. The fates of newly-made proteins. The fig leaf may not accurately represent the 3-dimensional structure of preprotein translocase!

First steps. During the late 1970's and early 1980's, temperature-sensitive *sec* mutations in the export of periplasmic and outer membrane preproteins were identified. In addition, *prl* mutants which could export proteins with severely defective leader sequences were obtained, and shown to be in the same genes as the sec mutants. Assembly of the M13 virus coat protein into the inner membrane was shown to require a typical leader sequence and membrane electrochemical potential, but (for reasons still not understood) to not employ these Sec proteins (Wickner, 1988). Physiological studies showed that translocation is not mechanistically coupled to translation (Randall, 1983); the problems of getting large, folded, polar proteins across a membrane would later be shown to be solved by chaperones and even unfoldases rather than by extruding the polypeptide across the membrane as it emerged from the ribosome.

In vitro translocation. While the uncatalyzed insertion of M13 procoat had been reconstituted, it was not until the mid-1980's that in vitro protein synthesis of larger preproteins was successfully combined with preprotein translocation into the lumen of inverted, sealed inner membrane vesicles (Chen & Tai, 1985; Muller & Blobel, 1984). P.C. Tai, whose group developed this reaction as did Muller and Blobel, found that translocation required ATP (Chen & Tai, 1985); thus, both energy currencies of the cell, ATP and $\Delta\mu_{H^+}$, were coupled to protein export (Geller et al., 1986). Purified precursor proteins such as proOmpA (the precursor of outer membrane protein A) could undergo ATP and $\Delta\mu_{H^+}$ translocation into membrane vesicles, but the mechanism was still unknown (Crooke et al., 1988).

The first of the components of translocation to be purified was the soluble chaperone SecB (Kumamoto et al., 1989; Weiss et al., 1988)). The overproduction of this protein was genetically engineered, allowing its isolation in mg amounts. The pure protein was shown to function in supporting in vitro translocation by stabilizing preproteins and targeting them to the plasma membrane. Similarly, the overproduction of the SecA protein was genetically engineered, allowing its isolation (Cabelli et al., 1988). Functional in vitro assays for this protein were developed: 1. It could be largely "stripped" from membranes by urea, and its readdition was essential to restore translocation to these membranes (Cunningham et al., 1989). 2. Membranes which had been inactivated for translocation by exposure to light and ATP were restored by the addition of pure SecA protein (Lill et al. 1989); this led to the finding that SecA is an ATPase, and that its "translocation ATPase" activity requires its coordinate interactions with membrane proteins (SecY/E; see below), acidic membrane phospholipids, and an authentic preprotein.

The isolation of the intact integral membrane protein for translocation was, in retrospect, dependent on first having the SecA and SecB proteins in hand. This is because SecA is exquisitely sensitive to detergent, and thus any fractionation scheme which tried to solubilize it with detergent is doomed. SecB is often required to an even greater extent in vitro than in vivo, for in vitro reconstitutions require a healthy excess of phospholipids during the detergent removal phase, and the resultant lipid-rich proteoliposomes adsorb un-chaperoned preproteins with maddening efficiency! The third key element was the insight, gained from earlier studies of membrane-embedded transport proteins, that detergent wasn't enough. To keep the most hydrophobic membrane proteins in solution and active, one must form small, oligomolecular patches of lipid bilayer in detergent/phospholipid mixed micelles. Furthermore, mimicking the cells' environment, the water concentration has to be lowered. This can be achieved by agents such as glycerol. From octylglucoside/lipid/glycerol extracts of E. coli inner membrane vesicles (Driessen & Wickner, 1990), a trimeric complex was isolated which is sufficient to reconstitute the integral domain of translocase (Brundage et al., 1990). This is termed the SecY/E protein, as two of the subunits are the SecY and SecE polypeptide, while the third is given the neutral name "band 1" while reverse genetics is used to explore its physiological role(s).

Purified SecY/E can be reconstituted into membranes, often with a proton pump such as bacteriorhodopsin, and mixed with pure SecA. The complex of SecA, a peripheral protein domain, with the membrane-embedded SecY/E (in the presence of acidic phospholipids) is termed "E. coli preprotein translocase." Its function in translocation is illustrated in Fig. 2. It will translocate authentic preproteins with a rate, and extent, which is comparable to that of the starting organelle. Translocation has also been reconstituted from the separated subunits of SecY/E (Akimaru et al., 1991).

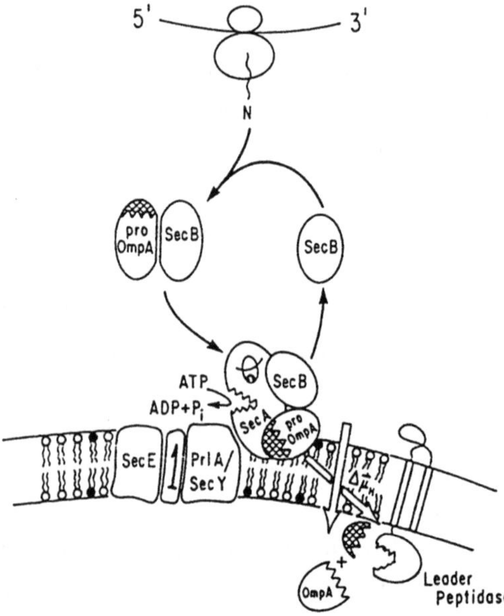

Figure 2. Preprotein translocase. See text for details. Modified, with permission, from Brundage et al. (1990).

Translocation mechanisms. How does translocase work? Proteins are guided to the membrane by a cascade of binding events, each entailing specific high-affinity recognition. SecB binds at the mature domain of exported proteins, perhaps due to their slow folding (Randall & Hardy, 1986). The SecB/preprotein complex then binds to the SecA subunit of translocase, guided by the affinity of SecA for SecB and that of SecA for the leader and the mature domain of preproteins. About half the preproteins examined do not require SecB chaperone and guidance functions for export; in these cases, direct recognition by SecA presumably confers sufficient specificity. Once the preprotein is bound, translocation proceeds by a 5-step catalytic cycle (Schiebel et al., 1991; see Fig. 3).

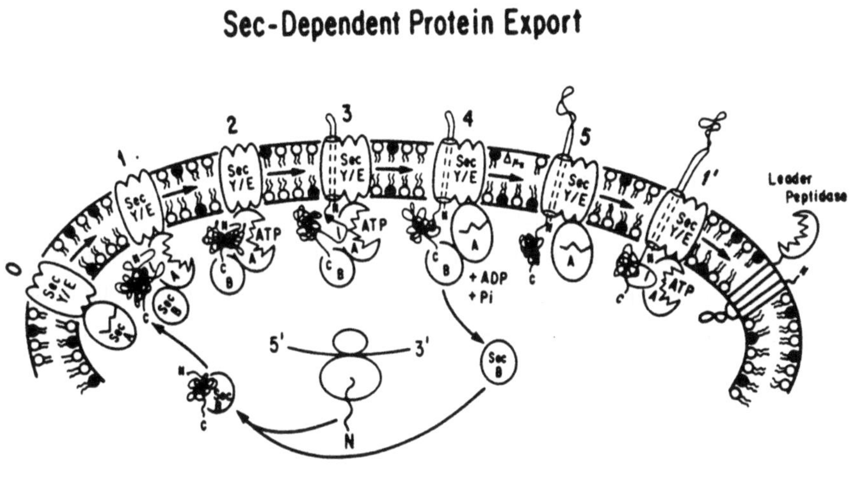

Figure 3. The 5-step catalytic cycle of <u>E. coli</u> preprotein translocase. Modified, with permission, from Schiebel et al. (1991).

The binding of preprotein activates SecA to bind ATP. The energy from the binding of ATP drives the first "loop" of 20-30 residues of the preprotein across the membrane. Recent experiments suggest that this may be accompanied by a major shift of the SecA protein itself into, or across, the membrane. In any case, a major portion of the pathway taken by the transiting protein as it crosses the membrane is comprised of SecY. After this limited insertion of 20-30 residues, ATP is hydrolyzed, the ADP and inorganic phosphate are released from SecA, and the preprotein is released from SecA. This released [from SecA], trans-membrane translocation intermediate is then rapidly driven forward in its translocation by the membrane electrochemical potential, $\Delta\mu_{H^+}$. At as-yet ill-defined points during translocation, the preprotein can again bind to SecA, be driven forward 20-30 residues by the binding energy of ATP, and be released from SecA upon ATP hydrolysis. The ATP-coupled mechanism may come into play during the translocation of unusually tightly folded domains or for a segment of the polypeptide chain which, for reasons not yet understood, cannot be driven to translocate by a membrane potential. Finally, at completion of translocation, folding in the periplasm may be catalyzed by SecD (Matsuyama et al., 1993) and F and by prolyl isomerase and disulfide isomerase (Bardwell et al., 1991). Cleavage of the leader peptide by leader peptidase is necessary to release the preprotein to continue into the periplasm or on to the outer membrane (Dalbey & Wickner, 1985).

Current questions. The progress in understanding translocation, cited above, is really just a beginning to understanding this basic biological phenomenon. What is the basis of recognition of the mature domain of preproteins by SecB and by SecA? How does translocase conduct amino acyl residues of all shape, polarity, and size across the membrane while keeping a proton-tight membrane? What is the stoichiometry and spatial arrangements of the subunits of translocase? Do SecB and SecA remain firmly bound to a single SecY/E throughout multiple cycles of preprotein translocation, or are they constantly shuttling on and off the membrane? How does the membrane potential drive translocation? It is clearly not simply a matter of electrophoresis, nor does translocase seem to function as a well-coupled proton/protein antiport. What are the roles of SecD, SecF, and FtsH, integral membrane proteins implicated in the export process by genetic studies? How is the transcription of SecA regulated by so many aspects of export, and what other levels of regulation may link export to other aspects of cell surface growth (such as lipid synthesis) or cell cycle control? The availability of pure translocase, of the subunit genes, and of natural and "designer" preproteins holds considerable promise that these questions can be addressed in the near future.

REFERENCES:

Akimaru, J., Matsuyama, S.-I., Tokuda, H., and Mizushima, S. (1991) Reconstitution of a protein translocation system containing purified SecY, SecE, and SecA from Escherichia coli. Proc. Natl. Acad. Sci. USA 88: 6545-6549.

Bardwell, J.C.A., McGovern, K., and Beckwith, J. (1991) Identification of a protein required for disulfide bond formation in vivo. Cell 67: 581-589.
Bieker, K.L., Phillips, G.J., and Silhavy, T. (1990) The sec and prl genes of Escherichia coli. J. Bioenerget. Biomembr. 22: 291-310.
Brundage, L., Hendrick, J.P., Schiebel, E., Driessen, A.J.M., and Wickner, W. (1990) The purified E. coli integral membrane protein SecY/E is sufficient for reconstitution of SecA-dependent precursor protein translocation. Cell 62: 649-657.
Cabelli, R., Chen, L.L., Tai, P.C., and Oliver, D.B. (1988) Secretory protein translocation into E. coli SecA protein is required for membrane vesicles. Cell 55: 683-692.
Chen, L. and Tai, P.C. (1985) ATP is essential for protein translocation into Escherichia coli membrane vesicles. Proc. Natl. Acad. Sci. USA 82: 4384-4388.
Crooke, E., Guthrie, B., Lecker, S., Lill, R., and Wickner, W. (1988) ProOmpA is stabilized for membrane translocation by either purified E. coli trigger factor or canine signal recognition particle. Cell 54: 1003-1011.
Cunningham, K., Lill, R., Crooke, E., Rice, M., Moore, K., Wickner, W., and Oliver, D. (1989) Isolation of SecA protein, a peripheral protein of the E. coli plasma membrane that is essential for the functional binding and translocation of proOmpA. EMBO J. 8: 955-959.
Dalbey, R.E. and Wickner, W. (1985) Leader peptidase catalyzes the release of exported proteins from the outer surface of the Escherichia coli plasma membrane. J. Biol. Chem. 260: 15925-15931.
Driessen, A.J.M. and Wickner, W. (1990) Solubilization and functional reconstitution of the protein-translocation enzymes of Escherichia coli. Proc. Natl. Acad. Sci. USA 87: 3107-3111.
Geller, B.L., Movva, N.R., and Wickner, W. (1986) Both ATP and the electrochemical potential are required for optimal assembly of pro-OmpA into Escherichia coli inner membrane vesicles. Proc. Natl. Acad. Sci. USA 83: 4219-4222.
Gorlich, D., Prehn, S., Hartmann, E., Kalies, K.-U., and Rapoport, T.A. (1992) A mammalian homolog of SEC61p and SECYp is associated with ribosomes and nascent polypeptides during translocation. Cell 71: 489-503.
Kumamoto, C.A., Chen, L., Fandl, J., and Tai, P.C. (1989) Purification of the Escherichia coli secB gene product and demonstration of its activity in an in vitro protein translocation system. J. Biol. Chem. 264: 2242-2249.
Lill, R., Cunningham, K., Brundage, L, Ito, K., Oliver, D., and Wickner, W. (1989) The SecA protein hydrolyzes ATP and is an essential component of the protein translocation ATPase of E. coli. EMBO J. 8: 961-966.
Matsuyama, S.-i., Fujita, Y., and Mizushima, S. (1993) SecD is involved in the release of translocated secretory proteins from the cytoplasmic membrane of Escherichia coli. EMBO J. 12: 265-270.
Muller, M. and Blobel, G. (1984) In vitro translocation of bacterial proteins across the plasma membrane of Escherichia coli. Proc. Natl. Acad. Sci. USA 81: 7421-7425.
Randall, L.L. (1983) Translocation of domains of nascent periplasmic proteins across the cytoplasmic membrane is independent of elongation. Cell 33: 231-240.
Randall, L.L. and Hardy, S.J.S. (1986) Correlation of competence for export with lack of tertiary structure of the mature species: a study in vivo of maltose-binding protein in E. coli. Cell 46: 921-928.
Schatz, P.J. and Beckwith, J. (1990) Ann. Rev. Genet. 24: 215-248.
Schiebel, E., Driessen, A.J.M., Hartl, F.-U., and Wickner, W. (1991) $\Delta\mu H+$ and ATP function at different steps of the catalytic cycle of preprotein translocase. Cell 64: 927-939.

Wickner, W. (1988) Mechanisms of membrane assembly: General lessons from the study of M13 coat protein and Escherichia coli leader peptidase. Biochemistry 27: 1081-1086.

Wickner, W., Driessen, A.J.M., and Hartl, F.-U. (1991) The enzymology of protein translocation across the Escherichia coli plasma membrane. Ann. Rev. Biochem. 60: 101-124.

Weiss, J.B., Ray, P.H., and Bassford, P.J. Jr. (1988) Purified secB protein of Escherichia coli retards folding and promotes membrane translocation of the maltose-binding protein in vitro. Proc. Natl. Acad. Sci. USA 85: 8978-8982.

Protease secretion by *Erwinia chrysanthemi* and *Serratia marcescens*

A signal peptide independent pathway in Gram-negative bacteria

J.M Ghigo, S. Létoffé, P. Delepelaire and C. Wandersman
Unité de Génétique Moléculaire, Institut Pasteur,
25 Rue du Docteur Roux, 75724 Paris CEDEX 15,
France

Introduction

The study of the pathways used by cells to translocate proteins from the cytoplasm to the cellular envelope or to the extracellular medium is still an open investigation field. Particular attention has been devoted to the mechanisms of protein secretion in Gram-negative bacteria, which requires that proteins cross both the cytoplasmic and outer membranes. The genetic reconstitution of these secretion systems in *Escherichia coli* led to the identification of at least two quite distinct pathways. One of them, called the General Secretory Pathway (GSP) is used by proteins made with typical cleavable amino-terminal signal peptides (Pugsley,1993). Initially these proteins cross the cytoplasmic membrane through a highly conserved process which involves the components of the general export pathway encoded by *sec* genes. During the subsequent steps, specific envelope factors are required to allow the translocation across the outer membrane.

Recently it was shown that a growing number of proteins, devoid of signal-peptides are not secreted through the GSP but utilize a unique secretion apparatus. The first protein of this group was the α-hemolysin (HlyA) produced by some uropathogenic strains of *Escherichia coli.* (Felmlee *et al.*, , 1985). Now it appears that this secretion pathway is widespread amongst Gram-negative bacteria and is responsible for the secretion of a large number of unrelated proteins (see for review Wandersman, 1992). Besides hemolysin-related toxins (see fig.3) or non homologous proteins such as colicins, our work on *Erwinia chrysanthemi* and *Serratia marcescens* proteases showed that a new family of metalloproteases can also used this pathway to reach the extracellular medium.

NATO ASI Series, Vol. H 82
Biological Membranes:
Structure, Biogenesis and Dynamics
Edited by Jos A. F. Op den Kamp

A new family of metalloproteases.

Erwinia chrysanthemi is a Gram-negative phytopathogenic bacterium which secretes four distinct proteases during the exponential growth (Wandersman *et al.*, 1987; Ghigo and Wandersman, 1992b). All of the genes whose products are involved in protease synthesis and secretion as well as a gene coding for a small intracellular inhibitor are clustered on a fourteen kb DNA region of the chromosome.(see fig.1)

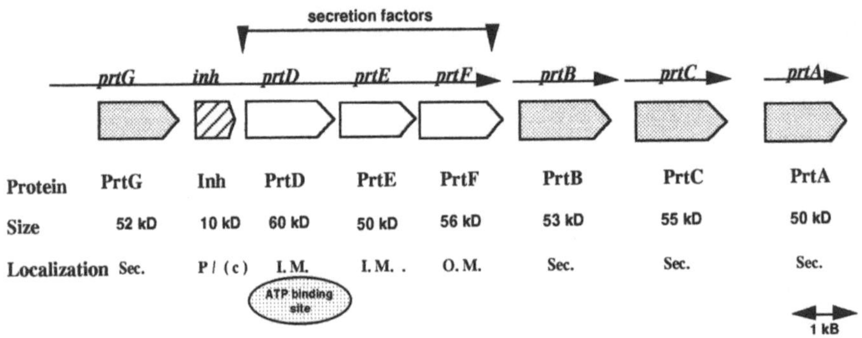

I. M.: inner membrane O . M.: outer membraneP : périplasm c: cytoplasm Sec: secreted proteins

Genetic organization of the region required for the specific synthesis and secretion of the A, B, C and G protease as well as for the synthesis of their specific intracellular inhibitor.

fig.1

When a DNA fragment encoding the proteases determinant is introduced in *E. coli* on recombinant plasmids it leads to the synthesis and secretion of proteases A, B, C and G (PrtA, B, C and G) The structural genes for PrtA, B and C are adjacent and form three transcription units whereas the region located upstream of *prtB* forms an operon of five genes which codes for PrtG, the inhibitor and three secretion factors PrtD, PrtE and PrtF, all absolutely required for the protease secretion.

The four proteases are synthesised as inactive zymogens which have short amino-terminal extensions named propeptides Cleavage of propeptides is not required for secretion since the zymogens are very efficiently secreted and activated in the extra cellular medium (Delepelaire and Wandersman, 1989). Furthermore,comparison of the primary amino acids sequence and the deduced protein sequence from DNA sequencing showed that there is no N-terminal signal-peptide sequence upstream of the propeptide. Instead, deletions of the 5' extremity of *prtB*, *prtA* and *prtG* showed that their secretion signal is located, as in the case of HlyA, in the C-terminal portion of the proteases (Mackman *et al.*, 1987 ; Delepelaire et Wandersman, 1990; Ghigo and Wandersman, 1992a, 1992b). Comparison of the *E. chrysanthemi* proteases with the metalloprotease of *S. marcescens* (PrtSM) and the alkaline protease of *Pseudomonas aeruginosa* (AprA) has shown that they share common features and thus belong to the same family: they are highly homologous (>50%), sensitive to the same protease inhibitor, synthesised as inactive zymogens and have a C-terminal secretion signal. Furthermore, an efficient secretion of PrtSM and AprA was obtained in *E. coli* expressing the *E. chrysanthemi* protease secretion factors (Létoffé *et al.*, 1991; Guzzo *et al.*, 1991).

A specific secretion apparatus.

The three *E. chrysanthemi* secretion factors PrtD, PrtE and PrtF are required for protease secretion (Wandersman *et al.*, 1987). Sequence data indicated the presence of six potential transmembrane hydrophobic segments in PrtD, one in PrtE and a putative signal sequence in PrtF (Létoffé *et al.*, 1990). Antibodies were obtained against PrtD, PrtE and PrtF which specifically localize these proteins in the bacterial envelope.

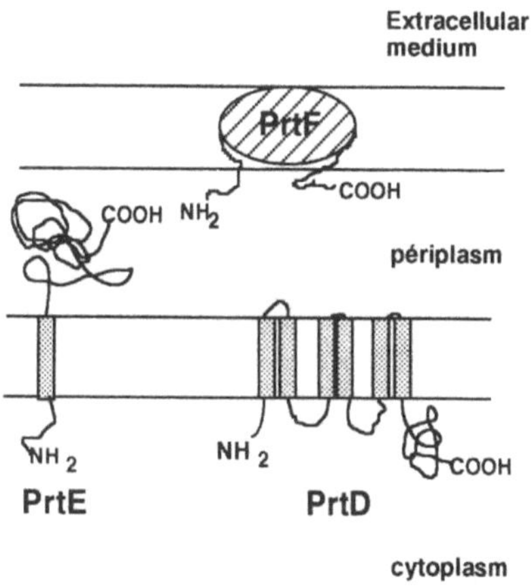

Fig.2

Fractionation of bacterial lysates was used to show that both PrtD and PrtE are cytoplasmic membrane proteins, whereas PrtF is an outer membrane protein. On the basis of sequence data and protease accessibility data on whole cells, spheroplasts and inverted cytoplasmic membrane vesicles we have proposed a model for the organisation of PrtD, PrtE and PrtF in the envelope (Delepelaire and Wandersman, 1991) (see fig.2)

In all cases where the secretion systems could be studied the specific secretion apparatus appeared to be composed of two inner membrane proteins and one outer membrane protein. Moreover, the genes for these proteins are often clustered with those coding for the secreted protein(s) themselves (see fig.3)

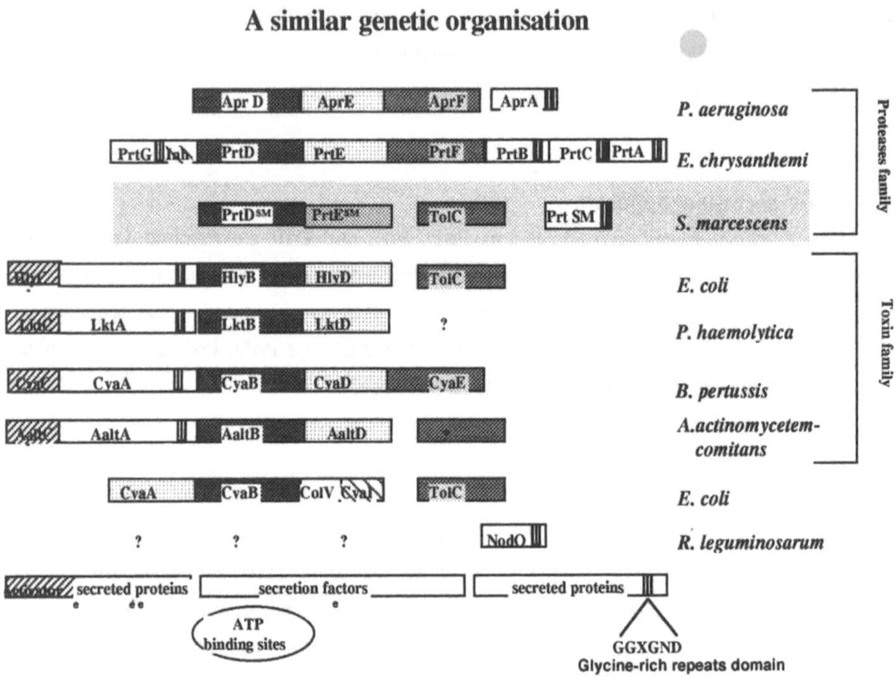

Fig.3

This genetic organisation is widely conserved in these transporters among which one component is a member of the ATP-Binding Cassette (ABC) membrane protein superfamily all involved in the translocation of various molecules acros biological membranes (Higgins, 1992). In the *E. chrysanthemi* transporter, PrtD is the ABC protein and its ability to be specifically labelled by 8-azido ATP suggest that its ATP-binding site is functional (P. Delepelaire, unpublished results).

DNA sequence comparison showed that *prtD* and *prtE* are homologous to *hlyB* and *hlyD* the only two previously described genes involved in HlyA secretion. On the basis of homologies between PrtF and the *E. coli* outer membrane protein TolC and on the fact that TolC mutants are defective in secretion of the α-hemolysin it was shown that the TolC protein is the outer membrane constituent of the hemolysin secretion system (Wandersman and Delepelaire, 1990).

Complementation between heterologous secretion systems.

The sequence homologies between the secretion apparati led us to undertaken cross-complementation experiments between various proteins secretion systems. These experiments allowed us to show that the HlyA transporter (HlyB, HlyD and TolC) can promote, albeit with low efficiency, the secretion of PrtB, PrtA, and PrtSM. On the other hand the *E. chrysanthemi* transporter allows the efficient secretion of PrtSM, AprA and ColV but not HlyA (Létoffé *et al.*, 1991; Wandersman, 1992)(see Table1).

COMPLEMENTATION BETWEEN HETEROLOGOUS SYSTEMS EXPRESSED IN *E. COLI*

Secretion factors	Secreted proteins	
HlyB, HlyD, TolC	HlyA	+++
	other hemolysins related to HlyA	+++
	Proteases	+
	Colicine V	+
PrtD, PrtE, PrtF	Proteases	+++
	Reconstituted system in *E.coli*	
	Colicine V	+++
	Toxines	—
AprD,AprE, AprF	Reconstituted system in *E.coli*	
	Proteases	+++
CyaB,cyaD,CyaE LktB,LktD, ?	Non functional systems in E. coli	
PrtD ^{SM}, PrtE ^{SM} and TolC	Proteases	+++
	Reconstituted system in *E.coli*	

Table1

We recently cloned and identified the genes for the two inner membrane components of the *S. marcescens* PrtSM secretion apparatus: *prtD*^{SM} and *prtE*^{SM}. They display a high level of homology with *prtD* and *prtE* respectively of *E. chrysanthemi.* Although the PrtSM secretion was efficiently reconstituted in *E. coli* a third gene, encoding the outer membrane component was not present on the DNA fragment cloned from *S. marcescens* and introduced in *E.coli* expressing PrtSM. However we were able to show using a *tolC* mutant that the TolC outer membrane protein of *E. coli* was able to enter in the composition of an efficient hybrid PrtSM secretion apparatus, as illustrated in Fig. 4. TolC enters thus in the constitution of several active transporters.

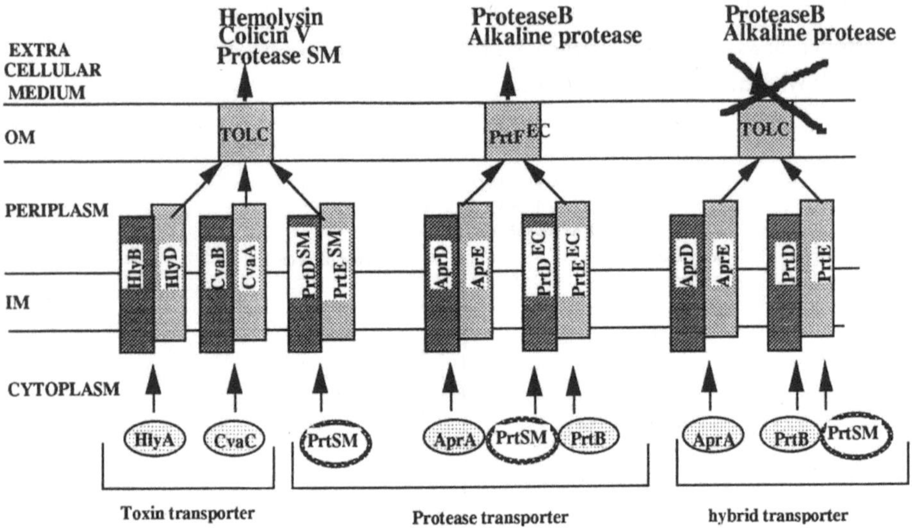

Fig.4

Interestingly, TolC cannot replace PrtF or AprF to secrete PrtB or AprA and conversely PrtF cannot replace TolC to promote the secretion of HlyA (SEE FIG.4). These results indicate that the outer membrane component in these apparati is not just an non-specific pore allowing transport across the outer membrane. Fig.4 also illustrates the fact that the substrates can therefore use the TolC protein if it is in contact with $PrtD^{SM}$ and $PrtE^{SM}$, but not if it is in contact with $PrtD^{EC}$ and $PrtE^{EC}$. This suggests that the specificity does not arise from a direct interaction between the signal and the outer membrane component (Létoffé *et al.*, submitted).

A C-terminal secretion signal.

Whereas the features of the N-terminal signal peptide are well defined it is still not true for the secretion information able to target proteins through the signal peptide independent secretion pathway. The search for such a signal has shown that as in the case of HlyA the secretion signal of PrtB, PrtA and PrtG of *E. chrysanthemi* and of PrtSM of *S. marcescens* is located at the C-terminal extremity of these proteins generally in the last 50 residues. In spite of the growing number of identified signals the primary amino acids sequence homologies are very low and no consensus can be found. However the conservation between the different secretion systems and the fact that they are at least partially interchangeable suggest that higher order complex motifs could be recognised on the C-terminal signals. This hypothesis has been extensively explored and several structural models were proposed in the case of the HlyA secretion signal (Hughes *et al.*, 1992). In the case of the proteases secretion signal, experiments are undertaken to precisely define the secretion signal which was shown to lie in the last 38 residues of PrtB (Delepelaire and Wandersman,1990)and in the last 33 residues of PrtG (J.M. Ghigo, unpublished results).

Although the extreme C-terminal part is essential for the secretion, other regions of the proteins could play a role in the secretion process.This is the case for the glycine rich repeated motifs, located upstream of the C-terminal secretion signal of almost all the proteins secreted by such a pathway (Welch, 1991) (see fig.4). The repeats are not part of the secretion signal *sensus stricto* since small secreted fusions containing C-terminal regions of HlyA, PrtA, PrtB, PrtG and PrtSM lack these repeats. Nevertheless, it has been clearly shown in the case of PrtB that its C-terminal secretion signal can promote the secretion of passenger proteins only if a C-terminal signal containing the repeats was used (Létofé and Wandersman, 1992). Furthermore we and others have showen that the C-terminal secretion signal display a wide tolerance with respect to the nature and the size of passenger polypeptides having the ability to be specifically secreted through these type of transporters and can thus be used as a secretion vector.

Conclusion

Our work on the *E. chrysanthemi* and *S. marcescens* metalloproteases led to the characterisation of bacterial transport apparati allowing the secretion of a new family of metalloproteases by the signal-peptide independent pathway. Besides the absence of an N-terminal signal peptide these proteases share with others proteins secreted by this pathway the following characteristics: i) a C-terminal targeting sequence ii) the presence of glycine-rich repeated hexapeptide GGXGXD in the C-terminal third of the proteins. iii) the lack of cystein residues. These proteins are secreted by homologous and partially interchangeable secretion systems composed by two inner membrane and one outer membrane proteins. We are now currently investigating the nature of the secretion signal and of its interaction with the secretion apparatus using genetic and biochemical approaches.

References

Delepelaire, P. and C. Wandersman. 1989. Protease secretion by *Erwinia chrysanthemi.* J. Biol. Chem. **264**: 9083-9089.

Delepelaire, P. and C. Wandersman. 1990. Protein secretion in Gram negative bacteria. The extracellular metalloprotease B from *Erwinia chrysanthemi* contains a C terminal secretion signal analogous to that of *Escherichia coli* a hemolysin. J. Biol. Chem. **265**: 17118-17125.

Delepelaire, P. and C. Wandersman. 1991. Characterization, localization and transmembrane organization of the three proteins PrtD, PrtE and PrtF necessary for protease secretion by the Gram-negative bacterium *Erwinia chrysanthemi*. Mol. Microbiol. **10**: 2427-2435.

Felmlee,T., S. Pellett, E.Y. Lee and R.A. Welch. 1985. *Escherichia coli* hemolysin is released extracellularly without cleavage of a signal peptide. J. Bacteriol. **163**: 88-93.

Ghigo,J. M. and C. Wandersman. 1992a. Cloning, nucleotide sequence and characterization of the gene encoding the *Erwinia chrysanthemi* B374 PrtA metalloprotease : a third metalloprotease secreted via a C-terminal secretion signal. Mol. Gen. Genet. **236**: 135-144.

Ghigo, J. M. and C. Wandersman. 1992b. A fourth metalloprotease gene in *Erwinia chrysanthemi*. Res Microbiol. **143**: 857-867.

Guzzo, J., F. Duong, C. Wandersman, M. Murgier and A. Lazdunski. 1991. The secretion genes of *Pseudomonas aeruginosa* alkaline protease are functionally related to those of *Erwinia chrysanthemi* proteases and *Escherichia coli* a-haemolysin. Mol. Microbiol. **5**: 447:453.

Higgins, C. F. 1992. ABC transporters from microorganisms to man. Ann Rev. Cell Biol. **8**: 67-113.

Hughes C., P. Stanley and V. Koronakis 1992. *E. coli* hemolysin interactions with Prokaryotic and Eukaryotic cell membranes. Bioassays.14:519-525.

Létoffé,S., P. Delepelaire and C. Wandersman. 1990. Protease secretion by *Erwinia chrysanthemi*: the specific secretion functions are analogous to those of *Escherichia coli* a-hemolysin. EMBO J. **9**: 1375-1382.

Létoffé,S., P. Delepelaire and C. Wandersman. 1991. Cloning and expression in *E. coli* of the gene encoding the *S. marcescens* metalloprotease : secretion from *E; coli* in the presence of the *E. chrysanthemi* specific protease secretion function. J. Bacteriol. **173**: 2160-2166.

Létoffé,S., J.M. Ghigo and C. Wandersman. 1993. Identification of two Components of the *Serratia marcescens* Metalloprotease Transporter. Protease^{SM} Secretion in *Escherichia coli* is TolC dependent. Submitted

Létoffé,S. and C. Wandersman. 1992.Secretion of CyaA-PrtB and HlyA-PrtB fusion proteins in *Escherichia coli:* involvment of the glycine-rich repeat domain of *Erwinia chrysanthemi* Protease B. J.Bacteriol. **174**:4920-4927

Mackman, N., K. Baker, L. Gray, R. Haigh, J. M. Nicaud and I.B. Holland. 1987. Release of a chimeric protein into the medium from *Escherichia coli* using the C-terminal secretion signal of haemolysin. EMBO J. **6**: 2835-2841.

Pugsley,A. P..1993. The complete general secretory pathway in Gram-negative bacteria. Microb. rev. **57**:50-108.

Wandersman, C., P. Delepelaire, S. Létoffé and M. Schwartz. 1987. Characterization of *Erwinia chrysanthemi* extracellular proteases: cloning and expression of the protease genes in *Escherichia coli*. J. Bacteriol. **169**: 5046-5053.

Wandersman, C. and P. Delepelaire. 1990. TolC, an *Escherichia coli* outer membrane protein required for hemolysin secretion. Proc. Natl. Acad. Sci. USA. **87**: 4776-4780.

Wandersman, C. 1992. Secretion across the bacterial outer membrane. Trends in Genet. **8**: 317-321.

Wandersman, C. and S. Létoffé. 1993. Involvement of lipopolysaccharide in the secretion of *Escherichia coli* a-hemolysin and *Erwinia chrysanthemi* proteases. Mol. Microbiol. **7**: 141-150.

Welch, R. A. 1991. Pore forming cytolysins of Gram negative bacteria. Mol. Microbiol. **5**: 521-528.

ScFv ANTIBODY FRAGMENTS PRODUCED IN *Saccharomyces cerevisiae* ACCUMULATE IN THE ENDOPLASMIC RETICULUM AND THE VACUOLE.

Leon G.J. Frenken, Eveline van Tuijl, J. Wil Bos, Wally H. Müller*, Arie J. Verkleij*, and C. Theo Verrips.
Department of Gene Technology and Fermentation,
Unilever Research Laboratorium,
P.O. Box 114,
3130 AC Vlaardingen,
The Netherlands

INTRODUCTION

The yeast *Saccharomyces cerevisiae* has a number of properties which makes it an important organism for the production of heterologous proteins for research, medical and industrial use. Yeast is particularly amenable to genetic manipulation and a number of auxotrophic markers and strong (inducible) promoters are known. It can secrete proteins into the culture medium and, in contrast to prokaryotes, it can also glycosylate proteins, which is of particular importance for the production of heterologous eukaryotic proteins, which are glycosylated by their naturel hosts. Yeast can grow rapidly on simple, inexpensive media to high cell densities, and knowledge on large scale fermentation and downstream processing is extensive. Furthermore, as it has been used in the brewing and baking industries for a long time, it has become a GRAS (Generally Regarded As Safe) organism. So far, it has been successfully used for the production of a number of single-chain heterologous proteins (Romanos *et al.*, 1992).

Antibodies are proteins of high potential importance for industrial application. However, many of these applications cannot be realized because of the high cost of antibodies if produced by culturing hybridoma cells. Alternative routes for the production of antibodies have been

*Department of Molecular Cell Biology, Utrecht University, P.O. Box 80.054, 3508 TB Utrecht, The Netherlands.

NATO ASI Series, Vol. H 82
Biological Membranes:
Structure, Biogenesis and Dynamics
Edited by Jos A. F. Op den Kamp

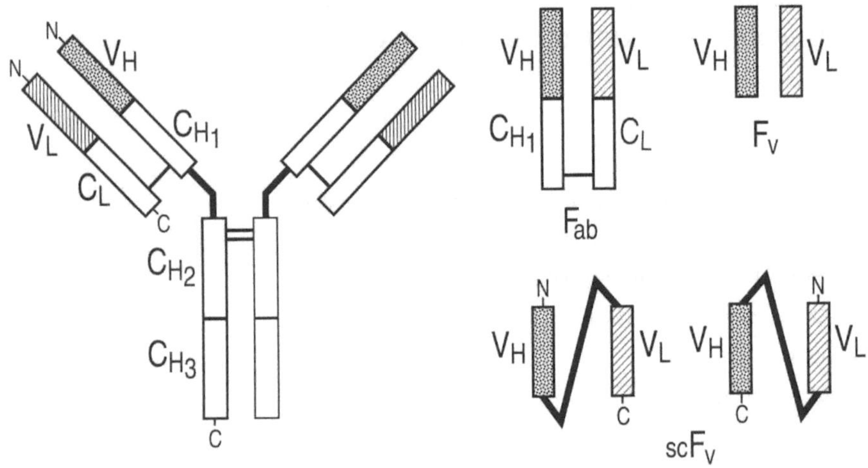

Figure 1. Schematic representation of an antibody and of Fab, Fv and scFv antibody fragments. In the scFv fragment, the V_H and V_L domains are connected via a peptide linker, which can either result in the V_H-L-V_L or in the V_L-L-V_H arrangement.

investigated, including the use of plants and microorganisms. Attempts to produce intact antibodies in yeast have resulted in disappointingly low yields (<500 µg/l) and the same holds for the much less complex Fab fragments (Fig. 1) (Horwitz *et al.*, 1988; Bowdish *et al.*, 1991). For a number of applications, however, the smaller Fv fragments are appropriate or even preferred. These fragments comprise only the variable regions of the heavy (V_H) and the light (V_L) chain, held together mainly by interactions between hydrophobic amino acids at the interface (Chothia *et al.*, 1985). Some Fv fragments may be completely stable, whereas others dissociate at high dilutions. A possibility to overcome this problem, is to link the fragments by a genetically encoded peptide linker, which results in a so-called single-chain Fv (scFv; Fig. 1). An additional advantage of this approach is the equimolar production of V_H and V_L fragments. Although much is known with respect to the production of this kind of fragments by *E. coli* (Plückthun, 1991), little is known about the production of scFv by yeast.

The aim of the present work has been to investigate the production of scFv fragments by *S. cerevisiae*. As a model system we used the scFv fragment of the anti-hen egg white lysozyme antibody D1.3 (scFv-LYS, Ward *et al.*, 1989). To achieve secretion of the scFv-LYS, it was either fused to signal peptides of yeast secretory proteins or to a heterologous protein which is secreted by yeast in high amounts. Surprisingly, in both cases the scFv accumulated intracellularly. The subcellular localization of the accumulating antibody fragments was investigated by immuno-electron microscopy and by using yeast strains deficient in vacuolar proteases.

PRODUCTION OF scFv FRAGMENTS

In recent years we developed a yeast expression system, using the inducible *GAL7* promoter, which has been successfully used for the production of a number of heterologous proteins. An example is the production of the α-galactosidase originating from the plant *Cyamopsis tetragonoloba*, which is produced at levels exceeding 0.5 g/l (Verbakel, 1991). Based on this system, an initial set of expression plasmids encoding scFv-LYS antibody fragments was constructed (Fig. 2). Since the signal sequence can have a profound effect on the secretion efficiency (Harmsen *et al.*, 1993; Sleep *et al.*, 1990) two different sequences were tested. Plasmids pUR4125 and pUR4126 encoded scFv-LYS having different arrangements, V_H-L-V_L and V_L-L-V_H, respectively, both preceded by the invertase signal sequence (SUC2). pUR4132 encodes the V_H-L-V_L arrangement preceded by the signal peptide from the prepro-α-factor (MFα1) (constructions will be described in more detail elsewhere, publication in prep). These plasmids were introduced into the yeast strain SU10 (MATα, *leu2*, *ura3*, *his4*, *cir*$^+$).

After growth for 24 to 48 h under inducing conditions (in the presence of galactose), the presence of scFv-LYS in the culture medium was checked by ELISA. No functional scFv-LYS could, however, be detected. Subsequently, the culture medium and different fractions of the yeast cells were subjected to Western blot analysis (Fig. 3). To this end, the cells were disrupted with glass beads and after centrifugation "soluble" and "insoluble" cell fractions were obtained. The latter comprise cell wall fragments and cell-organelles. As shown in Fig. 3, scFv-LYS fragments were only found in the insoluble cell fraction together with a number of degradation products. The processing of the scFv preceded by the SUC2 signal sequences is clearly more efficient when compared to the scFv preceded by the MFα1 prepro-sequence. The yield of the scFv-LYS from SU10(pUR4125) was estimated to be 3% of the total protein.

To determine whether the intracellular accumulation of scFv-LYS was yeast strain dependent, the above described expression plasmids were introduced into 4 different laboratory strains. The scFv-LYS was found to accumulate in the insoluble cell fraction in all these strains. Alternatively,

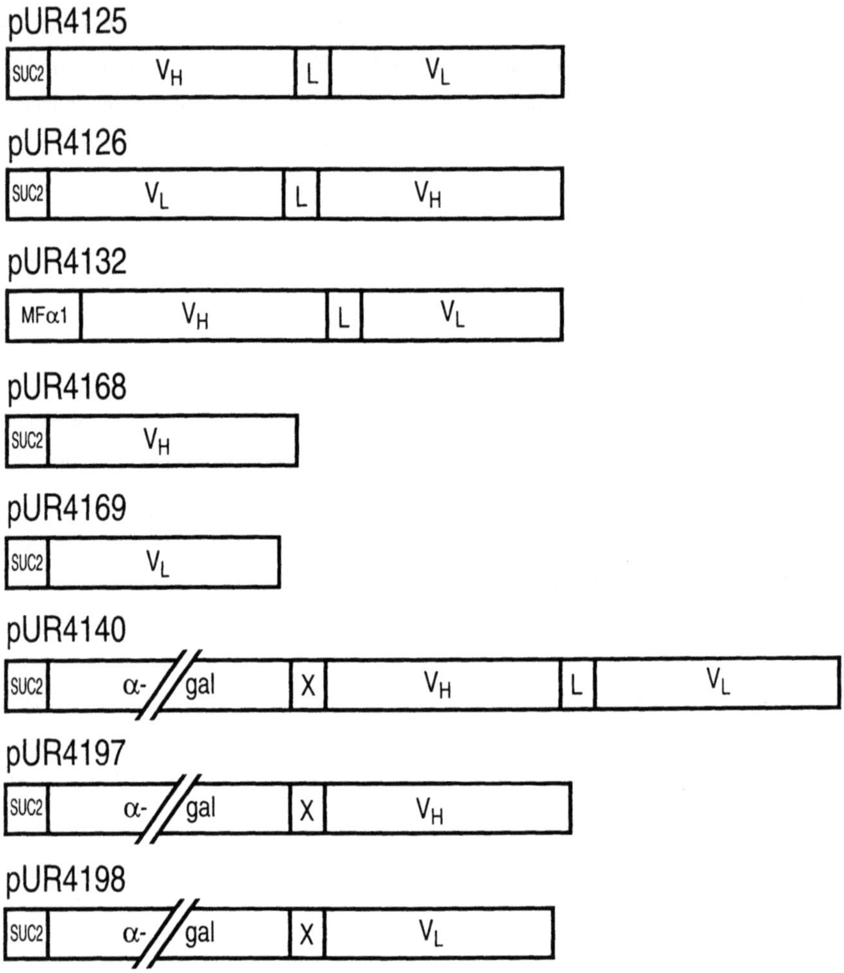

Figure 2. Schematic representation of the expression plasmids encoding different antibody fragments and fusion proteins.

as the intracellular accumulation of scFv-LYS could be due to some characteristics of the fragment itself, two other scFv fragments were tested. For this purpose, expression plasmids analogous to pUR4125 were constructed, encoding scFv fragments derived from an anti-Human Chorionic

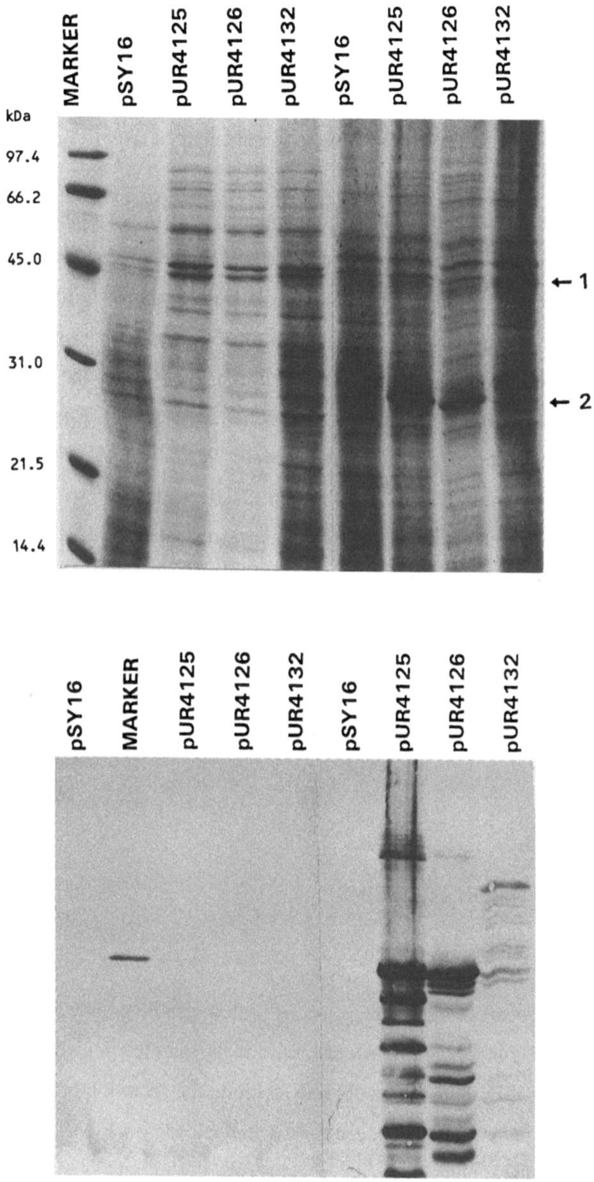

Figure 3. Coomassie Brilliant Blue stained gel (Top) and Western blot (Bottom) of SU10(pUR4125), SU10(pUR4126) and SU10(pUR4132). S = soluble cell fraction, and I = insoluble cell fraction. Arrows indicate scFv-LYS (1) and the MFα1-scFv-LYS fusion protein (2).

Gonadotropin (scFv-HCG, pUR4170) antibody and from an anti-Traseolide (a musk-like fragrance, scFv-TRAS, pUR4189) antibody. Western blot analysis of the culture media and the cell fractions of SU10 strains containing either of these plasmids, revealed that like the scFv-LYS fragments both, scFv-HCG and scFv-TRAS were found solely in the insoluble cell fraction (data not shown).

The scFv fragments comprise three different elements, i.e. *(i)* the variable region of the heavy chain, *(ii)* the variable region of the light chain, and *(iii)* the linker peptide.
To determine whether one of these segments is in particular responsible for the retention of the total scFv, they were tested separately. To this end three expression plasmids were constructed. Replacement of the $(GGGGS)_3$ peptide linker between the V_H and the V_L fragments by the linker -GSTSGSGKSSEGKG-, as described by Colcher *et al.* (1990), gave pUR4186. Plasmids pUR4168 and pUR4169 encode either the separate V_H or the V_L fragment, respectively. Introduction of these plasmids into yeast strain SU10 and Western blot analysis of the transformants, gave the same results as the previous constructs: the scFv-LYS having the modified linker sequence as well as the separate V_H and V_L fragments were detected only in the insoluble cell fraction (data not shown).

From these results it was concluded that the replacement of the linker peptide is not sufficient to overcome the secretion problems, and that both the V_H and the V_L fragments have intrinsic properties interfering with or blocking their secretion.

PRODUCTION OF FUSION PROTEINS

It is known that the secretion of a heterologous protein can be greatly enhanced by fusing the protein to another (homologous) protein which is normally secreted in high amounts (Ward *et al.*, 1990). To test whether this approach would also facilitate the secretion of antibody fragments by *S. cerevisiae*, three expression plasmids were constructed encoding scFv, V_H or V_L fused to the C-terminus of α-galactosidase (Fig. 2). As described above, this enzyme is readily secreted by yeast (Verbakel, 1991). Furthermore, is was shown by Schreuder *et al.* (1993), that a C terminal fusion of α-galactosidase to the C-terminus of α-agglutinin maintained full enzymatic activity when exposed at the cell surface. After introducing plasmids pUR4140, pUR4197 and pUR4198 into SU10, the transformants were grown under inducing conditions and subjected to Western blot analysis, using anti-Fv and anti-α-galactosidase antisera. Neither the antibody fragments (scFv, V_H

or V_L) nor the α-galactosidase (α-gal) could be detected in the culture medium or in the soluble cell fractions. Proteins reacting with the specific antisera could only be detected in the insoluble fractions of the yeast cells. Figure 4 shows Western blot analysis of the insoluble cell fraction of SU10(pUR4140), encoding the α-gal+scFv-LYS fusion. When using the anti-Fv serum, in addition to a band at the molecular weight (M_w) level expected for the fusion protein, several bands at lower M_w's were present, indicative of degradation products. When using the anti-α-galactosidase serum, again a band was found at the expected M_w level as well as some bands at lower M_w.

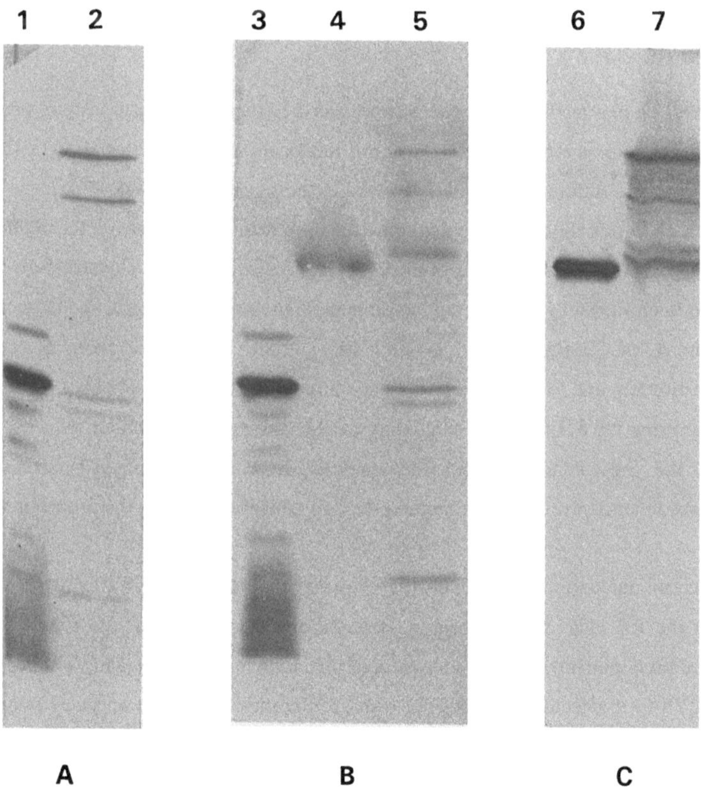

Figure 4. Western blot analysis of the insoluble cell fraction (Sample I) of SU10(pUR4140), expressing the fusion protein α-gal+scFv-LYS. (A) detection with anti-Fv serum, (C) detection with anti-α-galactosidase serum and (B) detection with both antisera. Lane 1 and 3, scFv-LYS (ex. *E. coli*); lane 2, 5 and 7, Sample I; lane 4 and 6, α-galactosidase (ex *S. cerevisiae*).

However, no bands were found below the M_w level of mature α-galactosidase, suggesting that the α-galactosidase portion of the fusion protein is more stable than the scFv part.

The finding that the antibody fragments and the fusion proteins are not secreted, but are retained in the insoluble cell fractions, suggests that their secretion is blocked somewhere between the endoplasmic reticulum (ER) and the secretion vesicles, or that they are targeted to a cell organelle. As a first step in elucidating the cause of the intracellular accumulation of the antibody fragments, we decided to investigate their subcellular localisation in more detail.

IMMUNO-ELECTRON MICROSCOPY

Since Western blot analysis gives only an impression of the total pool of proteins present in the cells, no information is available with respect to the localisation and distribution of the antibody fragments over the different organelles involved in the secretion pathway.
To obtain such information from immuno-electron microscopy, a preparation technique is needed which preserves both the antigenicity of the antibody fragments and the ultrastructure of the yeast cells. It has been shown by previous studies on microorganisms like bacteria and filamentous fungi (Voorhout *et al.*, 1989; Müller *et al.*, 1991), that the combination of cryofixation, freeze-substitution, and low-temperature embedding fulfils such requirements.

After growing the SU10 strains containing pUR4125 or pUR4140 under inducing conditions for 24 h, the cells were harvested and spheroplasts were prepared using zymolyase. The spheroplasts obtained were used for immuno-electron microscopy using the procedures described above.

The electron micrographs of SU10(4125) revealed that the scFv-LYS fragments were mainly located in the ER (Fig. 5). In contrast, the α-galactosidase moiety of the α-gal+ScFv fusion protein was predominantly present in a vacuolar like structure (Fig. 6). Thus, although fusion of the scFv-LYS to α-galactosidase does not result in the secretion of the antibody fragment it has a clear effect on the intracellular routing.

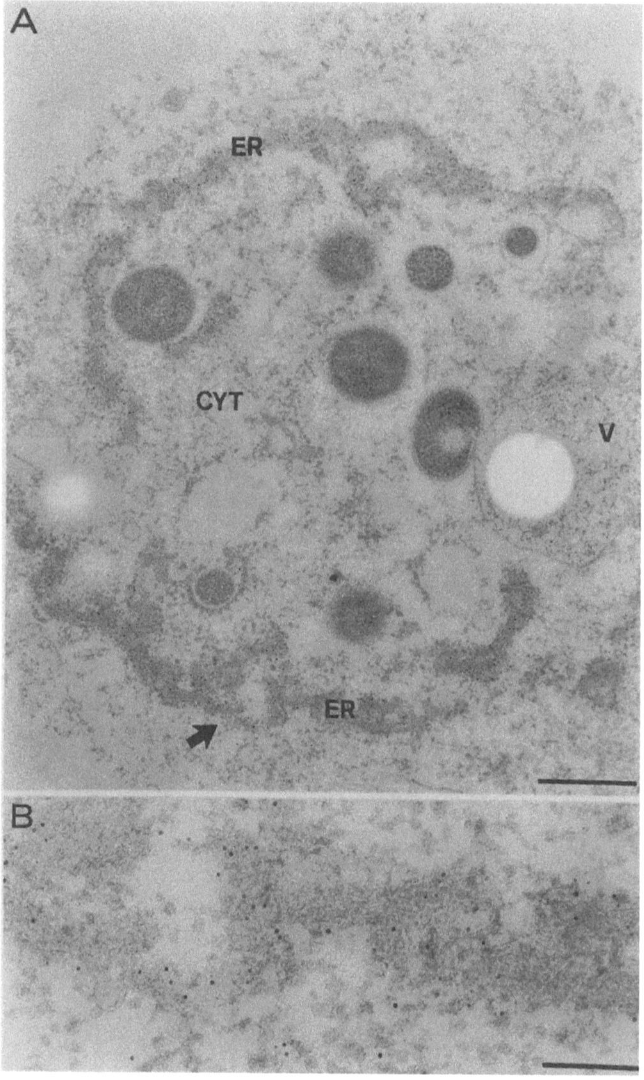

Figure 5. Electron micrograph of SU10(pUR4125) spheroplasts expressing scFv-LYS. The ScFv fragments were labelled with anti-Fv-LYS antiserum and the antigen antibody complex was visualized with goat anti-rabbit antibodies conjugated with 6 nm gold. A: Overview of pUR4125 yeast spheroplast, ER = endoplasmic reticulum, V = vacuole, Cyt = cytoplasm, bar = 500 nm. B: Higher magnification of area in figure A and indicated by the big arrow, bar = 150 nm.

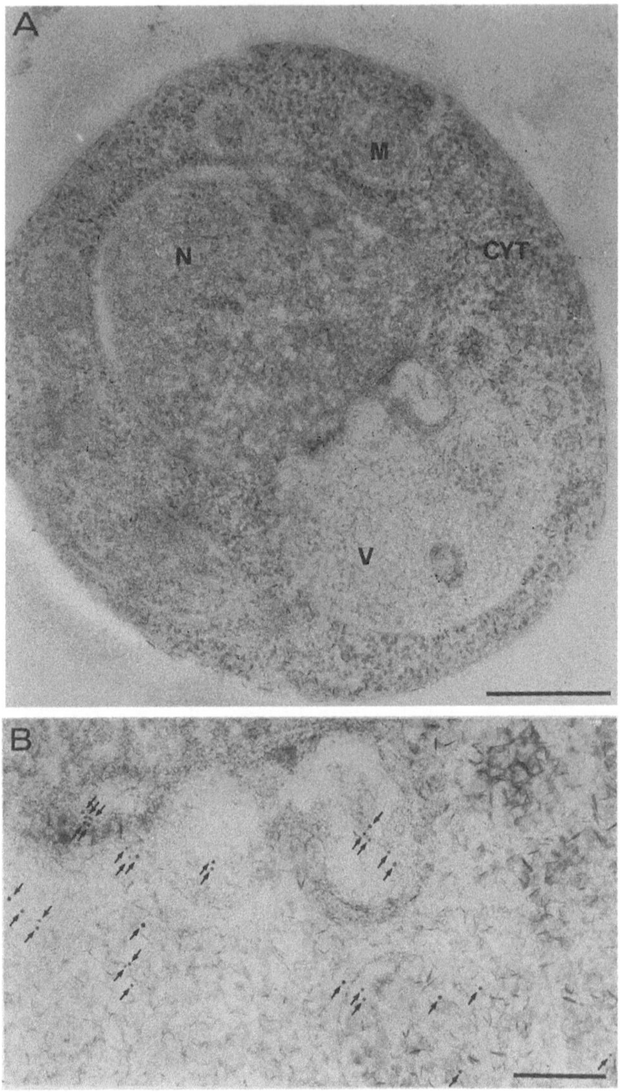

Figure 6. Electron micrographs of SU10(pUR4140) spheroplasts expressing the α-gal+scFv-LYS fusion protein. The fusion protein was labelled with antibodies against α-galactosidase, and the antibody antigen complex was visualized as described in Fig. 5. A) Overview of the yeast spheroplast pUR4140, N = nucleus, V = vacuole, M = mitochondrion, Cyt = cytoplasm, bar = 500 nm. B) Higher magnification of the vacuole in A). Arrows indicate gold particles, bar = 150 nm.

PROTEASE DEFICIENT YEAST STRAINS

Protease deficient mutants can be used to investigate the targeting of proteins (Takeshige *et al.*, 1992). For example, the *pep4-3* mutation for instance, results in a greatly reduced level of the vacuolar protease A (*PEP4*). Since protease A is involved in the activation of other vacuolar proteases e.g. protease B (*PRB1*) and carboxypeptidase Y (*PRC1*) (Woolford *et al.*, 1993), their levels are also reduced. In strains having additional mutations in the latter genes, a further reduction in protease activity is found.

To investigate whether the α-gal+scFv-LYS fusion protein indeed targets the vacuole, plasmid pUR4149 was introduced into the following strains: MT302 (MATa, *arg5-6, leu2-3, leu2-12, his3-11, his3-15,* ***pep4-3****, ade1*; Mellor *et al.*, 1983), and BJ2168 (MATa, *leu2, trp1, ura3-52,* ***prb1-1122, pep4-3, prc1-407****, gal2*; Jones *et al.*, 1991).

Western blot analysis of the culture medium and of the soluble and insoluble cell fraction showed that also in these strains the heterologous proteins were present only in the insoluble fractions (Fig. 7). Comparison of the band patterns of the insoluble fractions revealed a

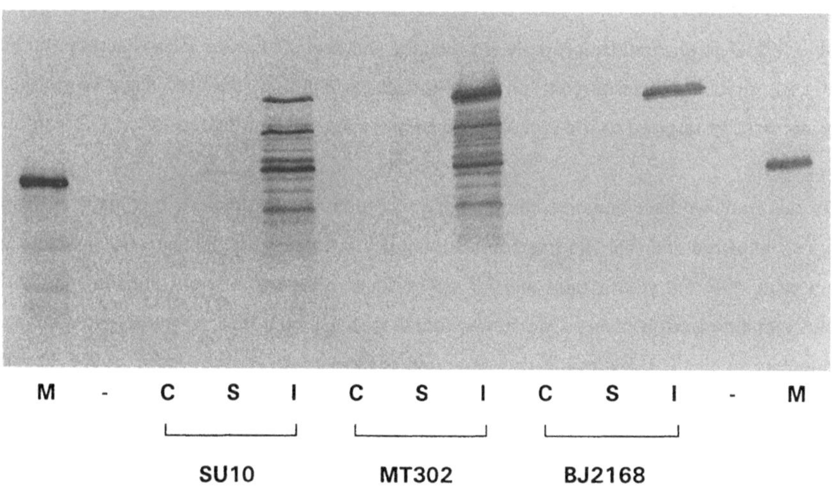

Figure 7. Western blot analysis of the culture medium (C), the soluble (S) and the insoluble (I) cell fraction of SU10(pUR4140), MT302(pUR4140) and BJ2168 (pUR4140), expressing the fusion protein α-gal+scFv-LYS. M = α-galactosidase (ex *S. cerevisiae*). For detection anti-α-galactosidase serum was used.

remarkable difference: whereas in SU10 a major band was found at the M_w level of mature α-galactosidase and a weaker band at the level of the fusion protein, the intensities found in the MT302 sample were different. In this strain, the band corresponding to the fusion protein was clearly the most intense band. In the case of BJ2168, only one band was found, representing the fusion protein.

From this result it is concluded that the reduced activities of the *PEP4*, *PRB1*, and *PRC1* encoded proteases, allow the stable accumulation of the fusion protein in the vacuole.

DISCUSSION

The first step in protein secretion by yeast is the translocation of the protein across the endoplasmic reticulum (ER) membrane, during which the signal sequence is processed. Subsequently, the proteins move from the ER to the Golgi apparatus and via secretory vesicles to the cell surface. Here the vesicles fuse with the plasma membrane of the young yeast bud and release their content into the surrounding medium (Pryer *et al.*, 1992). However, not all secretory proteins follow this route and by now it is well established that (heterologous) proteins which are unable to fold or oligomerize properly are retained and degraded in the ER (Klausner and Sitia, 1990). Alternatively, (heterologous) proteins having a partially unfolded state, might be recognized as abnormal and targeted to the vacuole were they are degraded (Roberts *et al.*, 1992; Takeshige *et al.*, 1992).

In this study we have demonstrated that the secretion of scFv antibody fragments by yeast is seriously impaired and that the fragments accumulate intracellularly. Similar results were found when using different yeast strains and for scFv fragments having different antigen specificities. From immuno-electron microscopy, it was found that the scFv-LYS fragments predominantly accumulate in the ER, indicating that secretion is blocked at an early stage of the secretion pathway.

Fusion of the scFv fragment to the C-terminus of α-galactosidase, a heterologous protein which is readily secreted by yeast, did not solve the secretion problems. Analysis of the α-gal+scFv-LYS fusion protein showed that the α-galactosidase moiety was more stable than the scFv fragment. Immuno-electron microscopy revealed that the α-galactosidase was predominantly present in a "vacuolar-like" organelle. On expression of the α-gal+scFv-LYS fusion protein in a mutant yeast strain, deficient in three vacuolar proteases, only intact fusion protein could be detected by

Western blot analysis, indicating that the scFv moiety is more stable in these mutants. It is therefore concluded that in the "wild type" strain the fusion protein is targeted towards the vacuole and that the vacuolar proteases are responsible for the degradation of the scFv moiety.

The finding that even the smaller V_H and V_L fragments could not be secreted, neither when they were preceded by the SUC2 signal sequence nor when they were fused to the C-terminus of α-galactosidase, indicates that they have intrinsic properties which are incompatible with the yeast secretion system.

A possible explanation might be the improper folding of these fragments, which seems to be underscored by the sensitivity of the antibody fragments to proteolytic degradation as found in the ER of the strain SU10. Alternatively, the secretion might be hampered by the formation of large aggregates, due to illegitimate interactions of the hydrophobic regions on the V_H and V_L fragments, normally involved in the binding of both fragments.

REFERENCES

Bowdish, K., Tang, Y., Hicks, J.B., and Hilver, D. (1991) Yeast expression of a catalytic antibody with chorismate mutase activity. J. Biol. Chem. **266:** 11901-11908.

Chothia, C., Novotny, J., Bruccoleri, R., and Karplus, M. (1985) Domain association in immunoglobulin molecules: The packing of variable domains. J. Mol. Biol. **186:** 651-663.

Colcher, D. *et al.* (1990) In vivo tumor targeting of a recombinant Single-Chain Antigen-Binding protein. J. Natl. Cancer Inst **82:**1191-1197.

Harmsen, M.M, Langedijk, A.C., Van Tuinen, E., Geerse, R.H., Raue, H.A. & Maat, J. (1993) Effect of a *pmrI* disruption and different signal sequences on the intracellular processing and secretion of *Cyamopsis tetragonoloba* α-galactosidase by *Saccharomyces cerevisiae*. Gene **125:** 115- 123.

Horwitz, A.H., Chang, C.P., Better, M., Hellstrom, K.E. and Robinson, R.R. (1988) Secretion of functional antibody and Fab fragment from yeast cells. Proc. Natl. Acad. Sci. U.S.A. **85:** 8678-8682.

Jones, E.W., (1991) Tackling the protease problem in *Saccharomyces cerevisiae*. In: Methods in Enzymology, **194:** 428-453.

Klausner, R.D., and Sitia, R. (1990) Protein degradation in the endoplasmic reticulum. Cell **62:** 611-614.

Mellor, J., *et al.*, (1983) Efficient synthesis of enzymatically active calf chymosine in *Saccharomyces cerevisiae*. Gene **24:** 1-14.

Müller, W.H., Van der Krift, T.P., Krouwer, A.J.J., Wösten H.A.B., Van der Voort, H.L.M., Smaal, E.B. & Verkleij, A.J. (1991) Localization of the pathway of the penicillin biosynthesis in *Penicillium chrysogenum*. The EMBO J. **10:** 489-495.

Plückthun, A. (1991) Strategies for the expression of antibody fragments in *Escherichia coli*. Methods **2:** 88-96.

Pryer, N.K., Wuestehube, L.J., and Schekman, R. (1992) Vesicle-mediated protein sorting. Annu. Rev. Biochem. **61:** 471-516.

Roberts, C.J., Nothwehr, S.F & Stevens, T.H. (1992) Membrane protein sorting in the yeast secretory pathway: evidence that the vacuole may be the default compartment. J. Cell Biol. **119:** 69-83.

Romanos, M.A., Scorer, C.A., and Clare, J.J. (1992) Foreign gene expression in yeast: a review. Yeast **8:** 423-488.

Schreuder, M.P., Brekelmans, S., Ende, H. v.d., and Klis, F.M. (1993) Targeting of a heterologous protein to the cell wall of *Saccharomyces cerevisiae*. Yeast **9:** 399-409.

Sleep, D., Belfield, G.P., and Goodey, A.R. (1990) The secretion of human serum albumine from the yeast *Saccharomyces cerevisiae* using five different leader sequences. Bio/technology **8:** 42-46.

Takeshige, K., Baba, M., Tsuboi, S., Noda, T. & Ohsumi, Y. (1992) Autophagy in yeast demonstrated with proteinase deficient mutants and conditions for its induction. J. Cell Biol. **119:** 301-311.

Verbakel, J.M.A. (1991) Heterologous gene expression in the yeast *Saccharomyces cerevisiae*. PhD Thesis, University of Utrecht, Utrecht, The Netherlands.

Voorhout, W., Leunissen-Bijvelt, J., Leunissen, J., Tommassen, J. & Verkleij, A.J. (1989) Immuno-gold labelling of *Escherichia coli* cell envelope components. In: Immuno-gold labelling in Cell Biology (ed. by A.J. Verkleij & J.L.M. Leunissen), pp. 292-304. CRC Press, Boca Raton, Florida.

Ward, E.S., Güssow, D., Griffiths, A.D., Jones, P.T., and Winter, G. (1989) Binding activities of a repertoire of single immunoglobulin variable domains secreted from *Escherichia coli*. Nature **341:** 544-546.

Ward, M., Wilson, L.J., Kodama, K.H., Rey, M.W., and Berka, R.M. (1990) Improved production of chymosin in *Aspergillus* by expression as a glucoamylase-chymosin fusion. Bio/technology **8:** 435-440.

Woolford, C.A., Noble, J.A., Garman, J.D., Tam, M.F., Innis, M.A., and Jones, E.W. (1993) Phenotypic analysis of proteinase A mutants. J. Biol. Chem. **268:** 8990-8998

PREPROTEIN BINDING BY ATP-BINDING SITE MUTANTS OF THE *BACILLUS SUBTILIS* SecA

J. van der Wolk, M. Klose,* R. Freudl* and A.J.M. Driessen
Department of Microbiology
University of Groningen
Kerklaan 30
9751 NN Haren
The Netherlands

The preprotein *translocase* of *Escherichia coli* (Wickner *et al.*, 1991) guards preproteins from the site of synthesis at the ribosome in the cytosol to the processed form to be released into the periplasm. SecA (Schmidt *et al.*, 1988) is the ATP-hydrolysing, peripheral subunit of the *translocase* (Brundage *et al.*, 1990; Wickner *et al.*, 1991), and plays an essential role in preprotein translocation (Lill *et al.*, 1989). The low endogenous ATPase activity of SecA is stimulated by interactions with acidic phospholipids, preproteins and the SecY/E protein (Lill *et al.*, 1989, 1990; Brundage *et al.*, 1990). This stimulated ATPase activity initiates translocation which is further driven by ATP hydrolysis and Δp (Schiebel *et al.*, 1991; Driessen, 1992). Biochemical studies suggest that SecA possess three ATP binding sites (Lill *et al.*, 1989; Oliver, 1993). Only one domain shows a significant level of sequence similarity to the Walker A- and B-motifs for a NTP-binding site (Walker *et al.*, 1982) (Fig. 1). Both regions are highly conserved among different bacterial and algal SecA homologues. To analyze the function of this putative ATP binding site, we started a site-directed mutagenesis approach and changed critical residues of the A-domain (Klose *et al.*, 1983; van der Wolk *et al.*, 1993) of the *Bacillus subtilis* SecA homolog (Overhoff *et al.*, 1991). Now we report on the localization of the B-domain and further characterized ATP- and preprotein-binding activities of the mutants.

---

*Institut für Biotechnologie 1, Forschungscentrum Jülich GmbH, Postfach 1913, 5170 Jülich, Germany

NATO ASI Series, Vol. H 82
Biological Membranes:
Structure, Biogenesis and Dynamics
Edited by Jos A. F. Op den Kamp

```
                              A-region                  B-region

Bacillus subtilis  96 NIAEMKTGEGKTLTSTLP  196 RPLHFAVIDEVDSILIDEARTPLII
Escherichia coli   98 CIAEMRTGEGKTLTATLP  198 RKLHYALVDEVDSILIDEARTPLII
Pavlova lutherii   94 KIAEMKTGEGKTLVAILP  194 NGFEFAIIDEVDSVLIDEARTPLII
Antithamnion sp    90 KIAEMKTGEGKTLVAMLT  190 RDFFFAIIDEIDSILIDEARTPLII
                          ****.*******.. *.       . . .*..**.**.***********
                            GxTGxGKT              RxxxhhhhDEADxhh
                                                      RxxxhhhhDEADxhh
```

Fig 1. Sequence alignment of the putative ATP-binding domains of SecA homologs.

EXPERIMENTAL PROCEDURES:

Bacteria, plasmids and materials. Wild-type and mutant *B. subtilis secA* genes were cloned into the expression vector pTRC99A under regulatory control of the *trc* promoter/*lac* operator (Klose et al., 1993). *E. coli* strain JM109 (*recA1, endA1, hyrA96, thi$^-$, hsdR17, relA1, supE44, λ$^-$, Δ(lac-proAB), (F', traD36, proAB, lacIq2ΔM15*) harbouring the plasmids pMKL40, pMKL20, and pMKL21 (Table 1) was used for purification of SecA proteins as described (van der Wolk *et al.*, 1993). pMKL18 contained the *E. coli* wild-type *secA* gene (Klose *et al.*, 1983). SecA growth complementation experiments were performed with *E. coli* strains MM52 (MC4100, *secA51^{ts}*), BA13 (MC4100, *secA13^{am}*, *supF^{ts}*, *zch::TN10*), MM54 (*geneX^{am}*, *supF^{ts}*) and JM105.1 (*sec51^{ts}*, *leu::Tn10*, *thi*, *rpsL*, *endA*, *sbcB15*, *hspR4*, Δ(*lac-proAB*), [F', *traD36*, *proAB*, *lacIq*Δ*M15*]) (*See* Klose *et al.*, 1993). ProOmpA and OmpA were isolated as described (*See* Driessen, 1992).

Proteolytic digestion. The conformation of SecA was probed by the sensitivity to Staphylococcal V8 protease (Shinkai *et al.*, 1991). Reaction mixtures (50 $\mu\ell$) contained: 60 μg of SecA, 50 mM TrisCl (pH 8.0), 50 mM KCl, 5 mM $MgCl_2$ and 1 mM DTT. ATP (4 mM), (pro-)OmpA (640 μg $m\ell^{-1}$), and/or *E. coli* phospholipid vesicles (800 μg ml^{-1}) were added as indicated. Controls received 0.48 M urea. After 10 min preincubation at 37°C, 450 ng of V8 protease was added and aliquots (10 $\mu\ell$) were withdrawn at different time intervals. Reactions were terminated with 5 $\mu\ell$ of blocking buffer (10 mM PMSF, 6 % SDS, 150 mM TrisCl [pH 6.8], 30% [v/v] glycerol, 15 mM DTT, and 0.03

% bromophenol blue) and heating for 5 min at 95°C. Samples were analyzed by SDS-PAGE.

Table 1. Plasmids

Plasmid	Amino acid substitution	*B. subtilis* *secA* allele	Reference
pMKL04	none	Wild-type	Klose *et al.*, 1993
pMKL20/pMKL200	$Lys^{101} \rightarrow$ Asn	K101N	Klose *et al.*, 1993
pMKL21/pMKL210	$Lys^{106} \rightarrow$ Asn	K106N	Klose *et al.*, 1993
pMKL4041	$Asp^{207} \rightarrow$ Asn	D207N	M.Klose, unpublished
pMKL4040	$Asp^{215} \rightarrow$ Asn	D215N	M.Klose, unpublished

ADP binding. Binding of ADP to SecA was assayed using the fluorescent analog 1,N^6-ethenoadenosine-5'-diphosphate (ε-ADP) (Molecular Probes, Eugene, OR). Solutions contained: 200 μg $m\ell^{-1}$ of SecA, 50 mM TrisCl (pH 8.0), 50 mM KCl, 5 mM $MgCl_2$, 1 mM DTT and 10 μM ε-ADP. Reactions received 5 mM of indicated nucleotides, 400 μg $m\ell^{-1}$ proOmpA and/or 0.3 M urea. Fluorescence polarization measurements at 20°C were performed with an SLM 4800C (Aminco, Urbana, IL) fluorimeter equipped with Glan-thompson polarizers. Excitation and emission was at 300 and 410 nm, respectively. The polarization (P) of ε-ADP fluorescence was calculated according to: $P = [(R_{vert}/R_{horz})-1] / [(R_{vert}/R_{horz})+1]$ where R_{vert} and R_{horz} are the ratios of parallel and perpendicular signals with the excitation polarizer placed in the vertical and horizontal orientation, respectively.

RESULTS:

Expression of B. subtilis secA A- and B-domain mutant genes in E. coli. Site-directed mutations were introduced in critical residues of the A- and B-domain of the putative catalytic ATP-binding site of the *B. subtilis* SecA (*See* table 1). Mutant proteins were expressed in various *E. coli* strains to test their ability to complement the growth-defects of conditional lethal *secA* mutations. After a shift to the nonpermissive temperature (42°C), full length, but non-functional SecA protein is synthesized in MM52 (Schmidt *et al.*, 1988), a truncated amino-terminal SecA fragment is synthesized in BA13 (Cabelli *et al.*, 1988), while due to a strong polar effect in the stop codon of *geneX*, no

SecA is synthesized in MM66 (Schmidt *et al.*, 1988). *E. coli* JM105.1 contains a single copy of the *lacI^q* gene on the F' episome (Klose *et al.*, 1993). When the temperature sensitive strains MM52, BA13 or MM66 were transformed with plasmids pMKL40, wild -type *B. subtilis* SecA protein was found to complement the growth defects of these strains at the non-permissive temperature, provided that the protein is not brought to high expression level (Table 2) (Klose *et al.*, 1993). With pMKL210 and pMKL440 only poor growth was observed, even at the permissive growth temperature (Table 2). Lys^{106} and Asp^{215} therefore seem to be crucial for complementation of the *E. coli secA* mutant strains by the *B. subtilis* (this paper) and *E. coli* (M. Klose, unpublished) SecA.

Table 2. Growth complementation of *E. coli secA* mutants

secA Allele	Plasmid	*E. coli* strain															
		MM52				JM105.2				BA13				MM66			
		30°C		42°C		30°C		42°C		30°C		42°C		30°C		42°C	
		G	I	G	I	G	I	G	I	G	I	G	I	G	I	G	I
	pTRC99A	+	+	-	-	+	+	-	-	+	+	-	-	+	+	-	-
B. subtilis																	
Wild type	pMKL40	+	+	+	-	+	+	+	-	+	+	+	-	+	+	+	-
K101N	pMKL200	+	+	+	-	+	+	+	-	+	+	+	-	+	+	+	-
K106N	pMKL210	(+)	-	-	-	(+)	-	-	-	(+)	-	-	-	(+)	-	-	-
D207N	pMKL441	+	+	+	-	+	+	+	-	+	+	+	-	+	+	+	-
D215N	pMKL440	(+)	-	-	-	(+)	-	-	-	(+)	-	-	-	(+)	-	-	-
E. coli																	
Wild type	pMKL18	+	+	+	+	+	+	+	+	+	+	+	+	+	+	+	+

E. coli strains were grown on LB plates supplemented with 0.5 % glucose (G) or 1 mM IPTG (I). Growth temperatures were 30 or 42°C; growth; (+), poor growth; -, no growth.

ATP enhances the V8 protease resistance of wild-type SecA, but not that of K106N SecA. Soluble *E. coli* SecA is sensitive to digestion by protease V8. ATP and the non-hydrolyzable analog AMP-PNP enhance the resistance of SecA against V8 digestion (Shinkai *et al.*, 1991), suggesting a conformational change upon binding of ATP. *B. subtilis* SecA proteins were purified and incubated with V8 protease. Wild-type SecA was rapidly digested by V8 protease, and was more resistant in the presence of ATP (Fig. 2A). K106N SecA was not protected against V8 protease digestion in the presence of ATP (Fig. 2B). These data suggest that Lys^{106} is involved in an ATP-dependent conformational change of SecA.

Effect of proOmpA on V8 protease sensitivity of wild-type and K106N SecA. B. subtilis SecA ATPase activity is stimulated by proOmpA and the protein binds to translocation intermediates of proOmpA (van der Wolk *et al.*, 1993). On the other hand, K106N SecA blocks the Δp-dependent chase of such intermediates possibly through tight binding. Addition of proOmpA to the *B. subtilis* SecA, preincubated with ATP, restored the sensitivity to V8 digestion to the level when no ATP was present (Fig. 2A). A strong effect was observed with the K106N SecA, suggesting that this mutant binds proOmpA with greater affinity than the wild-type. OmpA had only little effect on the V8 protease sensitivity (not shown). These results demonstrate that K106N SecA interacts with preproteins.

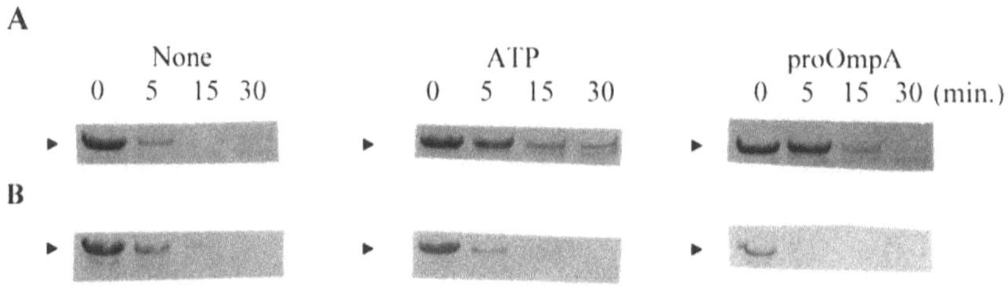

Fig 2. V8 protease sensitivity of wild-type (A) and K106N (B) SecA.

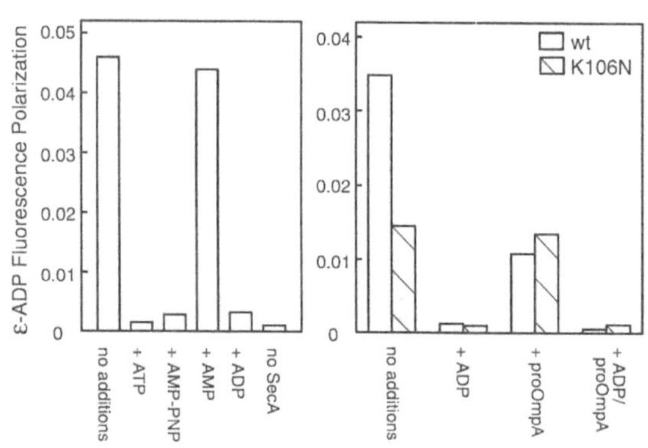

Fig 3. Binding of ε-ADP to wild-type and K106N SecA.

Binding of ε-ADP to wild-type and K106N SecA. ε-ADP was used to assay ADP binding to SecA through fluorescence polarization measurements (Fig. 3A). Bound ε-ADP was chased by an excess ATP, ADP or AMP-PNP, while AMP was ineffective. K106N SecA still binds ADP (Fig. 3B, hatched bars), although with lower specificity than the wild-type (open bars). ProOmpA elicits release of bound ε-ADP from the wild-type to the level found for the K106N SecA. ProOmpA had no effect on the binding of ε-ADP to K106N SecA, while further release required an excess ADP. These results suggest that binding of proOmpA to SecA lowers its affinity for ADP, while K106N SecA appears to be frozen in this low-affinity ADP-binding state.

DISCUSSION:

In a previous communication (van der Wolk *et al.*, 1983), we have shown that Lys^{106} is an essential residue of the A-domain of the catalytic ATP-binding site. The *in vivo* complementation assays now identify Asp^{215} as part of the B-domain. This site differs from the one suggested by Koonin and Gorbalenya (1992). Further work will involve the purification of this mutant in order to analyze its activities *in vitro.*

A conceptional model for the active site of SecA based on the structure of RecA which has been solved at 2.7 Å with bound ADP (Story and Steitz, 1992) is presented in Fig. 4. In RecA, the enzymatic mechanism of ATP hydrolysis involves activation of a water molecule by a glutamate residue for an in-line attack of the γ-phosphatepresented by the invariable lysine (Lys^{106} in SecA) of the A-domain. The position of the glutamate residue in SecA is unknown. The γ-phosphate is positioned such that Mg^{2+}, stabilized by the invariable aspartate (Asp^{215}) of the B-domain, bridges the oxygen atoms of the β- and γ-phosphates such that the phosphate oxygens, the Mg^{2+} ion and the Oγ of threonine of the A-domain (Thr^{107}) all lie in the same plane. In addition, the γ-phosphate is in contact with the so-called C-domain, not yet identified in SecA. The site for preprotein binding may be formed upon interaction between the β- and γ-phosphates and residues of the C-region that lie adjacent to the phosphate-binding loop. In analogy with the RecA model, ATP hydrolysis is expected to destroy interactions between the nucleotide and residues of the C-region, resulting in a change in conformation of the preprotein-binding regions causing release of the preprotein by SecA. ATP and ADP both stabilise the conformation of SecA; the conformation stabilized by the binding of ATP has a higher affinity for the preprotein as evident from cross-linking (Akita *et al.*, 1990) and dynamic light-scattering experiments (unpublished). Close proximity of the ATPase active site and the region likely to be involved in preprotein binding is suggestive of an allosteric mechanism by which the binding of the nucleotide affects the binding of the preprotein. Both domains may be highly interacting as signal peptides act as competitive inhibitors of SecA-translocation ATPase, while ATP antagonizes this inhibition (Cunningham and Wickner. 1989). Binding of the preprotein to SecA discharges the bound nucleotide possibly through lowering of the binding affinity (Shinkai *et al.*, 1991; this paper). This phenomenon is lost with the K106N mutant which binds ATP only with low affinity, maybe at an alternative low affinity binding site. We propose that the K106N SecA is prone to high-affinity preprotein binding. The inability to hydrolyse ATP (van der Wolk *et al.*, 1993) prevents the release of bound preproteins, thereby blocking preprotein translocation at the membrane translocation sites.

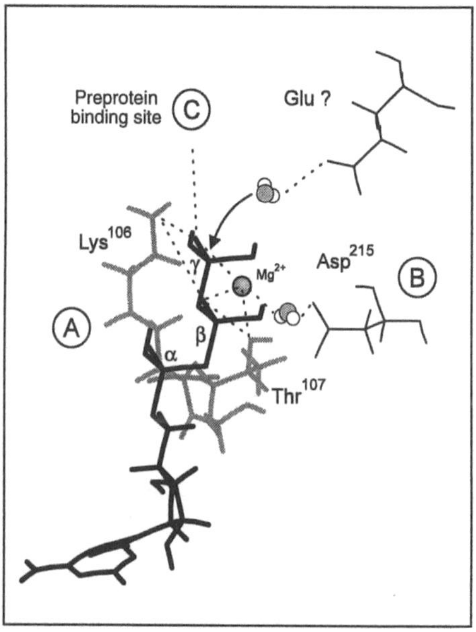

Fig 4. Model for the SecA ATP-binding site.

ACKNOWLEDGEMENTS:

This work was supported by the Netherlands Organization for Scientific Research (N.W.O.) and the Netherlands Foundation for Chemical Research (S.O.N.).

LITERATURE:

Akita M, Sasaki S, Matsuyama S-I, Mizushima S (1990) SecA interacts with secretory proteins by recognizing the positive charge at the amino terminus of the signal peptide in *Escherichia coli*. J Biol Chem 265:8164-8169

Brundage L, Hendrick JP, Schiebel E, Driessen AJM, Wickner W (1990) The purified

Escherichia coli integral membrane Protein SecY/E is sufficient for reconstitution of SecA-dependent precursor protein translocation. Cell 62:649-657

Cabelli RJ, Chen L-L, Tai PC, Oliver DB (1988) SecA protein is required for secretory protein translocation into *E. coli* membrane vesicles. Cell 55:683-692

Cunningham K, Wickner W (1989) Specific recognition of the leader region of precursor proteins is required for the activation of translocation ATPase of *Escherichia coli*. Proc Natl Acad Sci USA 86:8630-8634

Driessen AJM (1992) Precursor protein translocation by the *Escherichia coli* translocase is directed by the protonmotive-force. EMBO J 11:847-853

Klose M, Schimz K-L, van der Wolk J, Driessen AJM, Freudl R (1993) Lysine[106] of the putative catalytic ATP-binding site of the *Bacillus subtilis* SecA protein is required for functional complementation in *E. coli secA* mutants *in vivo*. J Biol Chem 268:4504-4516

Koonin EV, Gorbalenya AE (1992) Autogenous translation regulation by *E. coli* ATPase SecA may be mediated by an intrinsic RNA helicase activity of this protein. FEBS Lett 298:6-8

Lill R, Cunningham K, Brundage L, Ito K, Oliver DB, Wickner W (1989) SecA protein hydrolyzes ATP and is an essential component of protein translocation ATPase of *Escherichia coli*. EMBO J 8:961-966

Lill R, Dowhan W, Wickner W (1990) The ATPase activity of SecA is regulated by acidic phospholipids, SecY, and the leader and mature domains of precursor proteins. Cell 60:271-280

Oliver DB (1993) SecA protein: autoregulated ATPase catalysing preprotein insertion and translocation across the *Escherichia coli* inner membrane. Mol Microbiol 7:159-165

Overhoff B, Klein M, Spies M, Freudl R (1991) Identification of a gene fragment which codes for the 364 amino-terminal amino acid residues of a SecA homologue of *Bacillus subtilis*: further evidence for the conservation of the protein export apparatus in gram-positive and gram-negative bacteria. Mol Gen Genet 228:417-423

Schiebel E, Driessen AJM, Hartl F-U, Wickner W. (1991) $\Delta\tilde{\mu}_{H^+}$ and ATP function at dif ferent steps of the catalytic cycle of preprotein translocase. Cell 64:927-939

Schmidt MG, Rollo EE, Grodberg J, Oliver DB (1988) Nucleotide sequence of the *secA* gene and *secA*(ts) mutations preventing protein export in *Escherichia coli*. J Bacteriol 170:3404-3414

Shinkai A, Mei LH, Tokuda H, Mizushima S. (1991) The conformation of SecA, as re vealed by its protease sensitivity, is altered upon interaction with ATP, presecretory proteins, everted membrane vesicles, and phospholipids. J Biol Chem 266:5827-5833

Story RM, Steitz TA (1992) Structure of the recA protein-ADP complex. Nature 355: 374-376

van der Wolk J, Klose M, Breukink E, Demel RA, de Kruijff B, Freudl R, Driessen AJM (1993) Characterization of a *Bacillus subtilis* SecA mutant protein deficient in translocation ATPase and release from the membrane. Mol Microbiol 8:31-42

Walker JE, Saraste M, Runswick MJ, Gay NJ (1982) Distantly related sequences in the α- and β-subunits of ATP synthase, myosin, kinases and other ATP-requiring enzymes have a common nucleotide binding fold. EMBO J 1:945-951

Wickner W, Driessen AJM, Hartl F-U (1991) The enzymology of protein translocation across the *Escherichia coli* plasma membrane. Annu Rev Biochem 60:101-124

Internal Disulfides in the Diphteria Toxin A-Fragment Block Its Translocation to the Cytosol

Pål Ø. Falnes, Seunghyon Choe*, Inger H. Madshus, R. John Collier†, and Sjur Olsnes.
Department of Biochemistry
Institute for Cancer Research
Montebello
0310 Oslo
Norway

INTRODUCTION

A number of protein toxins from plants and bacteria efficiently kill eukaryotic cells by inhibition of protein synthesis. The toxin binds to receptors on the plasma membrane and is endocytosed, before the enzymatically active part of the toxin is translocated to the cytosol. Here it exerts its action by inactivating a crucial component of the cellular protein-synthesizing machinery.

The molecular mechanisms involved in the translocation process have been studied in most detail for diphtheria toxin, the main pathogenicity factor in clinical diphtheria (For reviews, see London, 1992, Sandvig and Olsnes, 1991). Diphtheria toxin is secreted as a polypeptide chain of 58 kD from *Corynebacterium diphtheriae*, and can easily be cleaved by trypsin into two fragments A (21 kD) and B (37 kD), joined by a disulfide bond. Diphtheria toxin entry into cells is initiated by binding of its B-fragment to specific surface receptors. Subsequently, the exposure to low pH in endosomes induces the translocation of the A-fragment to the cytosol. Translocation can also be induced at the level of the plasma membrane by exposing cells with surface-bound toxin to low pH. Low pH has been shown to induce unfolding and the exposure of hydrophobic regions in the toxin molecule, and this has lead to the speculation that unfolding of the toxin may be necessary for the translocation to occur.

---

*Molecular Biology Institute and Department of Chemistry and Biochemistry, University of California, Los Angeles, USA
†Department of Microbiology and Molecular Genetics, Harvard Medical School, Boston, USA

NATO ASI Series, Vol. H 82
Biological Membranes:
Structure, Biogenesis and Dynamics
Edited by Jos A. F. Op den Kamp

So far it has not been clear whether the A-fragment is translocated as a folded structure or if it must unfold before translocation can take place. The crystal structure of diphtheria toxin was recently solved (Choe et al., 1992), and this has allowed us to identify pairs of amino acids that are positioned sufficiently close to be able to form disulfide bonds if the residues are mutated to cysteines. In the present work we have introduced disulfides at 5 different locations in the A-fragment and demonstrated that the presence of disulfide linkages in the A-fragment strongly inhibits its translocation to the cytosol.

RESULTS

We constructed five double cysteine mutants of the diphtheria toxin A-fragment and termed them CC1 through CC5 (Fig. 1A). The disulfide bonds expected to be formed are

A

NAME OF MUTANT	MUTATIONS
CC1	K24C/D68C
CS1	K24S/D68C
CC2	N58C/S146C
CS2	N58C
CC3	K76C/G171C
CS3	K76C/G171S
CC4	D97C/V135C
CS4	D97C/V135S
CC5	G119C/N152C
CS5	G119S/N152C

B

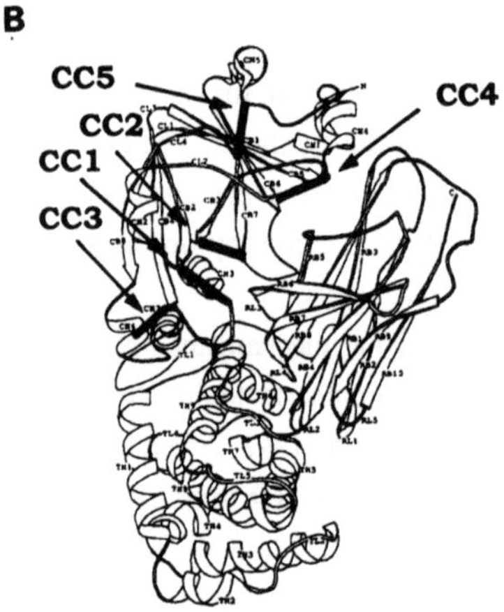

C

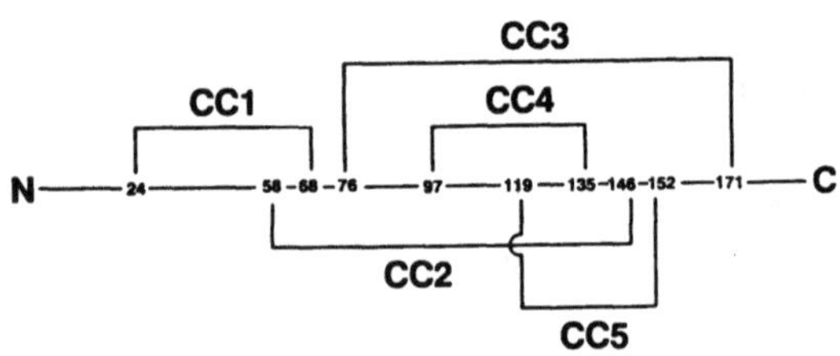

Fig. 1. Construction of toxins with internal disulfides in the A-fragment. A, list of constructs used. B, ribbon drawing of diphtheria toxin molecule (From Choe et al., 1992) with the genetically introduced disulfide bridges indicated, for simplicity all five drawn onto one molecule. C, linear representation of the location of internal disulfides in the A-fragment.

indicated on the diphtheria toxin crystal structure (Fig. 1B), and on a linear representaion of the diphtheria toxin molecule (Fig. 1C). Since the introduced mutations are not conservative they could reduce the biological activity of the A-fragment, irrespective of disulfide formation. To test this we also constructed five control mutants, in which one of the cysteines was replaced by serine. The five control mutants were termed CS1 through CS5 (Fig. 1A).

Proteins with internal disulfide bridges usually migrate more rapidly in SDS-PAGE than their reduced counterparts. To test if the double cysteine mutants formed intramolecular disulfide bonds under non-reducing conditions, we compared the migration rates on SDS-PAGE of the double cysteine mutants and their controls under reducing and non-reducing conditions. For the purpose of comparing the migration rates of A-fragments, an additional mutation (C186S) was also introduced to remove the cysteine originally present in the wild-type A-fragment, thereby avoiding the formation of disulfides between this residue and the introduced cysteines. The results showed that under reducing conditions each of the putative disulfide forming mutants CC1 through CC5 migrated at the same rate as its respective control mutant (Fig. 2A), whereas under non-reducing conditions (Fig. 2B) the double cysteine mutants in four out of five cases migrated faster than their respective controls, as expected if the four double cysteine mutants CC1, CC2, CC3, and CC5 all efficiently formed intramolecular disulfide bonds. In the case of CC4, a substantial part of the A-fragment migrated at the slow rate in the unreduced preparation, indicating that disulfide formation is less efficient in this mutant than in the other ones.

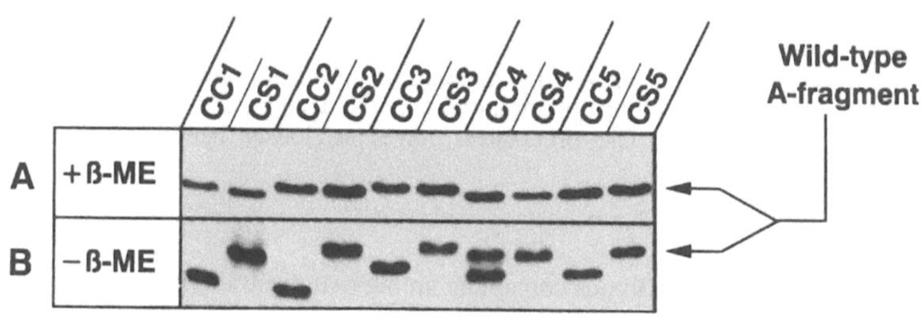

Fig. 2. Migration rates of mutant A-fragments (all with the additional mutation C186S) on SDS-PAGE in the presence (A) or absence (B) of 7% β-mercaptoethanol. The arrows indicate the positions where wild-type A-fragment would have migrated.

Entry of diphtheria toxin is initiated by binding of its B-fragment to specific surface receptors. Previous studies have shown that modifications of the A-fragment may interfere with the binding properties of the B-fragment. To test the possibility that the mutations here introduced in the A-fragment could affect the binding, full-length toxin was reconstituted from in vitro translated, [^{35}S]methionine labelled A- and B-fragments (Stenmark et al., 1992), and Vero cells were incubated with the mutant toxins, washed and analysed by SDS-PAGE. When reconstituted with B-fragment, all the 10 mutants were able to bind to cells with approximately the same affinity as wild-type toxin (Fig. 3, upper panel). Also, the binding was efficiently competed out by an excess amount of unlabelled wild-type toxin, showing that the mutants bind to specific diphtheria toxin receptors (data not shown).

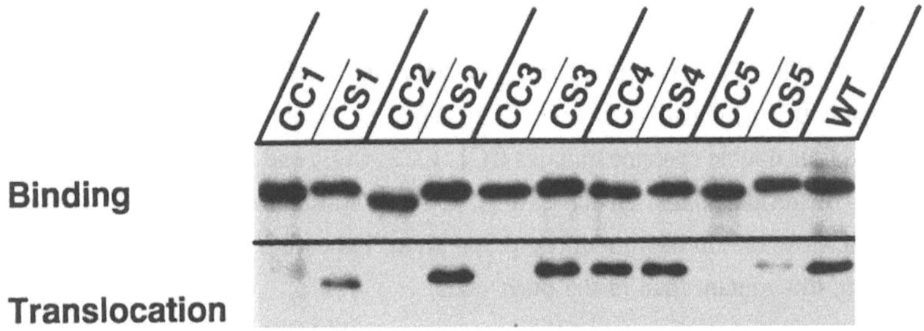

Fig. 3. Binding and translocation of mutant toxins. In vitro translated, [^{35}S]-methionine labelled mutant A-fragments were reconstituted with B-fragment. The binding ability of the mutant toxins was analysed by incubating Vero cells with the toxin, and then the cells were washed, lysed, and the TCA-precipitated material was analysed by SDS-PAGE and fluorography (upper panel). The translocated A-fragment was assayed by exposing Vero cells with bound toxin to pH 4.8, followed by removal of untranslocated material with pronase, and after lysis, cellular protein was precipitated with TCA and analysed by SDS-PAGE and fluorography (lower panel). The procedures have previously been described in detail (Moskaug et al., 1988)

To test the ability of the toxins containing an internal disulfide bond in the A-fragment to be translocated across the plasma membrane, cells with bound toxin were exposed to low pH to induce the translocation of the A-fragment to the cytosol. Subsequently, the cells were treated with pronase to remove untranslocated material remaining at the surface, and finally the protected material was analysed by SDS-PAGE and fluorography. The bands corresponding to protected A-fragment are shown in Fig. 3, lower panel. All the four mutants that were efficiently forming disulfide brigdes (CC1,CC2,CC3, and CC5) showed a dramatic reduction in their translocation efficiency as compared to their respective serine control mutants. The mutant CC4 which is inefficient in forming an intramolecular disulfide bond, showed no substantial reduction in its efficiency of translocation. One of the control mutants, CS5, displayed a reduction in its translocation efficiency, and in this case the mutations introduced must be specifically inhibitory for the translocation process, or they may have led to a conformational change in the protein. However, the results indicate that several different mutations can be introduced into the A-fragment without an effect on translocation efficiency, but that an intramolecular SS-bridge strongly inhibits translocation of the A-fragment to the cytosol.

DISCUSSION

On the basis of crystallographical data we constructed five double cysteine mutants of the diphtheria toxin A-fragment, each expected to form an intramolecular disulfide bond. As controls were constructed five corresponding mutants where one of the cysteines was replaced by serine. Four out of the five double cysteine mutants were efficiently forming an intramolecular disulfide bridge, and the translocation to the cytosol of all these mutant A-fragments was strongly reduced compared to their respective controls.

The diphtheria toxin A-fragment concists of two subdomains, and the catalytic site is located in the cleft between these subdomains. Two of the translocation-blocking disulfides (CC1 and CC3) are within one subdomain, one (CC5) is within the other, and one disulfide (CC2) is between the two subdomains. Therefore, these data suggest that extensive unfolding of its the tertiary structure is required for the translocation of the A-fragment to the cytosol, and to our knowledge this has never previously been shown for a protein toxin.

One of the mutants (CC4) was inefficient in disulfide formation, and for this mutant no block in translocation was observed. The stability of a disulfide can vary much depending on its local environment (Goldenberg et. al, 1993), and a disulfide bond that is not easily formed would be expected to be readily reduced. It could therefore be that some of the translocated material in the case of the CC4 mutant originates from A-fragments that have had their disulfide bridges reduced either at the cell surface or in the process of translocation. However, it is difficult to conclude from our translocation data whether or not all the translocated A-fragments were already reduced prior to binding to the cells.

Several examples from other translocation systems indicate that tight folding is generally inhibitory for translocation of proteins across membranes. The structure of many proteins can be stabilized by the binding of small ligands, leading to a block in translocation (Rassow at al., 1989, Chen and Douglas, 1987, Eilers and Schatz, 1986, Arkowitz et al., 1993). Also, when a heparin-binding growth factor is fused genetically to the N-terminus of the diphtheria toxin A-fragment, heparin blocks the translocation of this fusion protein to the cytosol (Wiedlocha et al., 1992).

In many cases, internal disulfides have been shown to block protein translocation. Reduction of internal disulfide bridges is necessary for the posttranslational translocation of preprolactin across microsomal membranes in vitro (Maher and Singer, 1986). The relatively small disulfide loop from Cys290 to Cys302 in the E.coli outer membrane protein OmpA blocks its translocation across the inner membrane in the absence, but not in the presence, of a protonmotive force (Tani et al., 1990). When the B subunit of cholera toxin was fused to Neisseria IgA protease, the presence of a disulfide in the toxin B subunit halts export of the fusion protein across the outer membrane of E.coli (Klauser et al., 1992). When the compactly folded protein bovine pancreatic trypsin inhibitor, containing 3 disulfides, is linked to a mitochondrial precursor protein, import into the matrix is blocked (Vestweber and Schatz, 1988).

In the translocation systems most extensively studied (mitochondrial import, ER import, and bacterial export), the translocation process is either coupled to translation, or newly synthesized proteins are often kept in a translocation-competent conformation by molecular chaperones. Protein toxins are quite different in this respect, since they have already aquired their native conformation when their translocation into cells is initiated. Conceivably, proteins in the membrane, possibly associated with the diphtheria toxin receptor, are catalysing the unfolding of the toxin prior to its membrane translocation. Also, treatment of the toxin with low pH causes unfolding and exposure of hydrophobic domains (For review, see London, 1992), and this unfolding could be sufficient to render the toxin competent for translocation.

Compared to most translocation systems, diphtheria toxin entry is quite distinct in another respect. By evolution this translocation system has been specifically designed to mediate the translocation of only one protein, *viz.* the A-fragment, and therefore it was not obvious that the requirements for unfolding observed in other systems would also apply to diphtheria toxin translocation. However, the mechanisms of entry of protein toxins into cells share many common features, and the observation in the present work with diphtheria toxin that unfolding of the A-fragment is required for its translocation may also be valid for other protein toxins.

ACKNOWLEDGEMENTS

We are grateful to Marianne Andreassen for excellent technical assistance with the cell cultures . P. Ø. F. is a fellow of The Norwegian Cancer Society. This work was supported by The Norwegian Cancer Foundation and The Jahre Fund.

REFERENCES

Chen,W-J, Douglas, MG (1987) The role of protein structure in the mitochondrial import pathway. Unfolding of mitochondrially bound precursors is required for membrane translocation. J. Biol. Chem. 262, 15605-15609

Choe, S, Bennett, MJ, Fujii, G, Curmi, PMG, Kantardjieff, KA, Collier, RJ, Eisenberg, D (1992) The crystal structure of diphtheria toxin. Nature 357, 216-222

Eilers, M, Schatz, G (1986) Binding of a specific ligand inhibits import of a purified precursor protein into mitochondria. Nature 322, 228-232

Goldenberg, DP, Bekeart, LS, Laheru, DA, Zhou, JD (1993) Probing the deteminants of disulfide stability in native pancreatic trypsin inhibitor. Biochemistry 32, 2835-2844

Klauser, T, Pohlner, J, Meyer, TF (1992) Selective extracellular release of cholera toxin B subunit by Escherichia coli: dissection of Neisseria Iga_{β}-mediated outer membrane transport. EMBO J. 11, 2327-2335

Moskaug, JØ, Sandvig, K, Olsnes, S (1988) Low pH-induced release of diphtheria toxin A-fragment in Vero cells. Biochemical evidence for transfer to the cytosol. J. Biol. Chem. 263, 2518-2525

London, E (1992) Diphtheria toxin: membrane interaction and membrane translocation. Biochim. Biophys. Acta 1113, 25-51

Maher, PA, Singer, SJ (1986) Disulfide bonds and the translocation of proteins across membranes. Proc. Natl. Acad. Sci. USA 83, 9001-9005

Rassow, J, Guiard, B, Wienhues, U, Herzog, V, Hartl, F-U, Neupert, W (1989) Translocation arrest by reversible folding of a precursor protein imported into mitochondria. A means to quantitate translocation contact sites. J. Cell Biol. 109, 1421-1428

Sandvig, K, Olsnes, S (1991) Membrane translocation of diphtheria toxin. In Sourcebook of Bacterial Protein Toxins (Alouf, JE, Freer, JH, eds.) pp 57-73, Academic Press Limited, London, UK

Stenmark, H, Afanasiev, BN, Ariansen, S, Olsnes, S (1992) Association between diphtheria toxin A- and B-fragments and their fusion proteins. Biochem. J. 281, 619-625

Tani, K, Tokuda, H, Mizushima, S (1990) Translocation of proOmpA possessing an intramolecular disulfide bridge into membrane vesicles of Escherichia coli. J. Biol. Chem. 265, 17341-17347

Vestweber, D, Schatz, G (1988) A chimeric mitochondrial precursor protein with internal disulfide bridges blocks import of authentic precursors into mitochondria and allows quantitation of import sites. J. Cell Biol. 107, 2037-2043

Wiedlocha, A, Madshus, IH, Mach, H, Middaugh, CR, Olsnes, S (1992) Tight folding of acidic fibroblast growth factor prevents its translocation to the cytosol with diphtheria toxin as vector. EMBO J. 11, 4835-4842

MOLECULAR CHARACTERIZATION, ASSEMBLY AND MEMBRANE ASSOCIATION OF THE GROEL-TYPE CHAPERONINS IN *SYNECHOCYSTIS* PCC 6803.

E. Kovács[1], I. Horváth[1], A. Glatz[1], Zs. Török[1], Cs. Bagyinka[2] and L. Vigh[1].
Institute of Biochemistry[1], Institute of Biophysics[2]
Biological Research Centre
H-6701 Szeged, POB 521
Hungary.

Chaperonins, a subclass of molecular chaperones, are a highly conserved family of proteins that are present in most, and probably all, living organisms. The level of their synthesis can be dramatically enhanced in response to stress, especially to heat stress (Georgopoulos, 1992). The GroE proteins of *Escherichia coli* are the best characterized molecular chaperones, at both the genetical and biochemical level, and have been used extensively to investigate interactions between this class of proteins and a number of target polypeptides (Lubben et al., 1989). There are two basic types of chaperonins of quite distinct sizes: the larger type known as GroEL contains subunits with an appropriate molecular mass of 60 kDa that are arranged in a complex structure comprising two rings of seven subunits stacked on each other. The second type of chaperonin (or co-chaperonin, GroES) is significantly smaller and contains subunits of about 10 kDa and, at least in bacteria, is thought to form a single ring of seven subunits. GroES and GroEL genes are essential for bacterial survival and they constitute an operon whose expression is enhanced during heat shock. It now appears that molecular chaperones exert their influence by stabilizing protein folding intermediates, thus partitioning them towards a pathway leading to the native state rather than forming inactive aggregated structures. As an implication of such effects, one of their main cellular functions is probably to provide protection under stress conditions (Gatenby et al., 1993).

In this study we shall concentrate on GroES and GroEL type chaperonins in a cyanobacteria, *Synechocystis* PCC 6803, which might be involved in conferring thermotolerance of the photosynthetic membranes. Since proteins related to GroES and GroEL have now been identified in numerous eukaryotic systems, we suppose, that the mechanism revealed for well-transformable model systems like cyanobacteria might also be relevant concerning thermoprotection in higher plants.

NATO ASI Series, Vol. H 82
Biological Membranes:
Structure, Biogenesis and Dynamics
Edited by Jos A. F. Op den Kamp

***Synechocystis* PCC 6803 contains two groEL genes**

Based on a previous purification and characterization of a 64 kDa heat shock protein (HSP) as a GroEL-type chaperonin (Lehel et al., 1992) we postulated that in contrast with findings described by Chitnis and Nelson (1991) a second groEL-type gene should also be present in the genome of *Synechocystis* PCC 6803. In fact, in addition to the earlier described cpn-60 gene possessing no groES in the neighbouring regions, recently we could demonstrate the existence of a groESL operon in this organism (Lehel et al., 1993/a). Similar to our findings, the presence of more than one copy of groEL genes has been reported for different prokaryotic organisms e.g. species of *Streptomyces* (Mazodier et al., 1991) and *Mycobacterium leprae*. (de Wit et al., 1992).

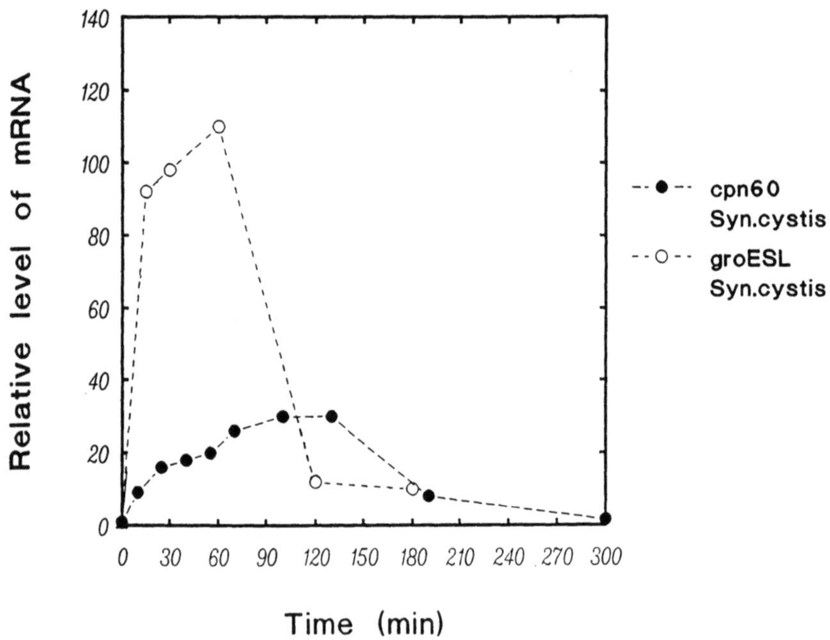

Figure 1. Transcription of cpn60 and groESL operon under heat stress. The transcript level was quantitated by dot blot analysis of total RNA obtained from cells at different times after shifting 30°C-grown cultures to 42°C and using cpn60 and groES probes.

The unexpected observation concerning the finding of multiple GroEL proteins supported the notion that they might fulfil physiologically distinct roles. In fact, the level of the bicistronic 2.2-kb transcript of the groESL operon increased 100-fold upon exposing *Synechocystis* cells to heat stress, whilst the transcript level of cpn-60 accumulated about 30-fold under identical conditions (Fig.1.). Although we have not yet made the characterization of the heat-inducible promoter of groESL, we revealed, however, a 9-bp inverted repeat (IR) separated by a 9-bp spacer immediately upstream of groES. It is noteworthy that the IR sequence of the stem-loop structure matches completely with that of described to precede the groESL operon of *Bacillus subtilis, Clostridium acetobutylicum*, and several other bacterial species (Schmidt et al., 1992, Narberhaus and Bahl, 1992). Introduction of point mutations within the IR preceding the dnaK coding sequence of *B. subtilis* led to a constitutively high level of expression of that gene, suggesting that a repressor is involved in the regulation of the heat shock response (Schön and Schuman, 1993).

Predicting coiled-coils and assembly of the two GroEL type chaperonins

Based on phylogenetics and cellular localization, J. Ellis divided the chaperonin group into GroEL and TCP-1 subclasses (Ellis, 1992). Recently it was suggested by Webb and Sherman (1992), that the above grouping could be further refined by analysis of the probability that a residue in a protein is part of a coiled-coil structure. These authors applied the algorithm designed by Lupas et al. (1991), which awards high scores to amphipathic helices containing patterns of hydrophillic and hydrophobic amino acids arranged in heptad repeats within a moving window of 28 residues. We adapted the original VAX Pascal program to PC (available upon request) and the analysis was extended to the two GroEL type proteins found in *Synechocystis* PCC 6803 (Fig.2). Obviously, both chaperonin proteins possess two common regions of structural similarity that may form coiled-coil domains but having strikingly differing score profile. Hence, the relative probability of predicted coiled-coil domain centred at amino-acid residue number 350 was more than twice as high for cpn60 as for GroEL. Since the predicted roles of coiled-coil domains are (i) involvement in binding of nascent polypeptides and (ii) in the oligomerization, as well, the above data strongly suggest differing functions of chaperonins encoded by lonely and groES-linked groEL-type genes.

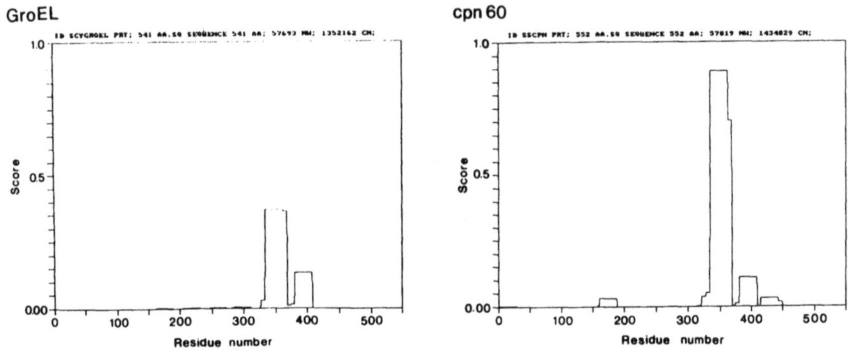

Figure 2. Analysis of the two chaperonins made by using Coils algorithm adapted to PC. Predicted regions that may form coiled-coil domains are indicated by bars: the scale on the y-axis shows a probability of 1.

We assumed, that the two GroEL-analog proteins, formed predominantly during high temperature stress are assembled into oligomeric species. Accordingly, cell extract derived from heat-shocked (42°C, 3 hours) cells have been resolved by sucrose density gradient centrifugation (Fig. 3/A). Samples were then taken from fractions of gradient centrifugation and analyzed by SDS-PAGE. Fraction 5, containing the peak quantity of chaperonins was further investigated by 2D-SDS-PAGE and visualized by Coomassie Blue staining (Fig. 3/B). Arrows indicate the locations of a larger and a smaller portions of 64 kDa components. If antibody prepared against purified *Synechocystis* GroEL was applied for these separated proteins, both of them was immunoreactive (data not shown). Next, the N-terminal amino acid sequence of the right-hand protein was determined: the analysis of the first 20 amino acids fully confirmed that the larger protein component corresponds to the product of the groEL gene. Moreover, only one N-terminal amino acid signal was detected from this spot, confirming the homogeneity of 2D-SDS-PAGE separated component. Although the N-terminal sequence determination of the smaller, 64 kDa immunoreactive component is not yet accomplished, it is predictably identical with the gene product of cpn60.

Our results demonstrate, that the gene products of both groEL-type genes assemble into oligomeric species. Since the assembly of the gene products of cpn-60 (L^1) and groEL (L^2)

theoretically may be either restricted (L^1_{14}, L^2_{14}, $L^1_7L^2_7$) or unrestricted ($L^1_nL^2_{14-n}$), understanding the exact organization of oligomeric (supposedly tetrademameric) species needs further studies.

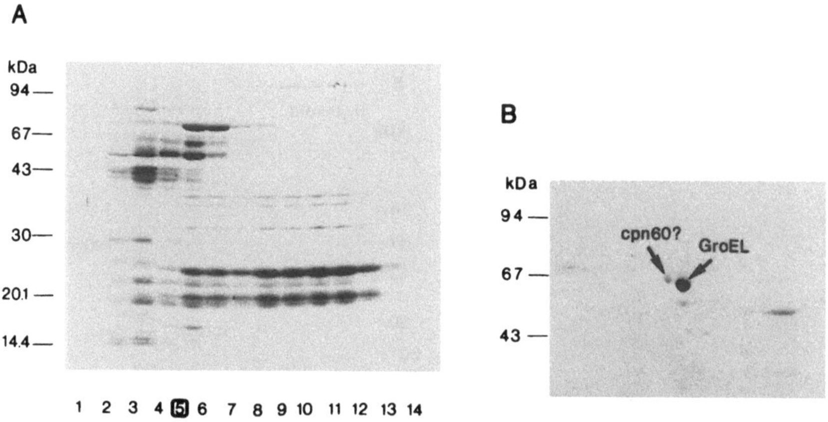

Figure 3. (A) Sucrose density gradient centrifugation of cell extracts from heat-shocked (3 hours, 42°C) *Synechocystis* cells. (B) Fraction 5, further analyzed by 2D-SDS-PAGE and visualized by Coomassie Blue staining.
Arrows indicate the position of GroEL and the predicted migration of cpn60.

Heat-induced thylakoid association of GroEL proteins

It is widely accepted, that in photosynthetic organisms exposed to high temperature, prior to impairment of other cellular functions, the photosynthetic apparatus and especially Photosystem II is damaged first (Berry and Björkman, 1980; Lehel et al, 1993/b). In line of this presumption, we proposed that if the chaperonin proteins played any protective role against heat-damage to the photosynthetic apparatus, they would probably accumulate in thylakoid membrane during heat treatment. In fact, whereas the total level of GroEL-analogues (being unresolved by 1D-SDS-PAGE) is increasing in the cytosol, heat-treatment results in a strikingly elevated level of association of GroEL-like proteins to thylakoid membranes, simultaneously (Fig. 4). Moreover, whilst the level of the cytosolic co-chaperonin, GroES (identified previously by Lehel et al., 1993/a) is also enhanced gradually upon heat-treatment, its membrane binding occurs at a discrete threshold of temperature, corresponding to the temperature regime at which the efficiency of Photosystem II became impaired. Detailed characterization of the heat-induced membrane binding of chaperonin proteins has recently been accomplished and under publication (Kovács et al, submitted).

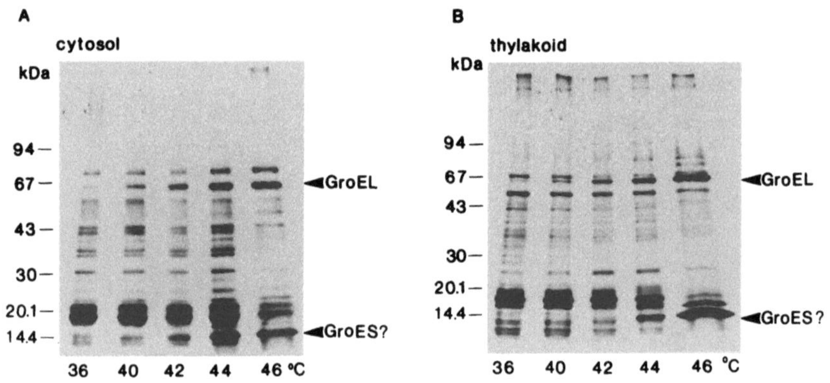

Figure 4. The temperature dependence of the level of GroEL and GroES proteins present in the cytosol (A) and bound to the thylakoid membranes (B) in *Synechocystis* PCC 6803.

It is known, that all organisms so far examined are able to acquire tolerance to a previously lethal high temperature by prior exposure to a non-lethal heat shock (Howarth, 1991). By measuring threshold temperatures at which the efficiency of Photosystem II is impaired, we have shown that heat pretreatment (42°C, 3 hours) affords significant protection to the photosynthetic membranes (Fig. 5). Based on the correlation between increase in heat stability of PS II and simultaneous association of GroEL-proteins with thylakoids demonstrated above, our finding might be explained by the high-temperature induced production of partially denatured proteins in accordance with the capability of chaperonins to prevent the formation of non-functional aggregates by binding to partially denatured proteins (Gatenby et al., 1993; Hartman et al., 1993). A recent study on mitochondrial malate dehydrogenase (MDH) not only demonstrates the chaperonin mediated reconstitution of heat shock effect, but underlines also the need of the concerted action of GroES and

GroEL. Whereas GroEL alone did not protect MDH against thermal inactivation but prevented its aggregation, reactivation could be achieved by addition of GroES.

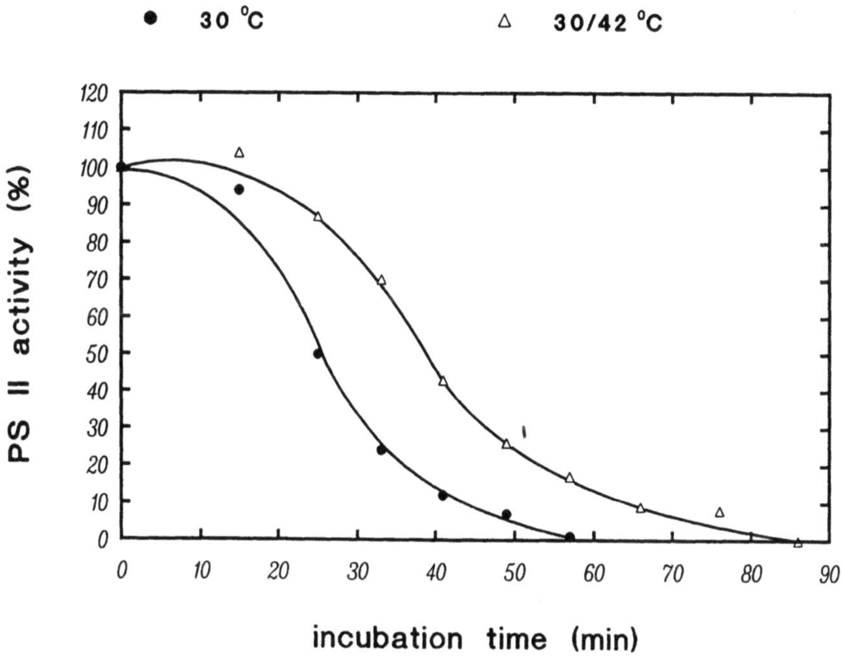

Figure 5. Time courses of the preservation of Photosystem II activities under heat-stress conditions (49°C) of *Synechocystis* cells, tested in cells grown at 30°C or after they were exposed to 42°C for 3 hours. 100% represents the photosynthetic rate found before heat-treatment, and measured at 30°C.

Conclusion and perspectives

In contrast to most bacterial species, *Synechocystis* PCC 6803 contains a groESL operon in addition to a separately arranged second groEL-type gene (cpn60). The two groEL-type genes revealed different expression kinetics upon heat shock. Variance in probabilities of forming coiled-coils within the two related chaperonins has also been demonstrated. Both of the gene products are able to assemble to oligomeric forms, whilst the precise subunit composition awaits further studies. The presented data on heat-induced GroEL-, GroES-thylakoid interaction might provide the first evidence on the possible implication of a chaperonin protein in the development of thermal stability of photosynthetic membranes. In order to gain further insights into regulation and roles of cpn-60 and groESL under normal and heat shock conditions, construction of chaperonin mutants defected in either of the groE-type genes is in progress in our laboratory.

Acknowledgements

This research was supported by grants from Hungarian Research Foundation (OTKA) to L.V. and I.H.

REFERENCES

Berry JA and Björkman O (1980) Photosynthetic response and adaptation to temperature in higher plants. Annu. Rev. Plant. Phys. 31:491-453

Chitnis PR and Nelson N (1991) Molecular cloning of the genes encoding two chaperone proteins of the cyanobacterium *Synechocystis* sp. PCC 6803. J. Biol. Chem. 266:58-65.

Ellis RJ (1990) The molecular chaperone concept. Seminars in Cell Biology 1:1-9

Gatenby AA, Donaldson GK, Baneyx F, Lorimer GH, Viitanen PV and van der Vies SM (1993) Participation of GroE heat shock proteins in polypeptide folding. Biocatalyst Design for Stability and Specificity, ACS Symposium Series (Himmel ME and Georgiou G eds.) American Chemical Society, Washington DC, 140-150

Georgopoulos C (1992) The emergence of the chaperone machines. TIBS 17:295-299

Hartman DJ, Surin BP, Dixon NE, Hoogenraad NJ and Hoj PB (1993) Substoichiometric amounts of the molecular chaperones GroEL and GroES prevent thermal denaturation and aggregation of mammalian mitochondrial malate dehydrogenase in vitro. Proc. Natl. Acad. Sci. 90:2276-2280

Howart CJ (1991) Molecular responses of plants to an increased incidence of heat shock. Plant, Cell and Environment 14:831-841

Kovács E, Lehel Cs, Gombos Z, Mustárdy L, Török Zs, Horváth I and Vigh L (1993) Heat stress induces association of the GroEL-analogue chaperonin with thylakoid membranes in cyanobacterium, *Synechocystis* PCC 6803. (submitted)

Lehel Cs, Wada H, Kovács E, Török Zs, Gombos Z, Horváth I, Murata N, and Vigh L (1992) Heat shock protein synthesis of the cyanobacterium *Synechocystis* PCC 6803. Plant. Mol. Biol. 18:327-336.

Lehel Cs, Los D, Wada H, Györgyei J, Horváth I, Kovács E, Murata N and Vigh L (1993/a) A second groEL-like gene, organized in a groESL operon is present in the genome of *Synechocystis* PCC 6803. J. Biol. Chem. 268:1799-1804.

Lehel Cs, Gombos Z, Török Zs and Vigh L (1993/b) Growth temperature modulates thermotolerance and heat shock response of cyanobacterium, *Synechocystis* PCC 6803.

Plant Phys. Biochem. 31:81-88
Lupas A, van Dyke M and Stock J (1991) Predicting coiled coils from protein sequences. Science 252:1162-1164.
Lubben TH, Donaldson GK, Viitanen PV and Gatenby AA (1989) Several proteins imported into chloroplasts form stable complexes with the GroEL-related chloroplast molecular chaperone. The Plant Cell 1:1223-1230
Mazodier P, Guglielmi G, Davies J and Thompson J (1991) Characterization of the groEL-like genes in *Streptomyces albus*. J. Bact. 173:7382-7386
Narberhaus F and Bahl H (1992) Cloning, sequencing and molecular analysis of the groESL operon of Clostridium acetobutylicum. J. Bact. 174:3282-3289
Schmidt A, Schiesswohl M, Völker U, Hecker M and Schumann W (1992) Cloning, sequencing, mapping and transcriptional analysis of the groESL operon from *Bacillus subtilis*. J. Bact. 174:3993-3999
Schön U and Schumann W (1993) Molecular cloning, sequencing and transcriptional analysis of the groESL operon from *Bacillus stearothermophilus*. J. Bact. 175:2465-2469
Webb R and Sherman LA (1992) Chaperones classified. Nature 359:485-486
de Wit TFR, Bekelie S, Osland A, Miko TL, Hermans PWM, van Solingen D, Drijfhout JW, Schöningh R, Janson AAM and Thole JER (1992) Mycobacteria contain two, groEL genes: the second *Mycobacterium leprae* groEL gene is arranged in an operon with groES. Molecular Microbiology 6:1995-2007

IDENTIFICATION AND CHARACTERIZATION OF NOVEL ATP-BINDING CASSETTE PROTEINS IN *SACCHAROMYCES CEREVISIAE*

Jonathan Leighton
Department of Biochemistry
Biocenter, University of Basel
CH-4056 Basel, Switzerland

INTRODUCTION

In recent years, a great deal of work has been focused on elucidating the mechanism by which nuclear-encoded proteins are imported into mitochondria from the cytoplasm (Glick and Schatz, 1991; Wienhues and Neupert, 1992). In contrast, despite some reports in the literature of effects of mitochondrial mutations on nuclear gene transcription, very little is known about how mitochondria communicate with the nucleus and how such signals might be transmitted (Forsburg and Guarente, 1989). Since members of the rapidly growing superfamily of transport proteins known as ATP-Binding Cassette (ABC) transporters appear to be present in all organisms (Higgins, 1992), we asked whether such transporters might also exist in mitochondria and play an as yet undiscovered role in communication with the nucleus or, more generally, in transport of substrates across the mitochondrial membranes.

ABC proteins are involved primarily in the transport of a wide range of biochemical substances across membranes. The ABC family includes the intensively studied Multidrug Resistance (MDR) P-Glycoprotein (Endicott and Ling, 1989) and Cystic Fibrosis Transmembrane Regulator (CFTR) (Riordan et al., 1989). Many ABC proteins are also found in bacteria, the presumed evolutionary ancestors of mitochondria, and it would therefore not be surprising if mitochondria had retained members of the family. ABC transporters have been found in several other eukaryotic organelles: the yeast vacuole (Ortiz et al., 1992), the mammalian peroxisome (Kamijo et al., 1990), and the mammalian endoplasmic reticulum, where a putative peptide transporter encoded by the MHC has been localized (Monaco, 1992). Until now, only a few ABC proteins have been identified in yeast, the best-known being STE6, which is responsible for the export of the mating pheromone a-factor (McGrath and Varshavsky, 1989; Kuchler et al., 1989).

NATO ASI Series, Vol. H 82
Biological Membranes:
Structure, Biogenesis and Dynamics
Edited by Jos A. F. Op den Kamp

Members of the ABC transporter superfamily share a similar overall domain organization, namely two hydrophobic domains which span the membrane several times, and two cytoplasmic domains which contain the ATP-binding sites. The ATP-binding sites are located within a highly conserved stretch of about 200 amino acids, the "ATP-Binding Cassette". In view of this marked sequence homology, we decided to adopt a PCR-based approach to identify new members of the family in yeast. Any new ABC genes identified would be further analyzed for mitochondrial localization or an interesting null mutant phenotype.

RESULTS

Degenerate oligonucleotide primers were synthesized corresponding to two regions within the ATP-Binding Cassette with particularly high sequence conservation (Figure 1). PCR was performed on genomic DNA from *Saccharomyces cerevisiae*, using a standard amplification protocol with a 45°C annealing temperature, and amplified DNA was cloned into the pUC18 vector. DNA sequencing of the cloned fragments revealed ten fragments with homology to the ABC superfamily, most being 300-400 base pairs in length. Only one fragment corresponded to a previously identified gene, ADP1, which had been identified during the sequencing of yeast chromosome III (Purnelle et al., 1991).

Upstream Oligonucleotides (Sense):

	BamHI	S	G	A/S/C	G	K	S	T
GGC	GGATTC	TCI	GGI	T/GC/GI	GGI	AAA/G	AGC/T	AC
GGC	GGATTC	TCI	GGI	T/GC/GI	GGI	AAA/G	TCI	IC

Downstream Oligonucleotides (Antisense):

	EcoRI	D	L	A	S	V/T	A/P	E	D
CGG	GAATTC	TC	IAG/A	IGC	A/GCT	IAC	IGC/G	T/CTC	A/GTC
CGG	GAATTC	TC	IAG/A	IGC	A/GCT	IGT	IGC/G	T/CTC	A/GTC
CGG	GAATTC	TC	IAG/A	IGC	IGA	IAC	IGC/G	T/CTC	A/GTC
CGG	GAATTC	TC	IAG/A	IGC	IGA	IGT	IGC/G	T/CTC	A/GTC

Fig.1.Oligonucleotides used for PCR. Degenerate oligonucleotides were designed corresponding to two regions within the ATP-binding cassette which showed particularly high homology among different ABC proteins. The corresponding amino acids are shown above the nucleotides. Restriction sites were incorporated for ease of cloning.

In order to gain insight into the functions of the corresponding gene products, we performed gene disruptions with the cloned PCR fragments by inserting selectable markers into unique restriction sites within the fragments. Disruptions were performed in haploid and/or diploid strains and confirmed by Southern analysis. In one case where haploid disruptants were not obtained, a diploid disruptant was sporulated and dissected to obtain disrupted haploid spores. Of the nine fragments other than ADP1, five were used successfully to disrupt the corresponding genes. One fragment was too small to be used, while the other three failed to yield transformants that could be confirmed by Southern analysis as disruptants in the intended genes. The relatively small size of the fragments may have been a limiting factor in obtaining disruptants.

Haploid disruptants were tested for the ability to grow under different conditions, including ethanol and glycerol as sole carbon sources as an indicator of mitochondrial respiration. Of the five disruptants obtained, one displayed a discernible phenotype: a dramatic impairment of growth on YPD. Three independent disruptions were performed on this gene, temporarily named ABC1. The first disruption involved the insertion of the entire Yip5 yeast integrating vector, containing URA3 as a selectable marker, into an internal Nhe1 restriction site contained within the PCR fragment. No haploid disruptants were obtained, while dissection of a disrupted diploid resulted in two slow-growing URA^+ spores per tetrad. As this disruption strategy leaves repeated sequences within the gene which can lead to reversal of the disruption if the cells are not maintained on selective medium, we decided to perform a stable disruption of the gene, this time using LEU2 as a selectable marker so that a URA3-bearing library could subsequently be used for cloning the gene using complementation of the slow-growth phenotype. Two diploids bearing a LEU2 insertion in ABC1 were dissected. One gave little or no phenotype in the LEU^+ spores, while the other yielded two LEU^+ spores per tetrad with a substantial growth rate reduction, though not as dramatic as in the previous disruption. The growth rate in liquid YPD medium was reduced 2.9-fold compared with wild-type. Interestingly, while the disrupted spores were scorable as LEU^+, they grew extremely poorly on synthetic plates lacking leucine, while addition of leucine to the plates resulted in a substantial improvement in growth rate. This observation hinted at a possible physiological role for the ABC1 gene product in amino acid metabolism or transport, and we therefore disrupted the gene a third time in a LEU^+ strain to allow direct comparison of wild-type and disrupted spores under the same conditions. In this third disruption, performed with the insertion of URA3 into the NheI site, sporulation and dissection of a disrupted diploid led to a very severe reduction in growth of the URA^+ spores on rich medium (Figure 2), and to a complete growth arrest on synthetic medium. This represents one of the strongest phenotypes observed upon disruption or mutation of an ABC transporter.

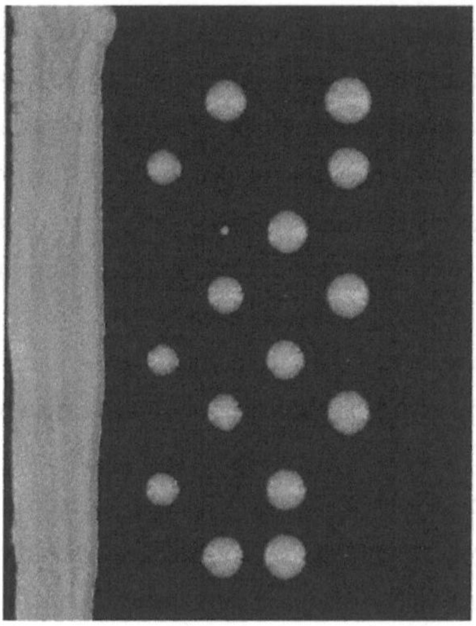

Fig. 2. Dissection of diploid ABC1 disruptant. A diploid yeast strain heterozygous for ABC1 bearing an internal URA3 disrupting marker was sporulated and dissected on YPD medium. Seven of eight tetrads exhibited a 2:2 segregation of two normal-sized colonies and two colonies with severely inhibited growth. The small colonies were confirmed to contain the URA3 marker.

The cloned PCR fragment was used to screen a yeast genomic library by hybridization, and a clone was obtained which complemented the slow-growth phenotype of the LEU2-bearing disruptant. Sequencing of a portion of the 11 kb insert revealed a 2082-base pair open reading frame which contained the PCR fragment. The predicted 694 amino acid protein is half the size of the yeast STE6 and human MDR1 proteins; it contains one hydrophobic and one hydrophilic domain, and has a hydropathy profile resembling that exhibited by each of the two homologous halves of MDR1 and STE6. It is likely that the protein forms a homo- or hetero-dimer, by analogy to other ABC transporter complexes.

DISCUSSION

The superfamily of ABC transporters is responsible for the transport of a wide range of substrates in species ranging from bacteria to humans, and the importance of this family is underscored by their involvement in some human diseases (Gartner et al., 1992). The ubiquity of the family is further demonstrated here by the generation of PCR fragments corresponding to potentially as many as nine new ABC genes in *S. cerevisiae*. This number surely underestimates the total number of these genes present, as the degenerate oligonucleotide primers did not encompass all the sequences observed in members of the ABC superfamily, and STE6, a known gene with which two of the primers had 100% homology, was not even obtained in this screen.

Identification of the functions of the corresponding genes will be a challenging task, in part because of the possibility of overlapping functions between different ABC transporters. The ABC1 gene identified here is more amenable to analysis because it exhibits a strong phenotype upon disruption. Subcellular localization of the protein is in progress and should provide further insight into its function. We are also continuing to search for a mitochondrial ABC transporter in the hope of learning more about how mitochondria communicate with the rest of the cell.

REFERENCES

Endicott JA and Ling V (1989) The biochemistry of P-glycoprotein-mediated multidrug resistance. Annu. Rev. Biochem. 58: 137-171

Forsburg SL and Guarente L (1989) Communication between mitochondria and the nucleus in regulation of cytochrome genes in the yeast Saccharomyces cerevisiae. Annu. Rev. Cell Biol. 5: 153-180

Gartner J, Moser H and Vallee D (1992) Mutations in the 70 kd peroxisomal membrane protein gene in Zellweger syndrome. Nature Genet. 1: 16-23

Glick BS and Schatz G (1991) Import of proteins into mitochondria. Annu. Rev. Genet. 25: 21-44

Higgins CF (1992) ABC transporters: from microorganisms to man. Annu. Rev. Cell Biol. 8: 67-113

Kamijo K, Taketani S, Yokota S, Osumi T and Hashimoto T (1990) The 70-kDa peroxisomal membrane protein is a member of the Mdr (P-glycoprotein)-related ATP-binding protein superfamily. J. Biol. Chem. 265: 4534-4540

Kuchler K, Sterne RE and Thorner J (1989) *Saccharomyces cerevisiae STE6* gene product: a novel pathway for protein export in eukaryotic cells. EMBO J. 8: 3973-3984

McGrath JP and Varshavsky A (1989) The yeast *STE6* encodes a homologue of the mammalian multidrug resistance P-glycoprotein. Nature 340: 400-404

Monaco JJ (1992) A molecular model of MHC class-I-restricted antigen processing. Immunol. Today 13: 173-179

Ortiz DF, Kreppel L, Speiser DM, Scheel G, McDonald G and Ow DW (1992) Heavy metal tolerance in the fission yeast requires an ATP-binding cassette-type vacuolar membrane transporter. EMBO J. 11: 3491-3499

Purnelle B, Skala J and Goffeau A (1991) The product of the *YCR105* gene located on the chromosome III from *Saccharomyces cerevisiae* presents homologies to ATP-dependent permeases. Yeast 7:867-872

Riordan JR, Rommens JM, Kerem BS, Alon N and Rozmahel R (1989) Identification of the cystic fibrosis gene: cloning and characterization of complementary DNA. Science 245: 1066-1073

Wienhues U and Neupert W (1992) Protein translocation across mitochondrial membranes. Bioessays 14: 17-23

MOLECULAR INTERACTIONS INVOLVED IN THE PASSAGE OF THE CYTOTOXIC PROTEIN α-SARCIN ACROSS MEMBRANES*

José M. Mancheño, María Gasset, Javier Lacadena, Alvaro Martínez del Pozo, Mercedes Oñaderra & José G. Gavilanes
Department of Biochemistry and Molecular Biology. Faculty of Chemistry. Complutense University. 28040 Madrid. Spain

α-Sarcin is a cytotoxic protein secreted by the mould *Aspergillus giganteus*. In vitro assays have revealed that this single polypeptide chain inactivates the ribosomes by cleaving a single phosphodiester bond in the larger ribosomal RNA (Wool et al., 1990). α-Sarcin inhibits the protein biosynthesis in a variety of cultured human tumour cell lines in the absence of any permeabilizing agent (Turnay et al., 1993; Olmo et al., 1993). No protein membrane receptors have been so far described for α-sarcin. Therefore, the cytotoxicity of α-sarcin would require some kind of interaction with the membrane lipids of the target cells. In this regard, α-sarcin strongly interacts with phosphatidylglycerol vesicles promoting their aggregation and lipid mixing (Gasset et al., 1989, 1990, 1991a, 1991b). These events may be involved in the passage of the protein across the membrane of the target cells, thus explaining the cytotoxicity of α-sarcin. The analysis of the membrane/α-sarcin interactions is the overall goal of our studies.

α-Sarcin binds to acid phospholipid vesicles at neutral pH and 0.1 M ionic strength (Table I). The protein exhibits a higher affinity for dimyristoylphosphatidylglycerol (DMPG) than for dimyristoylphosphatidylserine (DMPS) vesicles. DMPG/DMPS (1:1 molar ratio) vesicles present an apparent dissociation constant that corresponds to about the average affinity of each individual type of phospholipid. The observed differences may be related to intermolecular hydrogen bonding in phosphatidylserine (Boggs et al., 1986). No significant binding of the protein to the vesicles has been observed at high ionic strength (above 0.3 M NaCl). α-Sarcin does not bind to phosphatidylcholine (PC) vesicles. These results indicate that electrostatic interactions are primarily involved in the α-sarcin-vesicle binding.

The protein produces aggregation of the interacting vesicles. This has been deduced from light-scattering and absorbance measurements. Initial rates of aggregation

* This research has been supported by a research grant from the Spanish DGICYT #PB90-0007.

NATO ASI Series, Vol. H 82
Biological Membranes:
Structure, Biogenesis and Dynamics
Edited by Jos A. F. Op den Kamp

Table I.- Binding parameters of the phospholipid vesicle-α-sarcin interaction.

	n	K (μM)
DMPG	40	0.06
DMPS	37	3.00
DMPS:DMPG (1:1 molar ratio)	54	1.90

These analyses have been performed by measuring the free protein concentration in protein/vesicles mixtures, at temperatures above the corresponding phase transition (42 °C for DMPS and DMPG/DMPS, and 37 °C for DMPG). Ultrafiltration and ultracentrifugation have been used to separate the free protein. The obtained results have been analyzed by considering the equilibrium: protein(P) + "free lipid sites" = PLn (complex), being n, the number of phospholipid molecules affected by the protein, and K, an apparent dissociation constant.

have been measured by stopped-flow light-scattering analysis. The aggregation process is second order in phospholipid vesicles (Figure 1). This suggests that the formation of a vesicle dimer is the first step of the aggregation process. The initial aggregation rate is not decreased at high protein/lipid molar ratios (e.g. 6:10). Thus, coating of the vesicles by protein molecules does not result in inhibition of the aggregation. This would indicate that protein-protein interactions are involved in maintaining the vesicles aggregates.

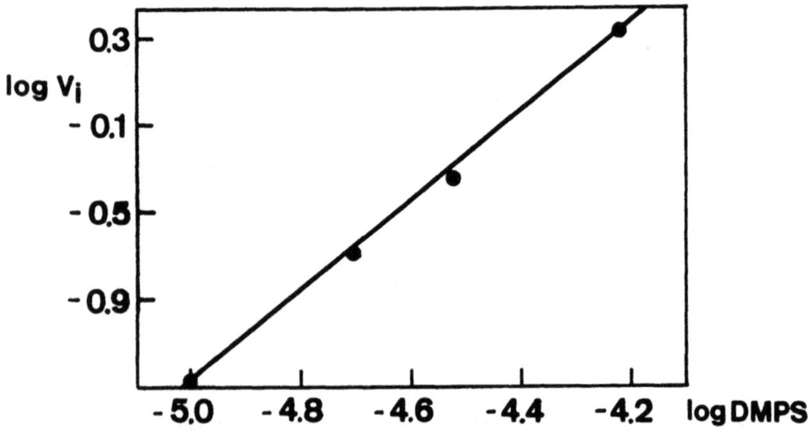

Figure 1.- Double logarithmic plot of the maximum initial rate of the aggregation of DMPS vesicles produced by α-sarcin (Vi, expressed as I per sec; I, scattered intensity) vs. molar concentration of phospholipid (DMPS).

The extent of the vesicles aggregation is decreased at high ionic strength. Less than 5% of the aggregation observed at 0.1 M NaCl is detected at 0.3 M NaCl. However, the aggregation is not fully reverted by increasing the ionic strength when the protein/vesicles system is under equilibrium. Increase of the ionic strength up to 0.4 M NaCl in protein-lipid mixtures immediately decreases both the absorbance at 360 nm and the 90° light-scattering of acid phospholipid vesicles aggregated by α-sarcin. These two spectroscopical parameters are decreased up to 50% of the initial values when the ionic strength of the sample is increased up to 0.4 M NaCl. Therefore, other interactions than the electrostatic ones are involved in maintaining the final α-sarcin-vesicles complexes.

In this regard, α-sarcin produces lipid mixing of acid phospholipid vesicles. Initial rates of this process have been studied by continuous recording the fluorescence increase at 530 nm in the classical assay that makes use of the resonance energy transfer between NBD-phosphatidylethanolamine (donor) and Rh-phosphatidylethanolamine (acceptor) (Struck et al., 1981) (Figure 2). Under these experimental conditions, the rates show a first-order dependence with the phospholipid concentration as would be expected for a fusion process of vesicle dimers in which the lipid mixing is the rate limiting step.

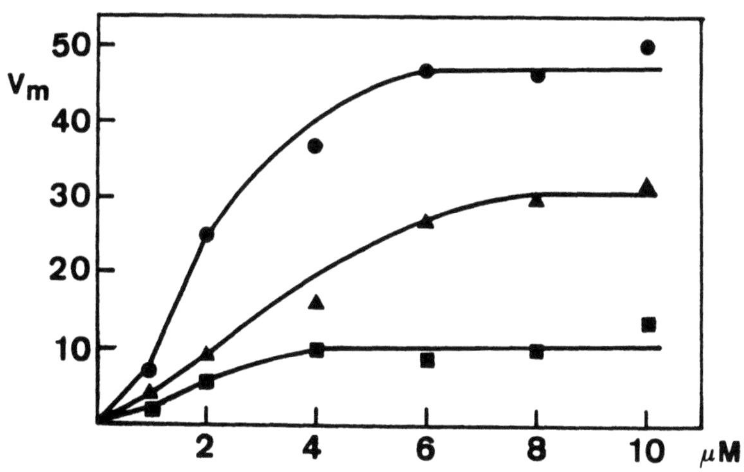

Figure 2.- Maximum rates (Vm, expressed as fluorescence emission variation per sec) of the fluorescence emission increase at 530 nm (for excitation at 450 nm) in a (1%) NBD-PE/(0.6%) Rh-PE resonance energy transfer system vs. α-sarcin concentration (μM). Fluorescence labelled vesicles were mixed with unlabelled vesicles in a 1:1 proportion. (■), 10 μM, (▲) 40 μM and (●) 70 μM total DMPG concentration. The fluorescence emission is expressed in arbitrary units.

The protein hydrophobically interacts with the bilayer. This is deduced from the existence of energy transfer from the two tryptophan residues of the protein to anthracene incorporated in the bilayer of vesicles (Figure 3). This approach has been

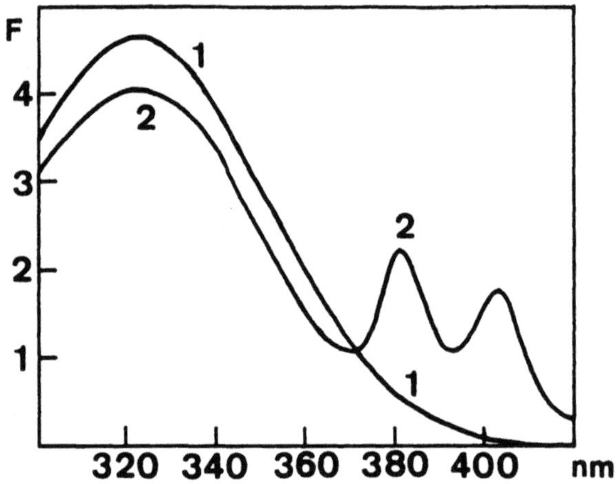

Figure 3.- Fluorescence emission spectra of α-sarcin (for excitation at 295 nm) in the presence of DMPG vesicles (1:40 protein/lipid molar ratio) without (1) and with (2) 1% weight anthracene incorporated into the lipid bilayer. Fluorescence is expressed in arbitrary units.

tested with synthetic peptides buried into the hydrophobic region of liposomes (Uemura et al., 1983). The regions around one or both tryptophan residues in α-sarcin (at positions 4th and 51st) would be close enough to the hydrophobic core of the bilayer to allow the energy transfer to anthracene. Moreover, the protein decreases the amplitude of the phase transition process of the phospholipid (Figure 4), the phase transition temperature value being not modified. This is also in agreement with a hydrophobic interaction, since it would indicate that the protein removes phospholipid molecules from participating in the phase transition.

The protein translocates across the bilayer of asolectin vesicles (Oñaderra et al., 1993). This has been demonstrated by using two different approaches. Trypsin-containing vesicles and yeast tRNA-containing vesicles have been prepared. Both types of vesicles have been incubated in the presence of α-sarcin. The protein is able to degrade encapsulated tRNA. This degradation is dependent on the concentration of α-sarcin externally added. Also, externally added α-sarcin is time-dependent degraded by the entrapped protease, in the presence of a saturating external concentration of trypsin inhibitor. A summary of these results is given in Table II.

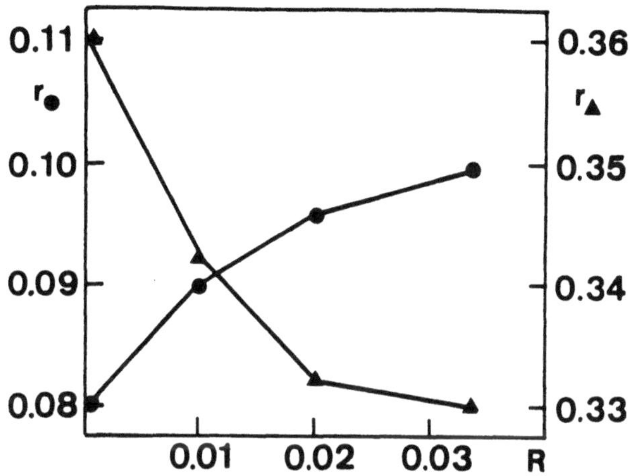

Figure 4.- Steady-state fluorescence anisotropy (r) variation of DPH-labelled DMPS vesicles (● ;22.5 °C) below and (▲;47.5 °C) above the phase transition temperature of the phospholipid vs. (R), α-sarcin/phospholipid molar ratio.

Table II.- Summary of the results obtained with trypsin-containing and tRNA containing vesicles in the presence of α-sarcin.

(A) % of the maximum degradation of encapsulated t-RNA at different (P/L) α-sarcin/phospholipid molar ratios.

(P/L)	0.005	0.010	0.015	0.020	0.030
(%)	19	40	58	80	100

(B) Time-course of the (%) degradation of externally added α-sarcin by encapsulated trypsin.

t (h)	4	8	16	24
(%)	15	20	70	94

(A) These assays have been performed at 1 mM lipid concentration by measuring the amount of soluble oligonucleotides produced by the action of α-sarcin, as described for the ribonuclease activity assay described for this protein (Martínez del Pozo et al., 1989). (B) The (%) degradation has been measured by considering the area under the absorbance peak from the densitometric scan of SDS-PAGE gels from the assay mixture.

α-Sarcin shows a significant degree of similarity with ribonuclease T1 (RNase T1), also a fungal ribonuclease (from *Aspergillus oryzae*), which tridimensional structure is well known (Pace et al. 1991). This similarity extends up to more than 40% of the amino acid residues in RNase T1. This similarity is also observed when considering secondary structures, the one predicted for α-sarcin from its amino acid sequence and the one deduced for RNase T1 from its tridimensional pattern. RNase T1 exhibits a characteristic barrel structure composed of ß-strands, which constitutes the nucleus of the molecule. These ß-strands would be also present in α-sarcin. Therefore, a similar tridimensional structure can be proposed for α-sarcin (Figure 5). The diagram given in this Figure contains all the periodic secondary structure elements predicted for α-sarcin arranged according to the tridimensional structure of RNase T1.

The polypeptide chain of α-sarcin contains 150 amino acid residues whereas that of RNase T1 has 104. The extra amino acid residues in α-sarcin lacks periodic secondary

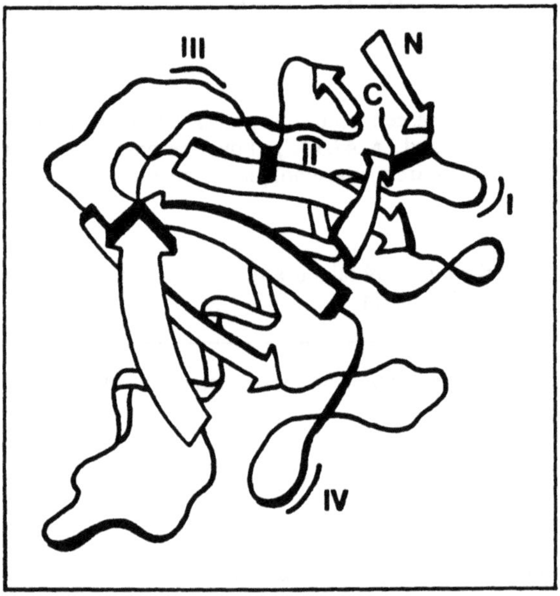

Figure 5.- Diagram of the proposed tridimensional structure of α-sarcin. (N) stands for the amino-terminal and (C) for the COOH-terminal ends of the protein molecule. The arrows represent the ß-strands. Loops are represented by irregular lines (I) to (IV) indicate the relative positions of positive charge enriched regions. The two disulfide bridges of the molecule are represented by gross segments.

structure other than ß-turns, and they would be located in external loops connecting periodic secondary structure elements. These loops contain a significant number of the positive charges of α-sarcin (16 among the 24 in the overall molecule). These positive charges are concentrated in four main regions (I to IV in Figure 5). This structural pattern may account for the ability of the protein in aggregating acid phospholipid vesicles. According to this hypothesis, the presence of the positive charges in those non-periodically organized loops would be responsible for the charge neutralization in the vesicles.

Reduction and carboxyamidomethylation of α-sarcin results in a chemically modified protein form that lacks periodic secondary structure, based on the observed circular dichroism spectrum in the peptide bond region. Denaturation of the protein is also deduced from the fluorescence emission analysis. This modified protein produces aggregation and fusion of acid phospholipid vesicles. These results are summarized in Figure 6.

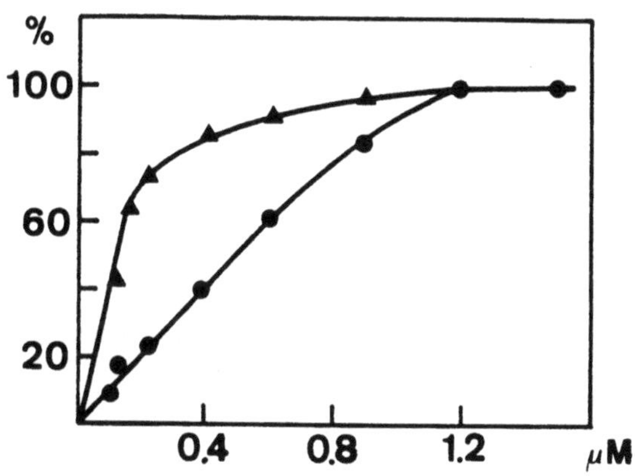

Figure 6.- Aggregation (●) and fusion (▲) of DMPS vesicles produced by chemically modified α-sarcin (μM). Values are expressed as percentages of the maximum effect. Aggregation has been measured from the absorbance variation at 360 nm. Fusion has been measured by using the above described resonance energy transfer assay.

According to these results, the native conformation of α-sarcin is not required in order to produce aggregation and fusion of phospholipid vesicles. These effects would arise from the presence of positive charge domains in the external loops of the molecule. This fact would facilitate charge neutralization in the vesicles. This interaction at level of the membrane would represent the first step in the cytotoxicity of α-sarcin. The second step, at the level of the ribosomes, would arise from the specific ribonuclease activity of the protein.

References:

Boggs, J.M., Chia, L.S., Rangaraj, G. & Moscarello, M. (1986) Interaction of myelin basic protein with different ionization states of phosphatidic acid and phosphatidylserine. Chem.Phys.Lipids 39: 165-184.

Gasset, M., Martínez del Pozo, A., Oñaderra, M. & Gavilanes, J.G. (1989) Study of the interaction between the antitumour protein α-sarcin and phospholipid vesicles. Biochem.J. 258: 569-575.

Gasset, M., Oñaderra, M., Thomas, P.G. & Gavilanes, J.G. (1990) Fusion of phospholipid vesicles produced by the antitumour protein α-sarcin. Biochem.J. 265: 815-822.

Gasset, M., Oñaderra, M., Martínez del Pozo, A., Schiavo, G., Laynez, J., Usobiaga, P. & Gavilanes, J.G. (1991a) Effect of the antitumour protein on the thermotropic behaviour of acid phospholipid vesicles. Biochim.Biophys.Acta 1068: 9-16.

Gasset, M., Oñaderra, M., Goormaghtigh, E. & Gavilanes, J.G. (1991b) Acid phospholipid vesicles produce conformational changes on the antitumour protein α-sarcin. Biochim.Biophys.Acta 1080: 51-58.

Martínez del Pozo, A., Gasset, M., Oñaderra, M. & Gavilanes, J.G. (1989) Effect of divalent cations on structure-function relationships of the antitumour protein α-sarcin. Int.J.Peptide Prot.Res. 34: 416-422.

Olmo, N., Turnay, J., Lizarbe, M.A. & Gavilanes, J.G. (1993) Cytotoxic effect of α-sarcin, a ribosome inactivating protein, in cultured Rugli cells. STP Pharma Sci. 3: 93-96.

Oñaderra, M., Mancheño, J.M., Gasset, M., Lacadena, J., Schiavo, G., Martínez del Pozo, A. & Gavilanes, J.G. (1993) Translocation of α-sarcin across the lipid bilayer of asolectin vesicles. Biochem.J. (in press).

Pace, C.N., Heinemann, U., Hahn, U. & Saenger, W. (1991) Ribonuclease T1: structure, function and stability. Angew.Chem. 30: 343-360.

Struck, D.K., Hoekstra, D. & Pagano, R.E. (1981) Use of resonance energy transfer to monitor membrane fusion. Biochemistry 20: 4093-4099.

Turnay, J., Olmo, N., Jiménez, J., Lizarbe, M.A. & Gavilanes, J.G. (1993) Kinetic study of the cytotoxic effect of α-sarcin, a ribosome inactivating protein from Aspergiullus giganteus, on tumour cell lines: protein biosynthesis inhibition and cell binding. Mol.Cell.Biochem. (in press).

Uemura, A., Kimura, S. & Imanishi, Y. (1983) Investigation on the interaction of peptides in the assembly of liposome and peptide by fluorescence. Biochim. Biophys.Acta 729: 28-34.

Wool, I.G., Endo, Y., Chang, Y.L. & Glück, A. (1990) In The ribosome, structure, function and evolution (A. Dahlber, R.A. Garret, P.B. Moore, D. Schlessinger & J.R. Warner Eds.) American Society of Microbiology. Washington DC.

OLIGOSACCHARYL TRANSFERASE AND PROTEIN DISULFIDE ISOMERASE: TWO KEY ENZYMES IN THE ENDOPLASMIC RETICULUM

Jack Roos, MaryLynne LaMantia and William J. Lennarz
Department of Biochemistry and Cell Biology
State University of New York at Stony Brook
Stony Brook, New York 11794-5215
USA

SUMMARY

The co-translational glycosylation of nascent polypeptide chains in the endoplasmic reticulum (ER) is catalyzed by oligosaccharyl transferase (OT). Recent studies in yeast by others have identified two proteins that are non-limiting components of OT. In an effort to identify other, limiting components of OT we have devised a sensitive assay for OT activity in lysed yeast protoplasts. This assay has been used to screen a collection of temperature sensitive yeast mutants. This approach has yielded several mutants that have defects in glycosylation, one of which may contain a limiting, defective component of OT. Efforts to characterize these mutants and to clone and sequence this putative OT subunit are in progress.

During earlier efforts to identify OT, hydrophobic peptides containing a glycosylation site for OT and a photoreactive, radiolabeled probe were developed. However, these probes were found to label a lumenal protein of the ER rather than OT. This protein was identified as protein disulfide isomerase (PDI). In yeast, deletion of the gene encoding this protein was lethal, demonstrating that PDI is encoded by an essential gene. C-terminal deletion and active site mutants of PDI were studied with respect to: 1) cell viability, 2) processing of secretory proteins, and 3) *in vitro* catalytic activity in disulfide bond formation. The results revealed that PDI does, in fact, catalyze disulfide bond formation *in vivo*. Furthermore, it is clear that the catalytic activity depends on the presence of the sequence -CGHC-. However, the essential role of PDI may be more closely related to its ability to bind and stabilize polypeptides than to its disulfide bond-forming function.

NATO ASI Series, Vol. H 82
Biological Membranes:
Structure, Biogenesis and Dynamics
Edited by Jos A. F. Op den Kamp

INTRODUCTION

Our laboratory is interested in two events that occur in the endoplasmic reticulum: One is the co-translational N-glycosylation of proteins destined for secretion or for routing to other membranes of the cell. The other is the folding and formation of disulfide bonds in these proteins. As shown in Figure 1, these two processes are catalyzed by enzymes that are associated with the endoplasmic reticulum in very different ways. The enzyme that catalyzes N-linked glycosylation, oligosaccharyl transferase, appears to exist as a complex of integral membrane proteins. The other enzyme, protein disulfide isomerase, is a resident protein of the lumen of the ER. In this chapter, progress on characterizing these two enzymes in yeast will be reported.

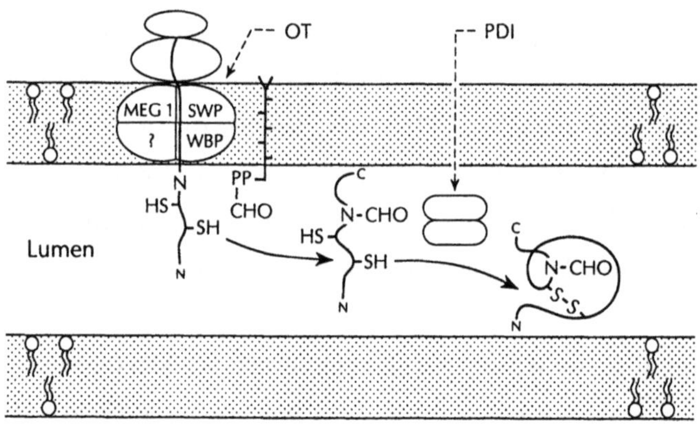

Figure 1. Protein N-glycosylation and folding in the ER.

Oligosaccharyl transferase. Since the discovery of the involvement of the long chain polyprenyl phosphate, dolichyl phosphate, in the pathway of assembly of the oligosaccharide chain of N-linked glycoproteins (for review, see Waechter and Lennarz, 1976), a great deal of progress has been made in understanding the individual steps of this complex process. As shown in Fig. 2, the overall process can be considered to occur in three phases.

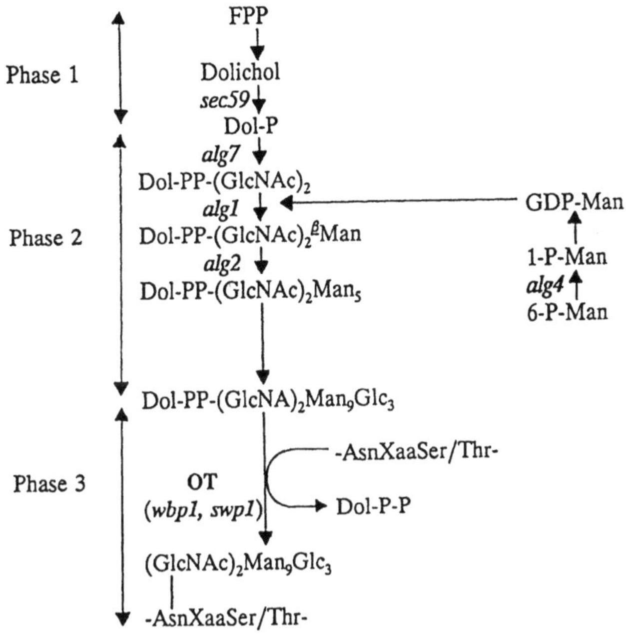

Figure 2. Potential site of defects in the glycoprotein synthetic pathway. Yeast mutants defective in specific steps of the pathway are indicated.

A number of the steps in the first phase, the synthesis of dolichol phosphate and several intermediate chain length polyprenoids that are important in polyprenylation of proteins, have been elucidated. However, the details of the late steps in formation of dolichol phosphate, as well as the regulation of its synthesis, are not well understood. The second phase is the multi-step lipid-linked oligosaccharide assembly process, whereby dolichol phosphate serves as a membrane anchor for the assembly of oligosaccharylpyrophosphoryldolichol. As shown, numerous steps in this process have been delineated in yeast as the result of the availability of *alg* mutants (asparagine-linked glycosylation mutants, Huffaker and Robbins, 1989), although at this stage, too, the details of some steps, and their topology and the regulation of their assembly, remain to be elucidated. Characterization of the enzyme involved in the final step, in which the oligosaccharide chain is transferred from the lipid anchor to -Asn-X-Ser/Thr- in protein has, for a number of years, eluded investigators. This step, catalyzed by oligosaccharyl transferase, has recently been characterized in part by Gilmore and co-workers (Kelleher et al., 1992; Silberstein et al., 1992), who reported that in canine pancreas microsomes oligosaccharyl transferase consists of a three component complex. Similarly, studies in yeast have provided evidence that oligosaccharyl transferase in yeast contains at least two protein components (te Heesen et al., 1992; 1993).

As an outgrowth of earlier efforts to determine the specificity of oligosaccharyl transferase, we prepared a variety of labeled -Asn-X-Ser/Thr- peptides to determine their efficiency as substrates. Using the most effective substrate, we have developed a simple assay that can be used on small cultures of lysed yeast protoplasts. The objective was to identify temperature sensitive yeast mutants that exhibited defects in one of the three phases of the dolichol-linked pathway: synthesis of dolichylphosphate, assembly of oligosaccharyl-pyrophosphorydolichol, or transfer of oligosaccharide to the acceptor protein or peptide. As is evident in Fig. 2, a defect leading to under production of any of the intermediates in phases 1 or 2 would result in impaired peptide glycosylation.

In this study we report the results of the use of this assay in the development of a screen for such yeast mutants and its application to a collection of temperature sensitive mutants. The validity of the screen is documented by identification of known mutants in phases 1 and 2 of the overall assembly process (Class 1 and Class 2 mutants). In addition, evidence for the identification of a mutant that exhibits a temperature sensitive defect in oligosaccharyl transferase activity (Class 3) is reported.

Temperature-sensitive mutants that are defective in glycopeptide formation. The initial objective of this screen was to identify yeast mutants defective in oligosaccharyl transferase (OT) activity. The simple assay employed in this screen is extremely sensitive due to the high affinity of the enzyme for the peptide substrate and its very high specific radioactivity (Welply et al., 1983; Geetha-Habib et al., 1990). Of 440 ts strains available to us from the Hartwell collection (Hartwell, 1967), 285 were assayed. Fifteen mutants exhibited either a significant (greater than 30%) reduction in peptide glycosylation upon shifting to the non-permissive temperature (36^0) or had less than 50% peptide glycosylation of the parental strain at 23^0. Arbitrarily, 7 mutants, shown in Table 1, were chosen to be further characterized; the remaining 8 mutants remain to be further characterized.

Table 1. Peptide glycosylation in lysates of mutants

Mutant	23^0	36^0
A364A (wt)	100[a]	100
59	24	43
64	37	71
148	46	61
163	9	9
212	1	1
265	19	29
283	64	19

[a]All values are expressed as % wt at activity 23^0.

Our first step in characterizing these mutants was to ascertain that the reduction in peptide glycosylation was the result of a lesion in an endoplasmic reticulum (ER) associated protein and not the result of a mutation elsewhere in the cell. To do this, microsomes were prepared from each mutant and equivalent amounts of microsomal protein were assayed for OT activity. The results of this enzyme assay, shown in Table 2, indicate that six of the seven mutants exhibit reduced levels of peptide glycosylation at the level of crude microsomes. Microsomes from mutant 212 could not be prepared because mutant 212 did not spheroplast under the conditions used in both the screen and in the microsome preparation. Thus, the apparent lack of peptide glycosylation in mutant 212 as determined by the screen was probably a result of the failure to deliver the peptide substrate to the cell. Further characterization of mutant 212 was abandoned at this step.

The second step in characterizing the remaining 6 mutants was to determine if the defect in peptide glycosylation was the result of a mutation in one gene, and, if it was, to determine if the mutated gene product gave rise to the phenotype of temperature sensitivity for growth at 37°. The results indicated that the lesions in mutant 163 and 283 arose from a single nuclear mutation that co-segregated with the temperature sensitivity for growth (data not shown). The remaining mutants, because their mutations failed to co-segregate with the temperature sensitivity, have either leaky mutations in essential genes or have mutations in non-essential genes.

Table 2. Peptide glycosylation assayed in microsomes from mutants

Mutant	Relative glycosylation activity
A364A (wt)	100[a]
59	23
64	31
148	14
163	20
212	n.d.
265	16
283	18

[a]All values are expressed as % wt activity at 23^0; 150 μg of microsomal protein was assayed.

Three classes of mutants are identified by this screen. Because the screen used to assay for a defect in OT activity utilizes endogenous oligosaccharylpyrophosphoryldolichol as the other substrate in the conversion of a ^{125}I-labeled glycosylatable peptide to a glycopeptide, it follows that any mutation that causes a block in synthesis of this substrate will also be detected. In fact, as shown earlier in Fig. 2, starting from the first step committed to dolichol synthesis, elongation of farnesyl-PP, a mutation detected by the screen may fall into one of three classes corresponding to the 3 phases in the early assembly process. Any mutation in the dolichol synthesis pathway (Class 1) will lead to a decrease in the endogenous pool of lipid-linked oligosaccharide and will be scored as a potential OT mutant. Furthermore, mutations in enzymes involved in the assembly of the lipid-linked oligosaccharide (the asparagine-linked glycosylation, or *alg*, mutants (Huffaker and Robbins, 1983)), will also be identified by this screen (Class 2). However, ConA-agarose is used to quantitate glycopeptide formation, and since ConA only binds with high affinity to

α-mannosyl residues, this screen will only detect alterations early in the oligosaccharide assembly pathway. Finally, Class 3 mutants are those with a defect in components of oligosaccharyl transferase. Using a combination of biochemical and genetic complementation experiments, the temperature-sensitive mutant 283 was shown to be allelic to *SEC59* (Class 1), where mutant 148 was shown to genetically interact with the *ALG1* gene product (data not shown). Similarly, using the same techniques, mutant 163 was shown to have a lesion in a novel protein in the glycoprotein biosynthetic pathway.

Mutant 163 is a candidate for OT. We found that one candidate, mutant 163, was not a Class 1 mutant or an already identified Class 2 mutant. Thus, mutant 163 was considered to be a strong candidate to be a Class 3 mutant. To confirm this, studies were carried out to characterize the lipid-linked oligosaccharide synthesized by mutant 163. As expected, mutant 163 is a Class 3 mutant because this strain synthesizes an oligosaccharide identical in size to that synthesized in a wild-type strain.

To further characterize this novel protein in the glycoprotein biosynthetic pathway, mutant 163 was transformed with a single-copy, centromere containing yeast genomic library. Transformants were selected at 23^0 on selective media plates and then replica plated and assayed for growth at 36^0. Of 8000 transformants that were screened, 5 transformants grew at 36^0. Plasmids from these transformants were isolated, electroporated into *E. coli*, and the DNA isolated from the bacteria was digested with *EcoRI*. The digests were then analyzed by agarose gel electrophoresis. The results revealed that the 5 transformants were rescued by 3 different plasmids, but all 3 possessed common 1.1 kb and 2.7 kb *EcoRI* fragments. Thus, one gene, situated within 3 overlapping partial digests of yeast genomic DNA, rescued the temperature-sensitive phenotype of mutant 163. This gene, designated *MEG1* (microsomal protein essential for glycosylation *1*), was used to transform mutant 163. Transformation of mutant 163 with pMEG1 generates tranformants that all grow at 36^0. The *in vitro* activity of transformed mutant 163(pMEG1) shows that the gene encoded on pMEG1 also rescues the glycosylation defect in mutant 163 (Table 3).

Table 3. OT activity of mutant 163 transformed with pMEG1

	Relative OT activity[a]
A364A	100
163	8
163(pMEG1)	92

[a]All cells were grown in SC-leu to assure retention of pMEG1.

Subclones of pMEG1 were generated and tested for complementation of the OT activity defect in mutant 163. pJR4, comprised of a 4.2 kb *EcoR1* partial fragment derived from pMEG1 and cloned into pSEYc68, was found to complement the OT activity defect in mutant 163. Preliminary sequence analysis of pJR4 revealed no sequence homology to any protein already identified, including those implicated as components of OT, i.e., Wbp1p and Swp1p in yeast and OST48, ribophorin I and ribophorin II in mammals. Furthermore, restriction analysis of pMEG1 revealed that the restriction pattern of pMEG1 was distinct from that of the *ALG7*, *WBP1*, or *SWP1* genes (data not shown). Finally, we found that transformation of mutant 163 with pSWP1 (a multicopy suppressor of *WBP1*) did not rescue the temperature sensitivity of mutant 163 (data not shown). All of this evidence is consistent with the hypothesis that *MEG1* encodes for a novel Class 3 protein, either a subunit of OT or a protein that affects the activity of OT. Further studies are focusing on the sequencing of *MEG1* and the precise function of the Meg1p protein, as well as its relationship to the *WBP1* and *SWP1* gene products.

Protein disulfide isomerase. The enzyme protein disulfide isomerase (PDI, encoded by the *PDI1* gene), so named for its catalytic activity observed *in vitro*, i.e. the ability to catalyze the formation, reduction and isomerization of disulfide bonds in a wide range of protein substrates, was discovered over thirty years ago (Goldberger et al., 1963). It is now known that disulfide bond formation occurs in the lumen of the endoplasmic reticulum (see Figure 1), where the redox state is more oxidizing than that of the cytosol (Hwang et al., 1992). Although there is substantial *in vitro* evidence suggesting that PDI functions in disulfide bond formation and rearrangement (Givol et al., 1965; Creighton et al., 1980; Koivu and Myllyla, 1987; Bulleid and Freedman, 1988), this role has not, in fact, been directly demonstrated *in vivo*.

In the present study, we have generated a series of *pdi1* mutant constructs and have tested their ability to support growth in a yeast strain in which *PDI1* has been completely deleted. We have also measured the catalytic activity of mutant PDI proteins *in vitro*. Mutations that altered both of the PDI active sites resulted in a protein that was devoid of catalytic activity *in vitro*. A yeast strain carrying this mutation exhibited a defect in disulfide bond formation, yet surprisingly, no growth defect was observed. We propose that the function of PDI in yeast viability is more closely related to its ability to bind, and perhaps assist in the folding of newly synthesized polypeptides in the lumen of the ER, rather than to its function as a catalyst in disulfide bond interchange reactions. Alternatively, PDI may exist as a subunit of a novel, unidentified protein complex that is essential for yeast viability.

Overall strategy for preparation of altered PDI. Analysis of disruptions created within the yeast *PDI1* gene provided preliminary evidence that limited deletions at the C-terminal end of PDI could be tolerated without affecting cell viability (LaMantia et al., 1991). To determine if functional domains of PDI could be delineated, we have prepared and analyzed 1) a more extended series of C-terminal deletions and 2) active site mutations. Growth phenotypes of strains carrying various *pdi1* alleles were assessed using the plasmid shuffling technique (Sikorski and Boeke, 1991). The chromosomal copy of *PDI1* was deleted from a wild-type yeast strain, and replaced with the *HIS3* nutritional marker. The strain survived the chromosomal deletion of *PDI1* because it expressed the wild-type *PDI1* gene in the plasmid, pML32 (*URA3 CEN4*). This strain, however, was not viable on media containing the counterselective agent 5-fluoroorotic acid (5-FOA), a drug which kills cells expressing the *URA3* gene product. Plasmids containing mutant forms of *PDI1* were transformed into the strain lacking *PDI1*, and the resulting transformants were tested for their ability to replace the plasmid containing wild-type *PDI1* (pML32). Productive "shuffling" was selected by replica plating the transformants onto media containing 5-FOA. Only those replacement plasmids encoding sufficient information to perform the essential function of *PDI1* allowed pML32 to freely segregate, thereby permitting growth on 5-FOA. The mutations analyzed included a series of C-terminal PDI deletions and three active site mutations.

Two features of the mutations, summarized in Figure 3, are noteworthy. First, each alteration destroys the integrity of one of the active sites by mutation of one of the Cys residues. The Cys mutated in each site is the second in the sequence -CGHC-, in contrast to previous studies, which have focused on the first Cys (Vuori et al., 1992). Second, each mutation changes an active site sequence normally present in *PDI1* to the corresponding sequence found in yeast *EUG1*, a multi-copy suppressor of the recessive lethality associated with a *PDI1* chromosomal deletion (Tachibana and Stevens, 1992). Cells that do not express *PDI1*, but instead overexpress *EUG1* are viable, but do not grow as well as wild-type cells.

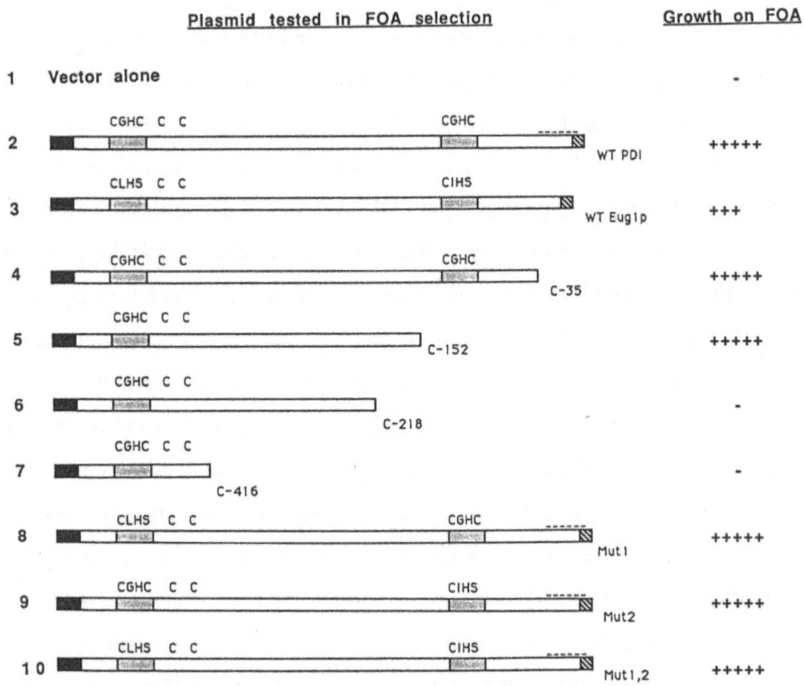

Figure 3. Effect of C-terminal deletions and active site mutations of *PDI1* on yeast viability.

Large C-terminal PDI deletions can be tolerated without affecting viability. It was found by this procedure that normal growth was exhibited by Δ*pdi1* C-35, a mutant in which 35 amino acid residues have been deleted from the C-terminus. This finding suggests that the ER-retention signal (-HDEL) and the highly acidic C-terminal domain present in all known PDI proteins (see Freedman et al., 1989) are not required for yeast viability. Further deletion of 152 C-terminal residues, including the second, C-terminal-most active site, also yielded a strain that exhibited normal growth. However, deletion of a total of 218 C-terminal residues resulted in cells that were not viable. The striking difference in growth characteristics between Δ*pdi1* C-152 and Δ*pdi1* C-218 demonstrates that there is a critical point, presently localized to within 66 amino acids, after which further C-terminal deletion results in a recessive lethal phenotype. We do not yet know whether the lethality associated with the C-218 mutation resulted from removal of an essential functional domain from PDI, or whether it was a consequence of improper folding and/or subsequent degradation of this truncated protein.

C-terminal deletions of PDI are secreted. The results of the deletion analysis above indicate that nearly 1/3 of the C-terminal amino acids of yeast PDI, including the ER-retention signal, can be removed without affecting viability. Therefore, we compared the cellular location of the C-terminal PDI deletions in the viable mutants with that of PDI in the wild-type strain. In an initial analysis, yeast strains were grown to mid-log phase, the cells and growth media were separated by centrifugation, and protein extracts were prepared from each. The samples (cells vs. media) were then subjected to Western blot analysis. The results of this analysis indicated that with both of the viable C-terminal PDI deletions, Δ*pdi1* C-35 and Δ*pdi1* C-152, the mutant PDI was secreted; only low levels of these C-terminally deleted proteins could be detected within the cell. In contrast, wild-type PDI was efficiently retained in the cells (specifically, within the ER lumen, data not shown). These results suggest that yeast PDI, like other lumenal proteins, utilizes a C-terminal four amino acid signal, -HDEL, for its retention in the ER.

The integrity of the PDI active sites is not required for viability. The results of the plasmid shuffle using strains carrying active site mutations were of particular interest. As indicated earlier, mutations altering either the first (Mut1) or second (Mut2) active site did not affect viability, a result which was not entirely surprising given the observation that strains, such as Δ*pdi1* C-152, carrying a complete deletion of the second active site, grew well. However, the FOA-growth pattern of the strain carrying a mutant form of *PDI1* in which both active sites were altered (Mut1,2) was unexpected. This mutant grew as well as wild-type, with no growth defect detectable on plates or in liquid culture. Thus, although *PDI1* encodes a protein necessary for viability in yeast, mutations that destroy both PDI active sites do not abolish the ability of this protein to perform some essential function.

The PDI mutant strain with both active sites altered accumulates the ER-form of vacuolar carboxypeptidase Y. Recently, two laboratories reported that depletion of PDI from yeast cells resulted in a defect in the transport and maturation of vacuolar carboxypeptidase Y (CPY), a disulfide-bonded protein (Günther et al., 1991; Tachibana and Stevens, 1992). Yeast cells depleted of PDI accumulated the ER form of CPY (p1CPY), which normally is only transiently detected in this compartment. Although it was possible that the observed transport defect was an indirect consequence of cell death, this result suggested a potential role for PDI in the transport of CPY out of the ER. That this defect was only partially suppressed by over-expression of Eug1p further suggested a PDI-specific role in CPY transport (Tachibana and Stevens, 1992). However, here again, secondary effects could not be excluded. We therefore examined the behavior of the double active-site mutant (Mut1,2) with respect to CPY transport, as this mutant does not exhibit any detectable growth defect. Wild-type and Mut1,2 cells were metabolically labeled with ^{35}S-methionine and -cysteine for 10 min, and were then chased for intervals of 0 to 60 min with excess unlabeled SO_4^{2-}, cysteine and methionine. Metabolically labeled CPY was then immunoprecipitated from extracts and analyzed by sodium dodecyl sulfate-polyacrylamide gel electrophoresis (SDS-PAGE) followed by autoradiography. Although Mut1,2 did not exhibit a detectable growth defect, an alteration in processing of CPY was observed. The results of this experiment revealed that the ER form of CPY (p1CPY) accumulated within the cell and was only very slowly transported through the secretory pathway and processed to the mature form (mCPY). The fact that there was no observable growth defect in the mutant cells suggested that the CPY transport defect was specifically caused by a defect in the PDI active sites, rather than a pleiotropic phenotype associated with cellular death.

The PDI mutant strain with both active sites altered exhibits a defect in disulfide bond formation. Because there was no direct evidence for PDI's proposed role in the formation of disulfide bonds *in vivo*, we asked if mutation of both of the active sites of PDI affected *in vivo* disulfide bond formation in CPY. Formation of the disulfide bonds in CPY was assessed by SDS-PAGE analysis of immunoprecipitated CPY under reducing and non-reducing conditions. Under non-reducing conditions, proteins that contain disulfide bonds often migrate faster in SDS-PAGE than they do when their cysteines are present as free sulfhydryl groups. Consequently, one can determine the extent of native disulfide bond formation by comparing the migration pattern of newly synthesized, disulfide-containing proteins in SDS-PAGE under reducing and non-reducing conditions (Scheele and Jacoby, 1982). Proteins whose free sulfhydryl groups have not been converted to disulfides upon termination of each pulse/chase time point are, at this point, alkylated with iodoacetamide, so that spontaneous formation of disulfides by air oxidation cannot occur (Gurd, 1972; Pollitt and Zalkin, 1983).

A comparison of the migration pattern of reduced and non-reduced CPY synthesized in the wild-type and Mut1,2 strains revealed that the mobility of the ER form (p1) differed significantly depending upon reduction prior to gel electrophoresis; CPY that had not been reduced migrated faster than reduced CPY. From this observation, it can be inferred that in wild-type cells CPY efficiently formed disulfide bonds in the ER. When synthesis of CPY in Mut1,2 cells was examined, it was found that under reducing conditions it migrated identically with CPY from wild-type cells, suggesting that modifications such as signal peptide cleavage, glycosylation and proteolytic maturation occurred normally. However, under non-reducing conditions, the accumulated ER form of CPY synthesized in Mut1,2 co-migrated with the corresponding reduced form of the protein, suggesting that disulfide bond formation had not occurred. Furthermore, only approximately 50% of the CPY synthesized in Mut1,2 had been subjected to the propeptide cleavage associated with transport to the vacuole after 60 min. Furthermore, in the absence of reductant, the CPY migrated as several broad bands located between the positions representing the mature, fully oxidized protein present in the wild-type strain, and the mature, reduced form. This suggested that little or no CPY synthesized in cells expressing the double active site mutation formed all of its native disulfide bonds.

viability, a result which was not entirely surprising given the observation that strains, such

Purification and measurement of the *in vitro* activity of wild-type and mutant yeast PDI proteins synthesized in E. coli. To obtain large quantities of PDI for enzymatic studies, wild-type *PDI1* and *pdi1* mutations were expressed and the proteins purified from E. coli. Using a BPTI reactivation assay we next asked if: 1) the C-terminal deletions leading to lethality *in vivo* were devoid of PDI catalytic activity *in vitro* and 2) active site mutations affected catalytic activity. In addition, an N-terminal PDI deletion mutant that lacked the first 97 residues of the mature PDI protein was analyzed. A diagram of the deletion mutants and their catalytic activity in reactivating BPTI relative to wild-type recombinant PDI is shown in Figure 4.

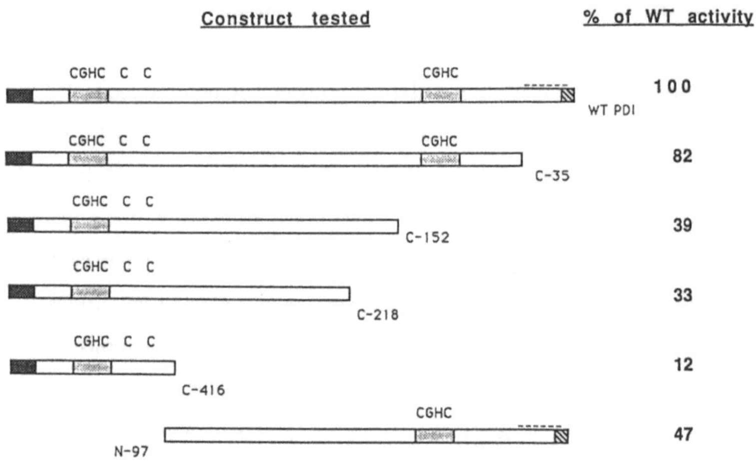

Figure 4. Comparison of the PDI catalytic activities displayed by recombinant wild-type and deletion mutant PDI proteins.

The results indicated that even the most severe deletions, which did not support growth *in vivo*, yielded mutant proteins that contained low, but detectable PDI activity *in vitro*. Of particular interest were the C-terminal deletions, C-152 and C-218. These two deletions exhibited dramatically contrasting growth phenotypes in the plasmid shuffling experiment; C-152, which encompasses residues 16-370 supported growth, whereas C-218, which encompasses residues 16-304, did not. Despite this difference, the PDI activity of these two truncated proteins measured at equimolar concentrations was nearly identical. Perhaps even more surprising was the finding that PDI activity, although low, was detectable in the most extensive deletion, C-416. This mutant protein contained only the first 86 residues of wild-type PDI (including the first active site), and yet it displayed low, but reproducible PDI activity (approximately 12% of wild-type PDI).

We next compared the catalytic activities of the active site mutants with wild-type PDI. The results in Figure 5 indicate that proteins containing mutations within either of the two active sites retained a little more than half (55-60%) of the activity observed in wild-type PDI, suggesting that in yeast PDI, as in human PDI (Vuori et al., 1992), each active site is independently functional, and contributes about one-half of the enzymatic activity observed *in vitro*. The mutant yeast protein with both active sites altered displayed virtually no catalytic activity (<5% of the wild-type PDI activity). This appeared consistent with the results obtained with human PDI, where mutations destroying the first cysteine in both active sites resulted in complete loss of detectable PDI activity (Vuori et al., 1992). Although the assays used to measure PDI activity and the residues mutated within each active site differed in their study and ours, the results from both were consistent with the proposal that the integrity of the -CGHC- active sites is essential for catalytic activity.

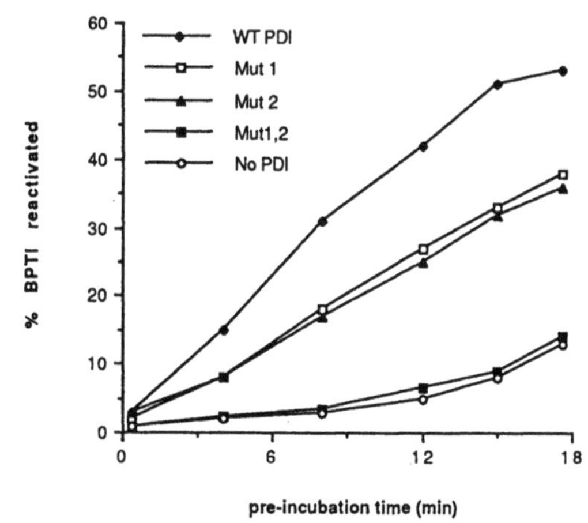

Figure 5. The integrity of the active sites of PDI is required for *in vitro* catalytic activity.

The results of this study lead us to conclude that catalysis of protein disulfide bond formation is not the essential role played by PDI in yeast viability. This conclusion is also consistent with the fact that Eug1p, although sustaining viability when over-expressed in cells devoid of PDI (Tachibana and Stevens, 1992), does not exhibit PDI catalytic activity (M.L.L. and W.J.L., unpublished data). The previous observations that 1) PDI transiently associates with newly synthesized proteins, 2) PDI stably associates with mutant, misfolded proteins, and 3) that the ER form of CPY remains stable in the Mut1,2 strain, but is not stable in cells depleted of PDI, all indicate that PDI may function directly in the protein folding pathway, perhaps acting via a peptide binding domain in a chaperone-like fashion. We speculate that the region in yeast PDI most similar to the mammalian PDI peptide binding domain may be important for this function, especially when this region is compared with Eug1p. Further studies should allow the testing of this idea, as well as the alternative possiblity that the essential role of PDI is as a component of an as-of-yet unidentified multi-protein complex. A precedent for this has been described in mammalian tissue, where PDI exists not only as a homodimer but as a component of prolyl hydroxylase and of the microsomal triglyceride transfer complex.

ACKNOWLEDGEMENT

This work was supported by grants GM33184 and GM33185 from the National Institutes of Health.

REFERENCES

Bulleid, N. J., and Freedman, R. B. (1988). Defective co-translational formation of disulfide bonds in protein disulfide-isomerase-deficient microsomes. Nature 335, 649-651.

Creighton, T. E., Hillson, D. A., and Freedman, R. B. (1980). Catalysis by protein-disulphide isomerase of the unfolding and refolding of proteins with disulphide bonds. J. Mol. Biol. 142, 43-62.

Freedman, R. B., Bulleid, N. J., Hawkins, H. C., and Paver, P. L. (1989). Role of protein disulfide-isomerase in the expression of native proteins. Biochem. Soc. Symp. 55, 167-192.

Geetha-Habib, M., Park, H. and Lennarz, W. J. (1990). *In vivo* N-glycosylation and fate of Asn-X-Ser/Thr tripeptides. J. Biol. Chem. 265, 13655-13660.

Givol, D., DeLorenzo, F., Goldberger, R. F., and Anfinsen, C. B. (1965). Disulfide interchange and the three-dimensional structure of proteins. Proc. Natl. Acad. Sci. USA 53, 676-684.

Goldberger, R. F., Epstein, C. J., and Anfinsen, C. B. (1963). Acceleration of reactivation of reduced bovine pancreatice ribonuclease by a microsomal system from rat liver. J. Biol. Chem. 238, 628-635.

Günther, R., Brauer, C. Janetzky, B., Forster, H.-H., Ehbrecht, E.-M. Lehle, L., and Kuntzel, H. (1991). The Saccharomyces cerevisiae *TRG1* gene is essential for growth and encodes a lumenal endoplasmic reticulum glycoprotein involved in the maturation of vacuolar carboxypeptidase. J. Biol. Chem. 266, 24557-24563.

Gurd, F. R. N. (1972). Carboxymethylation. Methods Enzymol. 25, 424-438.

Hardwick, K. G., Lewis, M. J., Semenza, J., Dean, N., and Pelham, H. R. B. (1990). *ERD1*, a yeast gene required for the retention of luminal endoplasmic reticulum proteins, affects glycoprotein processing in the Golgi apparatus. EMBO J. 9, 623-630.

Hartwell, L. (1967). Macromolecole synthesis in temperature-sensitive mutants of yeast. J. Bacteriol. 93, 1662-1670.

Huffaker, T. and Robbins, P. (1983). Yeast mutants deficient in protein glycosylation. Proc. Natl. Acad. Sci. USA 80, 7466-7470.

Kelleher, D., Kreibich, G. and Gilmore, R. (1992). Oligosaccharyltransferase activity is associated with a protein complex composed of ribophorins I and II and a 48kd protein. Cell 69, 55-65.

Koivu, J., and Myllyla, R. (1987). Interchain disulfide bond formation in types I and II procollagen. J. Biol. Chem. 262, 6159-6164.

LaMantia, M. L., Miura, T., Tachikawa, H. Kaplan, H. A., Lennarz, W. J., and Mizunaga, T. (1991). Glycosylation site binding protein and protein disulfide isomerase are identical and essential for cell viability in yeast. Proc. Natl. Acad. Sci. USA 88, 4453-4457.

Pollitt, S., and Zalkin, H. (1983). Role of primary structure and disulfide bond formation in ß-lactamase secretion. J. Bacteriol. 153, 27-32.

Scheele, G., and Jacoby, R. (1982). Conformational changes associated with proteolytic processing of presecretory proteins allow glutathione-catalyzed formation of native disulfide bonds. J. Biol. Chem. 257, 12277-12282.

Sikorski, R. S., and Boeke, J. (1991). *In vitro* mutagenesis and plasmid shuffling: from cloned gene to mutant yeast. Methods Enzymol. 194, 302-318.

Silberstein, S., Kelleher, D. and Gilmore, R. (1992). The 48-kDa Subunit of the Mammalian Oligosaccharyltransferase Complex Is Homologous to the Essential Yeast Protein WBP1. J. Biol. Chem. 267, 23658-23663.

te Heesen, S., Janetzky, B., Lehle, L. and Aebi, M. (1992). The yeast *WBP1* is essential for oligosaccharyl transferase activity *in vivo* and *in vitro*. EMBO J. 11, 2071-2075.

te Heesen, S., Knauer, R., Lehle, L. and Aebi, M. (1993). Yeast Wbp1p and Swp1p form a protein complex essential for oligosaccharyl transferase activity. EMBO J. 12, 279-284.

Tachibana, C., and Stevens, T. H. (1992). The yeast *EUG1* gene encodes an endoplasmic reticulum protein that is functionally related to protein disulfide isomerase. Mol. Cell Biol. 12, 4601-4611.

Vuori, K., Myllyla, R., Pihlajaniemi, T., and Kivirikko, K. I. (1992). Expression and site-directed mutagenesis of human protein disulfide isomerase in Escherichia coli. J. Biol. Chem. 267, 7211-7214.

Waechter, C. J. and Lennarz, W. J. (1976). The role of polyprenyl-linked sugars in glycoprotein synthesis. Ann. Rev. Biochem. 45, 95-112.

Welply, J., Shenbagamurthi, P., Lennarz, W. J. and Naider, F. (1983). Substrate recognition by oligosacchryl transferase: Studies on glycosylation of modified Asn-X-Thr/Ser tripeptides. J. Biol. Chem. 258, 11856-11863.

Division of the Golgi apparatus during mitosis in animal cells

Graham Warren
Imperial Cancer Research Fund
PO Box 123
44 Lincoln's Inn Fields
London WC2A 3PX

Introduction

In interphase cells, the Golgi probably exists as a single copy, comprising discrete stacks of cisternae linked to each other by tubules that connect equivalent cisternae in adjacent stacks (Rambourg & Clermont, 1990). By immunofluorescence microscopy the Golgi appears as a compact reticulum, located near to the nucleus, most often in the region of the centrioles (Ludford, 1924; Burke et al, 1982).

The division pathway for the Golgi apparatus has been worked out in detail for mitotic HeLa cells (Lucocq et al, 1987; Lucocq and Warren, 1987; Lucocq et al, 1989; Warren, 1993). Figure 1 gives an outline of this pathway. At the onset of mitosis, the Golgi apparatus fragments yielding discrete stacks (Colman et al, 1985) that initially surround the nucleus during prophase. These are presumably generated by the scission of the tubules linking the cisternae (Warren, 1993). During pro-metaphase and metaphase, the stacked cisternae vesiculate yielding several hundred clusters of Golgi vesicles (Lucocq & Warren, 1987). These shed vesicles so that by anaphase the entire cell cytoplasm is filled with up to ten thousand Golgi vesicles. During telophase these processes are reversed. Clusters grow by accretion of Golgi vesicles which then fuse to form several hundred discrete Golgi stacks (Lucocq et al, 1989). These Golgi stacks then congregate in the peri-centriolar region moving there probably along microtubules (Ho et al, 1985) where they fuse to form the interphase Golgi apparatus (Lucocq et al, 1989).

This process is thought in part to allow partitioning of the Golgi apparatus between the two daughter cells (Birky, 1983). Providing each mother cell is divided into two equally sized daughters, random partitioning of ten thousand vesicles will ensure that each daughter will receive half the vesicles with an error of only 1-2%. The

NATO ASI Series, Vol. H 82
Biological Membranes:
Structure, Biogenesis and Dynamics
Edited by Jos A. F. Op den Kamp

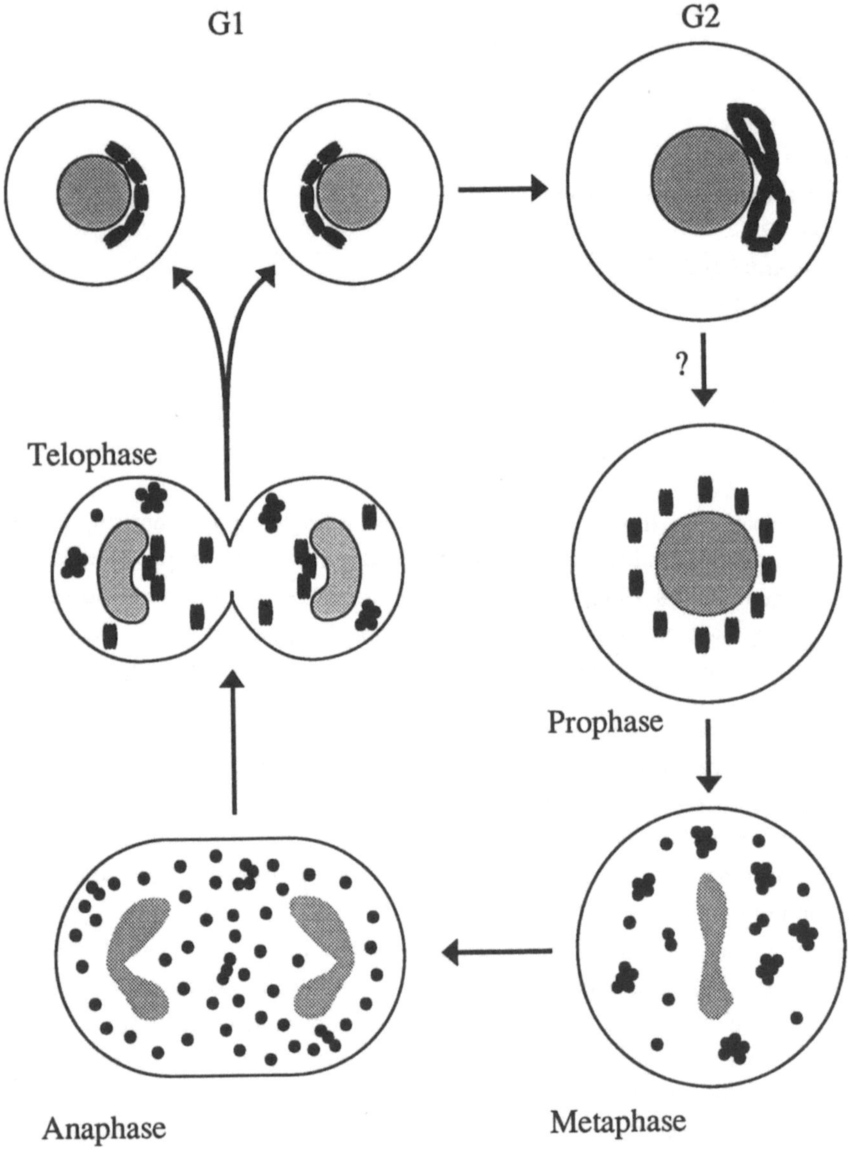

Figure 1. Schematic view of Golgi division during mitosis in HeLa cells.
The juxta-nuclear reticulum fragments during prophase and the resulting stacks vesiculate during pro-metaphase and metaphase forming Golgi clusters. These shed Golgi vesicles which become dispersed throughout the mitotic cell cytoplasm by anaphase. During telophase the clusters re-grow by accretion of vesicles which fuse to form Golgi stacks. These congregate in the peri-centriolar region and fuse to re-form a single copy Golgi in each daughter cell.

problem of partitioning can therefore be reduced to asking how vesiculation of the Golgi stack occurs.

We have put forward a simple hypothesis based on the known inhibition of membrane traffic during mitosis in animal cells (Warren, 1993; 1985; 1989). The model we have proposed makes three assumptions:

1. That transport vesicles continue to bud from the dilated cisternal rims but they are unable to fuse with the next cisterna in the stack.

2. That retention of Golgi enzymes is relaxed so that they can now enter the budding transport vesicles.

3. That cisternae separate from each other so that continued budding can occur.

An outline of this model is presented in figure 2.

Inhibition of Membrane Traffic

When animal cells enter mitosis there is a general inhibition of membrane traffic (Warren, 1993; 1985). With only one apparent exception (Kreiner and Moore, 1990) all vesicle-mediated steps are inhibited (figure 3). The inhibition of regulated secretion during mitosis suggests that the inhibited step is that of membrane fusion (Hesketh et al, 1984). The model for Golgi division demands that intra-Golgi transport is inhibited but this has proven difficult to demonstrate since transport from the endoplasmic reticulum (ER) to the Golgi is also inhibited (Featherstone et al, 1985). This means that proteins cannot be introduced into the early part of the Golgi to see if they are transported to the late part during mitosis.

This problem has recently been resolved by using glycosphingolipids as the markers for transport (Collins & Warren, 1992). The precursor ceramide is synthesised in the ER and is transported to the Golgi probably by a non-vesicular route. In the early part of the Golgi it is converted first to glucosylceramide and then lactosylceramide. Further modifications require vesicle-mediated transport to the later part of the Golgi stack where G_{A2} (GalNAc(β1,4) lactosylceramide) is synthesised. G_{A2} is not synthesised during mitosis showing that intra-Golgi transport is inhibited. Interestingly, during telophase, the resumption of glycosphingolipid transport within the Golgi stack occurs at about the same time as the resumption of protein transport (Souter et al, 1993). This argues that all of the inhibited steps are re-started

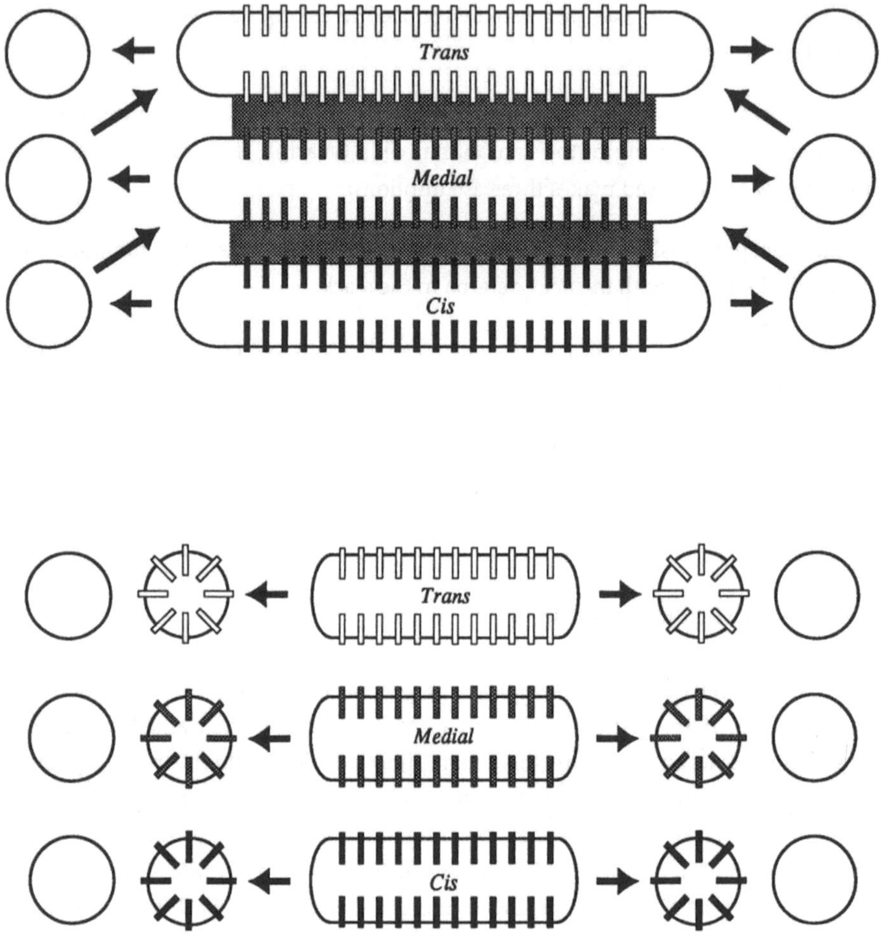

Figure 2. Model for the vesiculation of the Golgi stack.
Top: During interphase, transport vesicles bud from the dilated cisternal rims and fuse with the next cisterna in the stack towards the trans side.
Bottom: During mitosis the budding of transport vesicles continues but they can no longer fuse with the next cisterna in the stack. Relaxation of the process that retains Golgi enzymes in the stack coupled with the unstacking of cisternae ensures that complete vesiculation occurs.

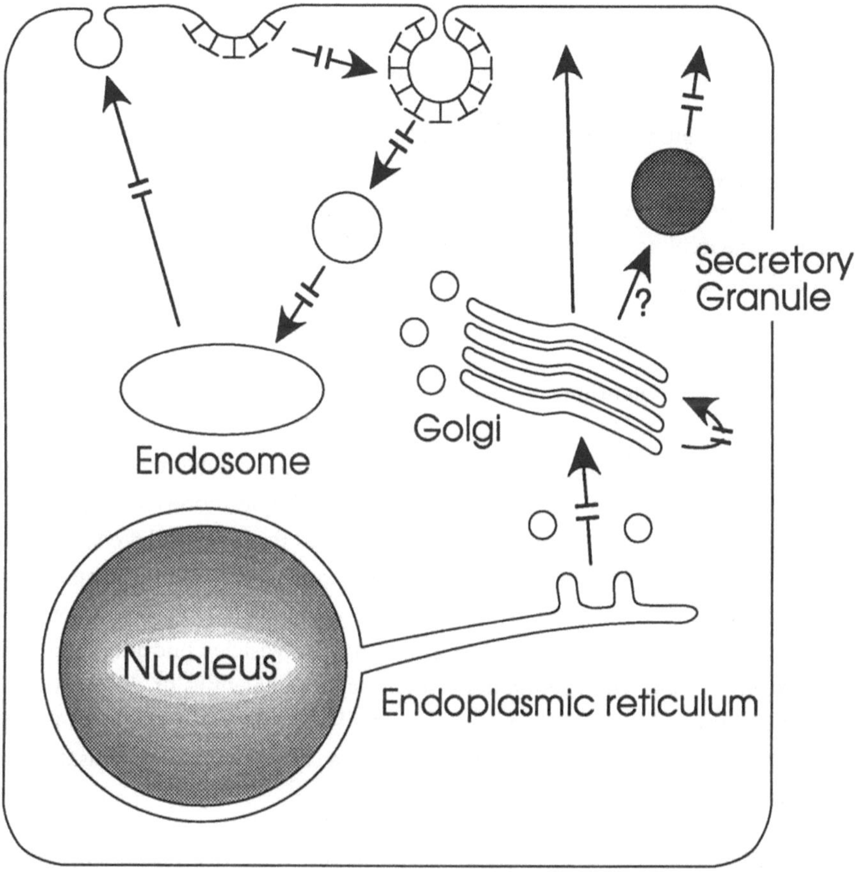

Figure 3. Steps inhibited during mitosis in animal cells.
Note that transport from the TGN to the cell surface is the only step that is not inhibited (13).

at the same time arguing for a common mechanism. It is also interesting to note that the Golgi apparatus reassembles very rapidly during telophase so that, by the time the proteins start arriving from the ER, the stack has already re-formed.

The inhibition of intra-Golgi transport has also been reconstituted in vitro (Stuart et al, 1993) using the cell-free assay described by Rothman and his colleagues. Inhibition is proportional to the activity of the mitotic cdc2 kinase and is fully reversible. The inhibited transport component is located in the cytosol though it has still to be identified.

Relaxation of Golgi retention

The retention signal for Golgi enzymes lies, surprisingly, in the membrane-spanning domain (Machamer, 1991; Nilsson et al, 1993a). A simple model to explain the retention mechanism is outlined in figure 4. It assumes that each Golgi enzyme is a homo-dimer formed by interactions between the lumenal domains. The spanning domains, however, are free to interact with those of their neighbours (or kin) forming long oligomers that are too big to enter the vesicles budding from the dilated cisternal rims. The neighbours can be the same enzymes or different ones and each cisterna will have different sets of hetero-oligomers (Nilsson et al, 1993b; 1993c). Kin recognition would therefore explain the compartmental organisation of the Golgi stack (Dunphy & Rothman, 1985).

The cytoplasmic tails play an accessory role in the retention mechanism (Nilsson et al, 1991). In their absence, some of the protein leaks to the cell surface. This suggests that modification of the cytoplasmic tails could provide a means, during mitosis, of breaking up the oligomers and allowing the enzymes to enter the budding transport vesicles. Phosphorylation is the obvious modification and experiments are underway to test this possibility (T. Nilsson, unpub. results).

Unstacking Golgi cisternae

The components that stack cisternae have proven to be surprisingly elusive. By extracting rat liver Golgi stacks with Triton X-100 followed by salt, we have been able to isolate a matrix that will specifically bind *medial* Golgi enzymes (Slusarewicz et al, 1993). This matrix is in the intercisternal space so by binding to the cytoplasmic tails of enzymes in adjacent cisternae it could explain cisternal stacking. In addition,

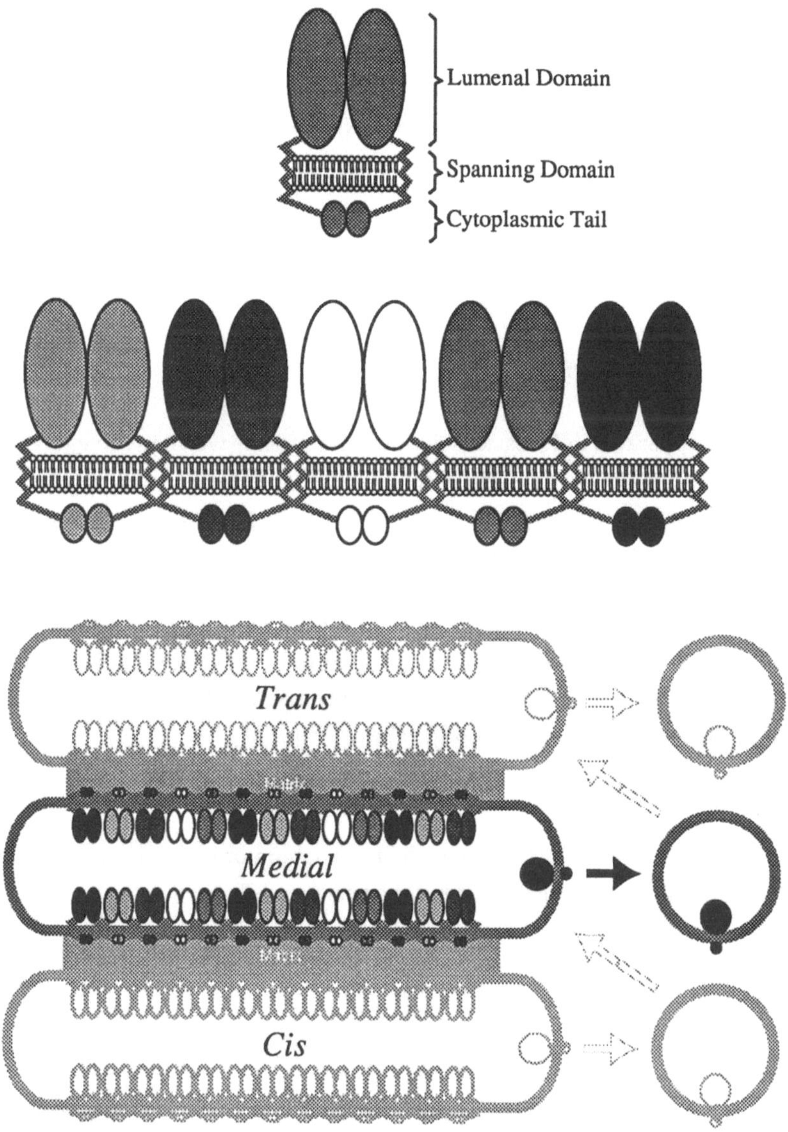

Figure 4. Retention of enzymes in the Golgi stack by kin recognition.
Each Golgi enzyme is assumed to be a homo-dimer (Top) with spanning domains that are free to interact with neighbouring enzymes of the same or different types (Middle). The linear oligomer so formed is anchored to an intercisternal matrix (probably via the cytoplasmic tails) thus preventing the enzymes from entering the budding vesicles (Bottom). Modification of the tails, perhaps by phosphorylation, could relax the retention mechanism by breaking up the oligomers.

it solves a problem associated with the retention of Golgi enzymes. No matter how long the linear hetero-oligomers are, it is difficult to see how they could be completely excluded from budding vesicles. If, however, these oligomers are anchored to a matrix, then full retention is explained.

The matrix contains only a dozen or so major proteins. Once these are cloned and sequenced it should be possible to ask how they are modified during mitosis to allow cisternal unstacking.

Summary and conclusions

The Golgi apparatus in HeLa cells divides by a process which involves breaking the organelle down to its smallest unit, the vesicle. Ten thousand vesicles become randomly dispersed throughout the mitotic cell cytoplasm thereby ensuring almost equal partitioning by a stochastic process. The molecular mechanisms underlying vesiculation are becoming clearer but the one outstanding problem is that of understanding the reassembly process. In a little over ten minutes, ten thousand vesicles associate and fuse to re-form the polarised Golgi stack that characterises the interphase state. An understanding of this process will help explain the rules that govern the assembly of complex organelles.

Acknowledgements

I would like to thank the many people in my laboratory who, over the years, have contributed to the work described in this review.

References

Birky CW (1983) The partitioning of cytoplasmic organelles at cell division. Int Rev Cytology 15: 49-89

Burke B, Griffiths G, Reggio H, Louvard D, Warren G (1982) A monoclonal antibody against a 135-k Golgi membrane protein. EMBO J 1: 1621-1628

Collins R, Warren G (1992) Sphingolipid transport in mitotic HeLa cells. J Biol Chem 267: 24906-24911

Colman A, Jones EA, Heasman J (1985) Meiotic maturation in Xenopus oocytes - a link between the cessation of protein secretion and the polarized disappearance of Golgi. J Cell Biol 101: 313-318

Dunphy WG, Rothman JE (1985) Compartmental Organization of the Golgi stack. Cell 42: 13-21

Featherstone C, Griffiths G, Warren G (1985) Newly synthesized G protein of vesicular stomatitis virus is not transported to the Golgi complex in mitotic cells. J Cell Biol 101: 2036-2046

Hesketh TR, Beaven MA, Rogers J, Burke B, Warren G (1984) Stimulated release of histamine by a rat mast cell line is inhibited during mitosis. J Cell Biol 98: 2250-2254

Ho WC, Allan VJ, vaMeer G, Berger EG, Kreis TE (1989) Reclustering of scattered Golgi elements occurs along microtubules. Eur J Cell Biol 48: 250-263

Kreiner T, Moore H-P (1990) Membrane traffic between secretory compartments is differentially affected during mitosis. Cell Regulation 1: 415-424

Lucocq JM, Warren G (1987) Fragmentation and partitioning of the Golgi apparatus during mitosis in HeLa cells. EMBO J. 6, 3239-3246

Lucocq JM, Berger EG, Warren G (1989) Mitotic Golgi fragments in HeLa Cells and their role in the reassembly pathway. J Cell Biol 109: 463-474

Lucocq JM, Pryde JG, Berger EG, Warren G (1987) A mitotic form of the Golgi apparatus in HeLa cells. J. Cell Biol. 104: 865-874

Ludford RJ (1924) The distribution of the cytoplasmic organs in transplantable tumour cells, with special reference to dictyokinesis. Proc Roy Soc Lond 97: 50-59

Machamer CE (1991) Golgi retention signals: do membranes hold the key? Trends in Cell Biology 1: 141-144

Nilsson T, Lucocq JM, Mackay D, Warren G (1991) The membrane spanning domain of β1,4 galactosyltransferase specifies *trans* Golgi retention. EMBO J 10: 3567-3575

Nilsson T, Slusarewicz P, Hoe, M, Warren G (1993a) Kin Recognition: A Model for the Retention of Golgi Enzymes. Tr Biochem Sci (in press)

Nilsson T, Pypaert M, Hoe MH, Slusarewicz P, Berger EG, Warren G (1993b) Overlapping distribution of two glycosyltransferases in the Golgi apparatus of HeLa cells. J Cell Biol 120: 5-14

Nilsson T, Hoe MH, Slusarewicz P, Rabouille C, Watson R, Hunte F, Watzele G Berger EG, Warren G (1993c) Kin recognition between *medial* Golgi enzymes in HeLa cells (submitted)

Rambourg A, Clermont Y (1990) Three-dimensional electron microscopy: structure of the Golgi apparatus. Eur J Cell Biol 51: 189-200

Slusarewicz P, Nilsson T, Hui N, Watson R, Warren G (1993) Isolation of a intercisternal matrix that binds *medial* Golgi enzymes (submitted)

Souter E, Pypaert M, Warren G (1993) The Golgi stack reassembles during telophase before arrival of proteins transported from the endoplasmic reticulum. J Cell Biol (in press)

Stuart R, Mackay D, Adamczewski J, Warren G, (1993) Inhibition of intra-Golgi transport *in vitro* by mitotic kinase. J Biol Chem. 268: 4050-4054

Warren G (1985) Membrane traffic and organelle division. Tr Biochem Sci 10: 439-443

Warren G (1989) Mitosis and membranes. Nature 342: 857-858

Warren G (1993) Membrane partitioning during cell division. Ann Rev Biochem (in press)

ORGANELLE INHERITANCE IN BUDDING YEAST

Teresa Nicolson and William Wickner
Department of Biological Chemistry
University of California, Los Angeles
School of Medicine
Los Angeles, California 90024
U.S.A.

Although a great deal of effort has been directed towards understanding the mechanism of replication and segregation of chromosomes during cell division, not much is known about organelle segregation. Inheritance of organelles is an important concern to the cell, since *de novo* synthesis does not appear to be a commonly used mechanism. Some important questions are: (1) What regulatory molecules are responsible for initiating the process of organelle inheritance? (2) How is maternally derived material targeted to the daughter cell? (3) If vesiculation and fusion are involved, do these mechanisms use the same types of molecules involved in other cellular fusion events such as protein trafficking in the secretory pathway?

Cytological studies in mammalian cells indicate that ER (endoplasmic reticulum) and Golgi apparatus segregation during mitosis involves complete fragmentation, dispersion, and reassembly of the organelle (Warren, 1989). During cell division, these organelles do not participate in their usual functions of protein trafficking, which involves intracellular budding and fusion of secretory pathway vesicles (Warren *et al.*, 1983). Instead, unrestricted budding of the ER and Golgi apparatus occurs during the onset of mitosis until the organelles are completely fragmented. It is thought that the mechanism of complete vesiculation may involve inhibition of fusion events normally driven by the secretory pathway machinery (Warren, 1985). As a result of the inhibition of fusion, vesicles derived from the ER and Golgi apparatus accumulate and can then disperse throughout the cytoplasm. Although there is evidence for inhibition of fusion of endocytic vesicles using mitotic cell extracts (Tuomikoski *et al.*, 1989), the above hypothesis remains largely untested.

Unlike mammalian cells, in budding yeast the ER and Golgi apparatus do not breakdown completely. The ER, which is mainly localized next to the plasma membrane, is partitioned into the bud soon after the bud has emerged (Preuss *et al.*, 1991). This is also the case with the Golgi apparatus, which is dispersed throughout the cytoplasm, numbering from one to ten stacks or bodies (Redding *et al.*, 1991). Inheritance of mitochondria in yeast also occurs early in the

NATO ASI Series, Vol. H 82
Biological Membranes:
Structure, Biogenesis and Dynamics
Edited by Jos A. F. Op den Kamp

cell cycle; mitochondria enter the bud as soon as it forms. Yaffe and coworkers have isolated recessive mutants defective in mitochondrial inheritance suggesting that segregation is an active process (Yaffe 1992). Transfer of mitochondria into the bud requires at least two genes, *MDM1* and *OLE1*, which encode a novel 51 kd cytoskeletal-like protein and fatty acid desaturase, respectively (Yaffe 1992). Another mutant, *mdm-11*, has a reversible temperature-sensitive phenotype. In the *mdm-11* strain, buds generated without mitochondria at the restrictive temperature inherit mitochondria from the mother cell when the strain is shifted back to the permissive temperature (Yaffe 1992).

In yeast, the counterpart to mammalian lysosomes is the vacuole. The yeast vacuole contains many degradative proteases, and in addition, participates in other cellular functions such as osmoregulation and ion homeostasis. The vacuole is usually the largest organelle in the yeast cell and may exist either as a single compartment or several compartments, depending on the strain (Weisman *et al.*, 1987). *In vivo* studies using specific dyes have shown that vacuoles undergo partial vesiculation or tubule formation and these structures are directed into the developing bud during early S phase (Weisman and Wickner, 1988). The amount of maternal material delivered via tubular/vesicular structures, called segregation structures, increases throughout S and G2 phase until the nucleus migrates into the bud neck (Gomez de Mesquita et al., 1991, Weisman et al., 1990). Two recessive mutants, *vac-1* and *vac-2*, are defective in partitioning material from the mother cell vacuole into the bud. The *vac-1* mutant mislocalizes vacuolar proteins to the cell surface. Genetic complementation demonstrates that the wild-type *VAC1* gene is identical to *VPS19*, a gene that is required for Golgi-to-vacuole protein transport. *vps 19*, along with seven other vacuolar protein sorting mutants, forms a morphologically related class of mutants which are defective in vacuole inheeritance. The *VAC1* gene encodes a 55 kd protein with zinc finger motifs (Weisman and Wickner, 1992). Unlike *vac-1*, *vac-2* is not a vps mutant and does not detectably mislocalize vacuolar proteins (Shaw and Wickner, 1991).

Recently, progress has been made in the biochemistry of vacuole segregation. *In vivo* studies reveal that vesicles derived from the mother cell vacuole fuse after they arrive in the bud. An *in vitro* assay has been developed which reflects this later vacuole fusion step in vacuole segregation. Vacuole-to-vacuole fusion requires ATP and cytosol, and is time and temperature dependent (Conradt *et al.*, 1992). Fusion may be assayed directly by microscope or assayed indirectly by fusion-dependent enzyme maturation. This latter assay involves mixing two types of vacuoles, one containing pro-enzyme (lacking proteinase A) and another type of vacuole containing proteinase A (lacking enzyme). If the vacuoles fuse and vacuolar contents mix, then proteinase A can process the pro-enzyme, resulting in a shift in molecular weight which can be detected by Western analysis, or enzyme activity which can be measured colorimetrically (Conradt *et al.*, 1992, Haas and Wickner, unpublished results).

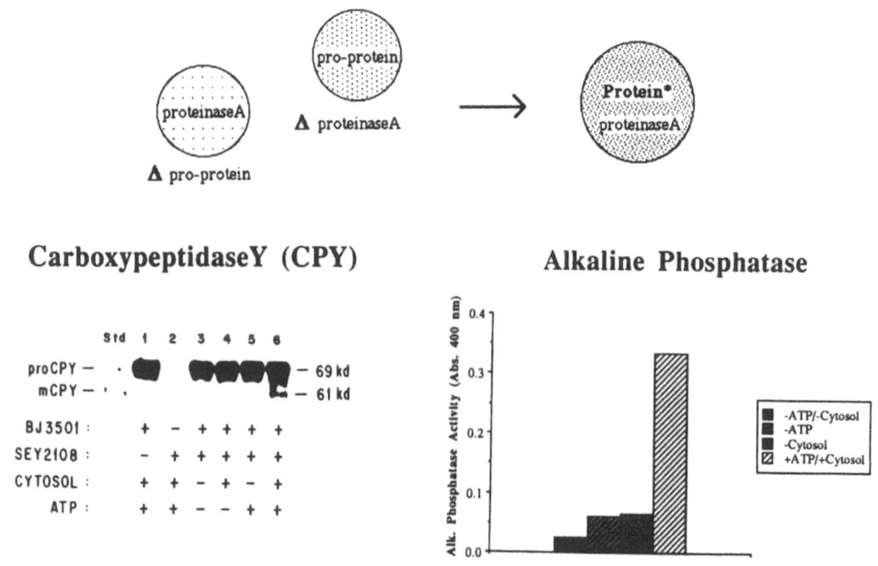

Figure 1. Vacuolar fusion can be measured by enzyme maturation assays. (Reprinted with permission from Barbara Conradt.)

By combining both genetic and biochemical approaches, we have begun to dissect the complex process of vacuole inheritance.

Mutants defective in vacuole inheritance.

The *vac-1* mutant was isolated from a screen of spontaneous mutants which secreted a soluble vacuolar enzyme (Weisman et al., 1990). The rationale for using this type of screen was that mutants which lack a vacuole in the bud would mislocalize proteins destined for that organelle. The problem with this approach is that it is unclear if the protein is directly involved in the inheritance process, or acts indirectly by causing the mislocalization of other factors which are necessary for inheritance. For this reason, a second screen of a temperature sensitive collection was initiated in which *vac-2* was isolated (Shaw and Wickner, 1991). This mutant does not mislocalize soluble vacuolar enzymes and thus, appears to be directly involved in vacuole inheritance.

The phenotype of these two mutants was determined by labeling fast growing cells with vacuole-specific fluorophores, such as FITC or CDC-FDA. An alternative method for labeling involves accumulating an endogenous fluorescent compound caused by an *ade-2* mutation in the strain background. Cells are grown into stationary phase in which they arrest in an unbudded state. When nutrients becomes limiting under these conditions, the cells attempt to synthesize adenine, but due to the *ade-2* block in the biosynthetic pathway, the vacuole accumulates an adenine precursor which becomes fluorescent. If cells are shifted into fresh media with adenine, they stop synthesizing the fluorescent compound and one can monitor the transfer of maternally derived fluorescent material into the bud.

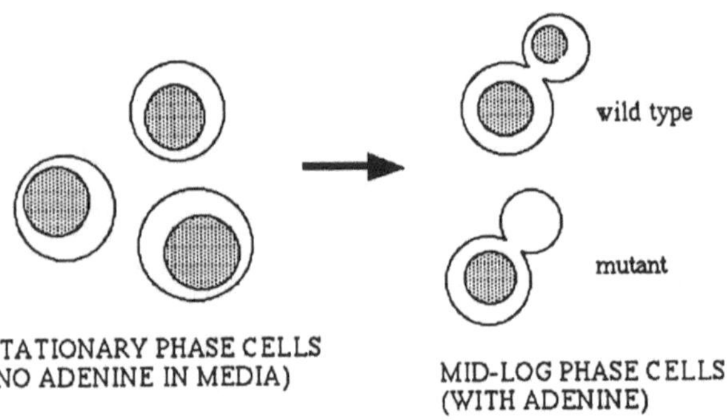

Figure 2. Inheritance of the *ade-2* fluorophore.

Such experiments were performed with *vac-1* and *vac-2*, and both proved to be defective in transfer of fluorescent material (Weisman et al., 1990, Shaw and Wickner, 1991). In *vac-2* cells, 73% of the buds did not inherit the maternal *ade-2* fluorophore, and in *vac-1* cells, the defect in inheritance was even more severe.

Semi-intact cell assay with *vac-1* and *vac-2*.

The inheritance process appears to involve several stages. Initiation of the process begins sometime during early S phase, as soon as the bud emerges (Gomez de Mesquita *et al.*, 1991). The mother cell vacuole forms a segregation structure by pinching off a string of connected vesicles or by creating a tubular structure. The differences in the types of structures appear to be strain dependent, and tubular formations may vary widely, ranging from short, thick structures to long, thin structures (unpublished observations). To better understand the role of *vac-1* and *vac-2* in segregation structure formation, Conradt *et al.* developed a semi-intact cell assay in which ATP and cytosol are required for structure formation. Semi- intact *vac-1* and *vac-2* cells can form structures in the presence of wild type cytosol and ATP, however, in the presence of their own respective cytosols, structure formation is greatly reduced.

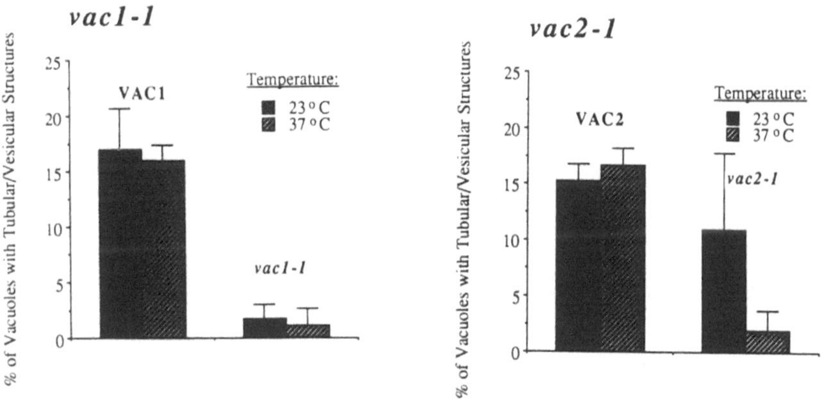

Figure 3. Segregation structure formation is defective in semi-intact cells of *vac-1* and *vac-2*. (Reprinted from Conradt *et al.*, 1992.)

Later steps in the inheritance process include targeting or delivery of maternally derived material via the segregation structures, then assembly or fusion of this material in the bud as it matures. Isolated vacuoles from *vac-1* and *vac-2* are also capable of fusion, when supplied with wild type cytosol. However, in the presence of their own respective cytosols, *vac-1* and *vac-2* vacuoles completely fragment.(Conradt *et al.*, 1992). Thus, the *in vitro* results suggest that the problem or defect of these mutants lies within the vacuole, not the cytosol.

Inhibition of *in vitro* vacuole fusion.

Vacuole-to-vacuole membrane fusion requires cytosolic components and an ATP regenerating system. In order to gain some insight into which types of molecules are involved, various inhibitors have been tested for their effects on fusion. First, trypsin was used to pre-treat both cytosol and vacuoles to determine if proteinaceous factors are required for fusion. As shown in figure 4., both membrane bound and soluble proteins are necessary for fusion.

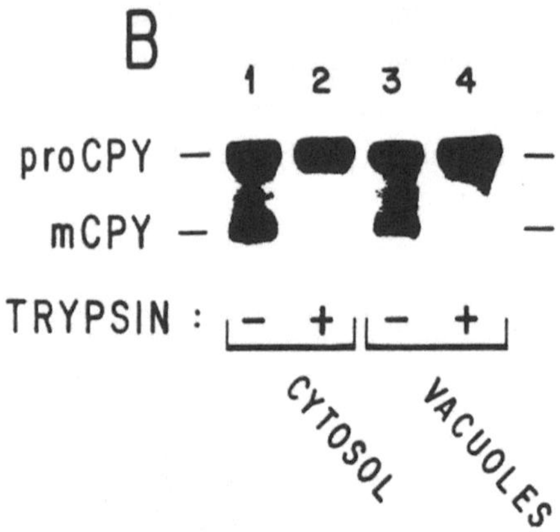

Figure 4. Sensitivity of vacuoles and cytosol to trypsin. (Reprinted from Conradt *et al.*, 1992.)

Subsequently, more specific inhibitiors were tested in the fusion assay. Okadaic acid and microcystin, potent inhibitors of phosphatases, had dramatic effects on fusion (Conradt *et al.*, 1992). Half-maximal inhibition of CPY maturation was achieved at submicromolar concentrations. These results suggest that phosphorylation/dephosphorylation events are important for fusion. Other inhibitors which cause a dramatic decrease in fusion include mastoparan and GTP-gamma-S (Haas *et al.*, unpublished results). These molecules interact with G proteins using different mechanisms, however, the net result is the same--G proteins are kept in their activated states. How continual activation prevents fusion is unclear, however, it is not suprising that G proteins are involved in vacuole membrane fusion, since many other membrane fusion events in the cell, such as those which are required for protein trafficking, involve G proteins.

Obviously, we are just beginning to learn how organelle inheritance works. Future efforts will directed towards isolating more of the factors involved in vacuole inheritance. Fractionation of cytosol and vacuole membranes should yield proteins which are required in all stages of inheritance: (1) initiation of segregation structure formation (2) delivery of material to the bud via segregation structures (3) coalescence or fusion of maternally derived material inside the bud. Staging the *in vitro* assay by dividing it into discrete steps with distinct and/or overlapping requirements will greatly aid the isolation and purification of components involved in vacuole inheritance. In addition, complementing the biochemical studies with genetic studies of more mutants defective in vacuole inheritance will also be important for gaining insight into the molecular basis of this complex process.

REFERENCES

Conradt, B., Shaw, J., Vida, T., Emr, S., and Wickner, W. (1992) *In vitro* reactions of vacuole inheritance in *Saccharomyces cerevisiae*. JCB 119:1469.

Gomez de Mesquita, D., Hoopen, R., and Woldringh, C. (1991) Vacuolar segregation to the bud of *Sacchromyces cerevisae* : an analysis of morphology and timing in the cell cycle. Journal of General Microbiology 137:2454.

Preuss, D., Mulholland, C., Kaiser, C., Orlean, P., Albright, C., Rose, M., Robbins, P., and Botstein, D. (1991) Structure of the yeast endoplasmic reticulum: localization of ER proteins using immunofluorescence and immunoelectron microscopy. Yeast 7:891.

Redding, K., Holcomb, C., and Fuller, R. (1991) Immunolocalization of Kex2 protease identifies a putative late golgi compartment in the yeast *Saccharomyces cerevisiae*. JCB 113:527.

Shaw, J., and Wickner, W. (1991) vac-2: a yeast mutant which distinguishes vacuole segregation from golgi-to-vacuole protein targeting. EMBO 10:1741.

Tuomikoski, T., Felix, M., Doree, M., and Gruenberg, J. (1989) Inhibition of endocytic vesicle fusion *in vitro* by the cell-cycle control protein kinase cdc2. Nature 342:942.

Warren, G. (1989) Mitosis and membranes. Nature 342:857.

Warren, G. (1985) Membrane traffic and organelle division. TIBS 10:439.

Warren, G., Featherstone, C., Griffiths, G., and Burke, B. (1983) Newly synthesized G protein of vesicular stomatitis virus is not transported to the cell surface during mitosis. JCB 97:1623.

Weisman, L., and Wickner, W. (1992) Molecular characterization of VAC1, a gene required for vacuole inheritance and vacuole protein sorting. JBC 267:618.

Weisman,L., Emr,S., and Wickner, W. (1990) Mutants of Saccharomyces cerevisiae that block intervacuole vesicular traffic and vacuole division and segregation. PNAS 87:1076.

Weisman, L., and Wickner, W. (1988) Intervacuole exchange in the yeast zygote: a new pathway in organelle communication. Science 241:589.

Weisman, L., Bacallao, R., and Wickner, W. (1987) Multiple methods of visualizing the yeast vacuole permit evaluatin fo its morphology and inheritance during the cell cycle. JCB 105:1539.

Yaffe, M. (1992) Organelle inheritance in the yeast cell cycle. TICB, in press.

FUSION ACTIVITY OF INFLUENZA VIRUS TOWARDS TARGET MEMBRANES: pH REQUIREMENTS AND EFFECT OF DEHYDRATING AGENTS

João Ramalho-Santos[1] & Maria da Conceição Pedroso de Lima[2]

Center for Cell Biology and Center for Neurosciences of Coimbra
University of Coimbra
3049 Coimbra Codex
Portugal

INTRODUCTION

Lipid enveloped viruses infect their target cells by a membrane fusion event often mediated by one or two virally encoded glycoproteins. The apparent simplicity of the mechanism involved makes these viruses important tools in the study of membrane fusion.

Among lipid enveloped viruses the fusion activity of influenza virus is one of the best characterized. This virus penetrates its target cells through receptor-mediated endocytosis, and fusion takes place between the viral envelope and the membrane of an intracellular acidic compartment, the endosome (for extensive reviews see Wiley & Skehel, 1987; Hoekstra & Kok, 1989; Stegmann et al., 1989; Hoekstra, 1990; White, 1990). Fusion activity of influenza virus is triggered by a well defined low pH-induced conformational change in the viral fusion protein. This protein, the hemagglutinin (HA[3]), is functionally organized in trimers, and each monomer is composed of two subunits linked via disulfide bonds (Wiley & Skehel, 1987; Hoekstra & Kok, 1989). The low-pH-induced conformational change involves the exposure of the N-terminal hydrophobic peptide of the HA2 subunit and the unfolding of the HA trimers, both of which have been implicated in the membrane fusion activity of influenza virus (Skehel et al, 1982; Doms & Helenius, 1986; White & Wilson, 1987).

Although these aspects are relatively well known, the overall molecular mechanism responsible for membrane fusion remains obscure. It should be noted that most of the data regarding the conformational change in the HA protein have been obtained with the purified water-soluble HA fragment, generated via bromelain treatment of intact influenza virus HA (Doms et al., 1985;

[1]Department of Zoology, University of Coimbra.
[2]Department of Biochemistry, University of Coimbra.

[3]Abbreviations: ANS: 1-aminonaphtalene-8-sulfonate; a.u.: arbitrary units; $C_{12}E_8$: octaethylene-glycol dodecyl ether; PC: phosphatidylcholine; PS: phosphatidylserine.

NATO ASI Series, Vol. H 82
Biological Membranes:
Structure, Biogenesis and Dynamics
Edited by Jos A. F. Op den Kamp

Ruigrok et al., 1986; Wharton et al., 1986; White & Wilson, 1987). It is not known whether the data thus obtained reflect the actual behaviour of the protein "in vivo". Indeed, fusion is not induced by the water-soluble HA fragment (Doms & Helenius, 1988). It should also be noted that the fusion activity of influenza virus may vary towards different types of target membranes, suggesting that the fusion mechanism does not depend solely on viral properties.

In this paper we describe work regarding the fusion activity of influenza virus (A/PR/8/34 strain) towards cultured PC-12 cells and model membranes. Fusion was monitored by the octadecylrhodamine (R18) assay (Hoekstra et al., 1984). Accordingly, the fluorescent probe R18 was introduced in the viral envelope at a selfquenching concentration. Fusion of the virus with a target membrane results in probe dilution, with concomitant relief of R18 selfquenching. The fluorescence scale was calibrated such that the initial fluorescence of R18 labeled virus and cell suspension was set at 0% fluorescence. The value obtained by lysing the virus and cellular membranes after each experiment with detergent (in this case $C_{12}E_8$) was set as 100% fluorescence (maximum probe dilution). Control experiments were employed to rule out nonspecific probe transfer, that would eventually mimic fusion. These experiments consisted in making sure that no fluorescence dequenching in the presence of target membranes took place when the virus was rendered inactive, by low pH and enzymatic or glutharaldehyde pretreatment.

Since fusion activity of influenza virus involves exposure of hydrophobic domains (the N-terminal peptide of the HA2 subunit) following the HA conformational change, influenza virus-dependent changes in ANS fluorescence were also studied. ANS is a fluorescent probe sensitive to changes in the polarity of solutions (Stryer, 1968). A decrease in polarity results in an increase in the quantum yield of ANS fluorescence and a shift of the emission maximum towards the blue. This property of ANS has been used to study conformational changes in proteins that involve exposure of hydrophobic segments to the external medium (Stryer, 1968; Yoshimura et al., 1987). For example, the exposure of hydrophobic domains, followed by changes in ANS fluorescence, has been correlated with low-pH fusion activity of the cytosolic protein clathrin (Yoshimura et al., 1987). The importance of hydrophobic interactions in the fusion activity of influenza virus stressed by the results obtained (see below), prompted us to study the effect of dehydrating agents on the process, using liposomes as target membranes and the R18 assay to follow membrane fusion.

RESULTS AND DISCUSSION

Two different experimental approaches were used to determine the effect of pH on influenza virus fusion activity towards PC-12 cells at 37°C . When influenza virus was added directly to a cell suspension already adjusted to the desired pH, the fusion activity (as monitored by

R18 dequenching) was optimal at pH 5.2 (Fig. 1A), virus-cell fusion being lower at pH values below 5.2. A short lag phase, prior to measurable dequenching, was detected in all the experiments. This delay could be attributed to either virus-cell binding and/or conformational changes required to activate the virions at low pH (see below). When parallel experiments were carried out with virus already bound to the cells at neutral pH, the results were markedly different (Fig. 1B). Here, the initial rate and extent of fusion were higher than in experiments without prebinding. Although the existence of prebinding could account for increased initial kinetics and extent of fusion observed for all pH values, it was interesting to note that, in this case, acidification below pH 5.2 did not result in any decrease in fusion activity. In fact, at pH 4.5, the extent of dequenching after 1 min was significantly larger than at pH 5.2, whereas after 5 min the extents were similar. These results may be explained by the kinetic analysis, which invokes an increase in the fusion and inactivation rate constants with decreasing pH (see Ramalho-Santos, 1993). The results also suggest that the lag phase observed without prebinding is mostly due to the time it takes for the virus to bind to the cells, since it could not be detected when prebinding took place.

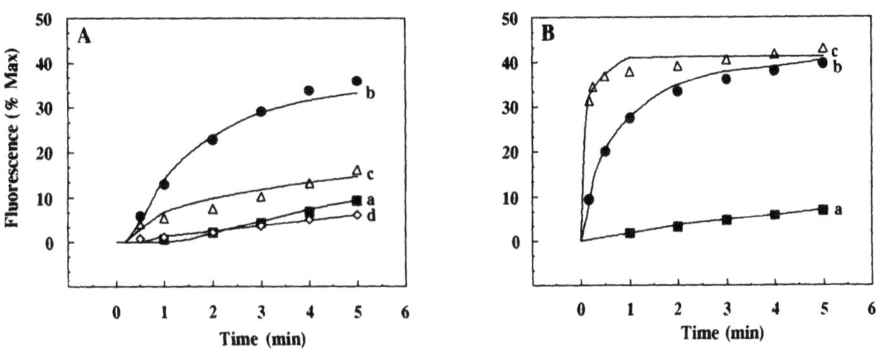

FIGURE 1- pH effect on influenza virus fusion activity towards PC-12 cells in the absence and presence of viral prebinding. Influenza virus (1 µg/ml viral protein) was added to 5 x 10^6 PC-12 cells in a final volume of 2 ml and R18 dequenching was monitored for 5 min at 37°C. **A**- The pH of the cell suspension was adjusted previously to pH 5.8 (**a**), 5.2 (**b**), 4.5 (**c**) and 4.0 (**d**). **B**- After 5 min at pH 7.4 the pH was lowered to pH 5.8 (**a**), 5.2 (**b**) and 4.5 (**c**), and R18 dequenching monitored for 5 min. When the pH was adjusted to 4.0 the dequenching was only slightly quicker and more extensive than the one obtained at pH 4.5 (not shown). The values calculated with a mass action kinetic model (see Ramalho-Santos et al., 1993) are also presented for pH 5.8 (■), 5.2 (●), 4.5 (△) and 4.0 (◇).

It is well known that when influenza virus is preincubated at low pH and 37°C in the absence of target membranes it looses most of its fusion activity (Sato et al., 1983; Stegmann et al., 1986). Since, under these preincubation conditions, the hydrophobic peptides exposed following the

HA low-pH-induced conformational change cannot interact with a target membrane, it is believed that the fusion proteins aggregate, forming non-functional clusters (Junankar & Cherry, 1986). This inactivation process, although generally extensive, is dependent on preincubation time, temperature and pH (as is also the case of the fusion process itself). The results presented above indicate that, in the absence of virus-cell prebinding (Fig. 1A), some inactivation takes place before the virus is able to bind to the cells, and this inactivation is more pronounced the lower the experimental pH. Virus-cell prebinding at neutral pH (Fig 1B) apparently protects the virus from inactivation.

It should be noted that the extent of influenza virus inactivation may vary with different types of target membranes, suggesting, as in the case of fusion, that the inactivation mechanism does not depend solely on viral properties (Düzgünes et al., 1992; Hoekstra & Lima, 1992). On the other hand, some data point to the fact that even virus bound to a target membrane may be subject to inactivation (Nir et al., 1990). It should also be noted that, in contrast to the general assumption, viral inactivation has been suggested to be somewhat reversible (Düzgünes et al., 1992).

As depicted in the Introduction we have also studied the effect of influenza virus on the fluorescence of ANS, a probe used to detect hydrophobic microenvironments in proteins (Stryer, 1968). Results displayed in Fig. 2 show that lowering the pH of the medium from 7.4 to 5.0 in the presence of influenza virus results in a drastic increase in ANS fluorescence and a simultaneous blue shift in the emission maximum of this fluorescent probe. Apparently this reflects a decrease in polarity caused by the exposure of the hydrophobic fusion peptide of the viral HA. When the same experiment is repeated with inactivated influenza virus (virus that has been preincubated at 37°C and pH 5.0 for 1 h) the changes in ANS fluorescence upon lowering the pH are much less dramatic. This was to be expected, since inactivated influenza virus has already undergone the low-pH conformational change, during the inactivation process. In this case the hydrophobic segments, exposed in the absence of target membranes, are thought to aggregate, rendering the virus unable to fuse (Junankar & Cherry, 1986).

However, a decrease in pH still results in some change of ANS fluorescence (increase in quantum yield, blue shift in the emission maximum). These changes indicate an increase in the hydrophobicity of the viral envelope and are probably due to protonation of negatively charged amino acids on the viral surface proteins, or even to some conformational change unrelated to viral fusion activity. This was confirmed by the observation that Sendai virus was shown to exhibit the same behaviour as inactivated influenza virus (Fig. 2). Sendai virus is believed to enter target cells by directly fusing with the plasma membrane at pH 7.4 (Hoekstra & Kok, 1989; Lima et al., 1992). The hydrophobic regions in the viral fusion protein, that are thought to be responsible for fusion activity, are already exposed at neutral pH, and therefore this virus apparently does not require any specific low-pH-triggered conformational change to be rendered active, as is the case with influenza virus (Hoekstra & Kok, 1989). Since Sendai virus and inactivated influenza virus have the same effect on ANS fluorescence when the pH is lowered to 5.0, the recorded changes are most likely not related to any biologically relevant process.

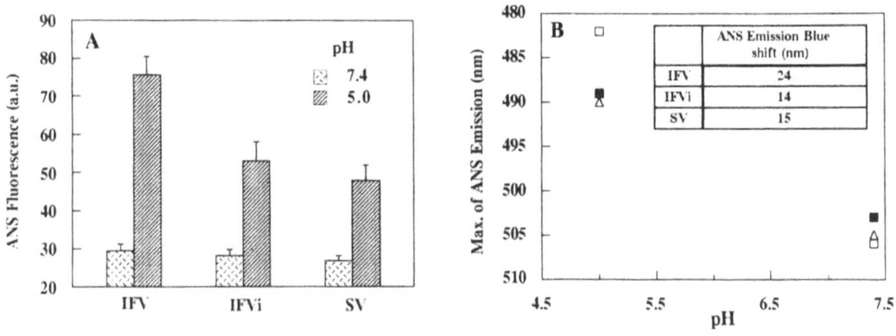

FIGURE 2- Effect of pH on lipid enveloped virus-mediated changes in ANS fluorescence. Influenza virus (IFV), inactivated influenza virus (IFVi- virus preincubated for 1 h at 37 °C and pH 5.0) and Sendai virus (SV) were added to buffer containing 23 μM ANS at pH 7.4 or 5.0 and at 37°C. The virus concentration was 15 μg/ml of viral protein. After 30 min fluorescence emission scans (450-560 nm) were taken with excitation set at 350 nm. **A**- Maximum ANS fluorescence (arbitrary units). The values presented are average ± standard deviation of at least three experiments. **B**- Emission maximum of ANS fluorescence (nm). □- IFV, ■- IFVi, Δ-SV.The calculated blue shift that took place in the emission maximum of ANS is also shown (insert).

The results obtained with ANS suggest that hydrophobic interactions play an important role in the infection of cells by influenza virus. Therefore we have studied the effect of dehydrating agents on the viral fusion activity. In this case we have used liposomes (large unillamelar vesicles) composed of PS and PS/PC 1:4 (molar ratio) as target membranes. Fusion was monitored by the R18 assay described above and, as reported previously, was more extensive with PS than with PS/PC liposomes. It is interesting to note that all the dehydrating agents tested (dimethylsulfoxide- DMSO; dimethylsulfone- DMSO2; poly(ethylene glycol) of 1500 and 3000 average molecular weight- PEG 1500 and PEG 3000) affected the initial rate of virus liposome fusion (calculated in the first 60 s following the onset of fusion), but had no effect on the extent of the process (monitored after 5 min of fusion).

When using DMSO/DMSO2 (Fig 3A), the initial rate of the fusion process increased linearly with the concentration of dehydrating agent, the enhancing effect being much greater in the case of DMSO2. This suggests a greater ability of DMSO2 as compared with DMSO to interact with water molecules through hydrogen bond formation. In turn, this effect may be explained by i) the number of hydrogen bound water molecules is double in the case of DMSO2, ii) the sulfur atom in DMSO2 has a geometric arrangement (approximately tetrahedral) more favourable for hydrogen bond formation, and iii) the SO bonds are more polarized in DMSO2 than in DMSO.

PEG is known to bind water molecules via hydrogen bond interactions, ca. 3 water molecules being bound per monomer, and has been shown to promote fusion in many systems involving liposomes (Yamazaki & Ito, 1990; Burgess et al., 1992) and enveloped viruses (Nir at al., 1986; Hoekstra et al., 1989). We have found (Fig. 3B) that only PEG 3000 is able to stimulate influenza virus fusion activity at low polymer concentrations, probably due to its higher dehydrating ability as compared with PEG 1500 (Yamazaki & Ito, 1990). At high polymer concentrations, both PEGs slow down the fusion process, possibly by acting as molecular spacers between the viral envelope and liposome bilayer. This effect is not seen with DMSO and DMSO2, probably because these are much smaller molecules.

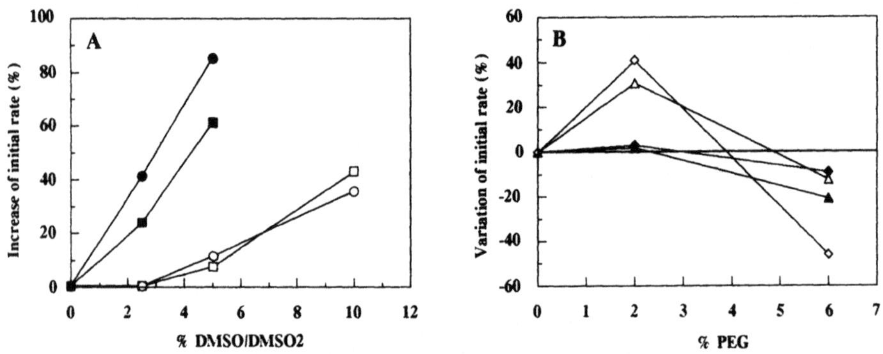

FIGURE 3- Effect of DMSO, DMSO2 and PEG on the initial rate of influenza virus fusion activity towards liposomes. Influenza virus (1 µg of viral protein) was added to target liposomes (100 nmoles of lipid) in a final volume of 2 ml and R18 dequenching monitored for 5 min at pH 5.0 and 37°C (Excitation- 560 nm; Emission- 590 nm). The initial rate was calculated in the first 10-20 s of the fusion process. Experiments were carried out either with PS (□,■) or PS/PC (1:4) (○,●) liposomes in the presence of either **A**-DMSO (open symbols) and DMSO2 (closed symbols) or **B**- PEG 3000 (open symbols) and PEG 1500 (closed symbols).

It is interesting to note that a greater enhancement of influenza virus fusion activity is obtained with both DMSO2 and PEG 3000 when PS/PC, rather than PS liposomes, are used as target membranes. This may be due to the fact that PC is more heavily hydrated than PS (Nir at al., 1986), and therefore the lowering of the free water content from a PC-containing bilayer may reflect the removal of a more important fusion barrier as compared with the less hydrated pure PS membrane.

In the course of these studies, we have conducted control experiments to rule out nonspecific probe transfer. We were able to completely abolish dequenching when the virus was inactivated (preincubated for 1 h at pH 5.0 and 37°C) in the presence of glutharaldehyde (5% v/v). However, we have found that virus that has only been low-pH inactivated (in the absence of glutharaldehyde), although completely unable to fuse with PS/PC liposomes, is still able to fuse with PS liposomes. In this case low concentrations of dehydrating agents (2.5% DMSO2, 5% DMSO), which increase the initial rate of fusion of intact virions (Fig. 3A), have no effect on the fusion activity of low-pH pretreated influenza virus (Table I). Dehydrating agents only stimulate the fusion of inactivated virus (in this case enhancing both initial rate and fusion extent) at high concentrations (Table I). These results suggest that the increase of influenza virus fusion activity at low concentrations of DMSO and DMSO2 reflects an enhancement in a biologically relevant process.

TABLE I: Effect of DMSO and DMSO2 on the fusion activity of low-pH preincubated influenza virus towards PS liposomes.

Dehydrating agent	initial rate (% max. fluor. increase/min)	fusion extent (% max fluor. increase)
None	4.3	18.4
DMSO2 (2.5 %)	3.7	17.4
DMSO (5 %)	4.3	17.6
DMSO2 (5 %)	13.0	29.0
DMSO (10 %)	9.0	26.4

Influenza virus was preincubated at pH 5.0 and 37°C for 1 h before being added to PS liposomes. Fusion was monitored by R18 dequenching at 37°C and pH 5.0. In each experiment 1 µg of viral protein and 100 nmoles of liposomes (final volume 2 ml) were used. The initial rate of the fusion process was calculated in the first 10-20 s following viral adition. The fusion extent was calculated following 5 min of dequenching.

Overall, our results therefore suggest that local dehydration may be important as a key element in the biological membrane fusion activity of the lipid-enveloped influenza virus, in agreement with previous results obtained in the case of PEG-stimulated Sendai virus fusion activity (Nir at al., 1986; Hoekstra et al., 1989).

ACKNOWLEDGEMENTS

This work was supported by JNICT, Portugal and by a NATO Collaborative Research Grant, CRG 900333.

REFERENCES

Burgess, S. W., McIntosh, T. J. & Lentz, B. L. (1992) Biochemistry 31: 2653-2661.
Doms, R. W. & Helenius, A. (1986) J. Virol. 60: 833-839.
Doms, R. W. & Helenius, A. (1988) In: Molecular mechanisms of membrane fusion (Ohki, S., Doyle, D., Franagan, T. D., Hui, S. W. & Mayhew, E., eds), pp. 385-398, Plenum Press, New York & London.
Doms, R. W., Helenius, A. & White, J. (1985) J. Biol. Chem. 260: 2973-2981.
Düzgünes, N., Lima, M. C. P., Stamatatos, L., Flasher, D., Alford, D., Friend, D. S. & Nir, S. (1992) J. Gen Virol. 73: 27-37.
Hoekstra, D., DeBoer, T., Klappe, K. & Wilschut, J. (1984) Biochemistry 23: 5675-5681.
Hoekstra, D. & Kok, J. W. (1989) Biosc. Rep. 9: 273-305.
Hoekstra, D., Klappe, K., Hoff, H. & Nir, S. (1989) J. Biol. Chem. 264: 6786-6792.
Hoekstra, D. (1990) J. Bioenerg. Biomemb. 22: 121-155.
Hoekstra, D. & Lima, M. C. P. (1992) In: Membrane interactions of HIV: Implications for pathogenesis and therapy in AIDS (Aloia, R. C. & Curtain, C. C., eds.) pp. 71-97, Wiley-Liss, New York.
Junankar, P. R. & Cherry, R. J. (1986) Biochim. Biophys. Acta 854: 198-206.
Lima, M. C. P., Ramalho-Santos, J., Martins, M. F., Carvalho, A. P., Bairos, V. A. & Nir, S. (1992) Eur. J. Biochem. 205, 181-186.
Nir, S., Klappe, K. & Hoekstra, D. (1986) Biochemistry 25: 8261-8266.
Nir, S., Düzgünes, N., Lima, M. C. P. & Hoekstra, D. (1990) Cell Biophys.: 17: 181-201.
Ramalho-Santos, J., Nir, S., Düzgünes, N., Carvalho, A. P. & Pedroso de Lima, M. C. (1993) Biochemistry 32: 2771-2779.
Ruigrok, R. W. H., Wrigley, N. G., Calder, L. J., Cusack, S., Wharton, S. A., Brown, E. B. & Skehel, J. J. (1986) EMBO J. 5: 41-49.
Sato, S. B., Kawasaki, K. & Ohnishi, S. (1983) Proc. Natl. Acad. Sci. U.S.A. 80: 3153-3157.
Skehel, J. J., Bayley, P. M., Brown, E.B., Martin, S. R., Waterfield, M. D., White, J. M., Wilson, I. A. & Wiley, D. C. (1982) Proc. Natl. Acad. Sci. U.S.A. 79: 968-972.
Stegmann, T., Hoekstra, D., Scherphof, G. & Wilschut, J. (1986) J. Biol. Chem. 261: 10966-10969.
Stegmann, T., Doms, R. W. & Helenius, A. (1989) Annu. Rev. Biophys. Biophys. Chem. 18: 187-211.
Stryer, L. (1968) Science 162: 526-533.
Wharton, S. A., Skehel, J. J. & Wiley, D. C. (1986) Virology 149: 27-35.
White, J. (1990) Annu. Rev. Physiol. 52: 675-697.
White, J. M. & Wilson, I. A. (1987) J. Cell Biol. 105: 2887-2896.
Wiley, D. C. & Skehel, J. J. (1987) Ann. Rev. Biochem. 56: 365-394.
Yamazaki, M. & Ito, I. (1990) Biochemistry 29: 1309-1314.
Yoshimura, T., Maezawa, S. & Hong, K. (1987) J. Biochem. 101: 1265-1272.

STRUCTURE AND REGULATION OF A GENE CLUSTER INVOLVED IN CAPSULE FORMATION OF *Y.PESTIS*

A.Karlyshev, E.Galyov[1], O.Smirnov[2] V.Abramov[2] and V.Zav'yalov[2]
Department of Microbiology
University of Reading
Reading, Berks, RG6 2AJ
United Kingdom

1. Introduction.

It has been known since 1894 that, the causative agent of plague disease, gram-negative bacterium *Yersinia pestis*, is capable of forming large capsules (Kitazato, 1894). The capsule substance is highly immunogenic for mice and has been under extensive study as a potential component of synthetic vaccine against disease. Chemical analysis of capsule (F1 antigen) has led to the conclusion that it consists of protein (Amies, 1951), lipid and polysaccharide (Glosnicka & Gruszkiewicz, 1980). We have cloned the gene cluster determining synthesis of the capsule protein subunut and shown that this subunit of protein origin can be secreted in *E.coli* cells with the formation of capsule. The gene, encoding capsule subunit, as well as genes involved in translocation of the protein across bacterial outer membrane have been sequenced. These data together with those of subcloning, mutagenesis and complementation analysis have led to the conclusion that the *f1* gene cluster is organised in a similar way to the gene clusters involved in extracellular secretion and assembly of some fimbriae subunits of *E.coli* and other gram-negative bacteria (Karlyshev et al., 1992a). It has been also shown that the *f1* gene cluster contains gene responsible for the regulation of capsule formation.

[1]Present address: Institute of Cellular and Molecular Biology, University of Umea, S90187 Umea, Sweden

[2]Institute of Immunology, Lyubuchany 142380, Chekhov District, Moscow Region, Russia

NATO ASI Series, Vol. H 82
Biological Membranes:
Structure, Biogenesis and Dynamics
Edited by Jos A. F. Op den Kamp

2. The structure of the *f1* gene cluster.

The vaccine strain EV of *Y.pestis* contains 3 plasmids: 10, 70 and 100 kb. The two large plasmids were isolated from total plasmid DNA by sucrose gradient centrifugation. Fragments of partial digestion of the isolated DNA were inserted into cosmid pHC79 and, after transduction into cells of *E.coli* HB101, colonies producing F1 antigen and capsule were selected. Radial immunodiffusion with cell lysates and agglutination of intact cells in the presence of anti-F1 antibodies were used during the selection step. The presence of capsule was cofermed by microscopy. Subcloning of hybrid cosmid DNA from positive colonies resulted in construction of pFS2 plasmid which contains 9 kb insert and still expresses F1 and capsule in *E.coli* cells. Subsequent subcloning and complementation analysis as well as deletion and insertion mutagenesis allowed us to localize the *f1* gene cluster to a 5.8 kb region of the DNA insert. These data together with the results of nucleotide sequencing of the entire region revealed the organization of the *f1* gene cluster shown in Fig.1. It has been found that overall organization of this region has striking similarity to the gene clusters involved in the secretion of some fimbria subunits (Karlyshev et al., 1992a). Moreover, the proteins involved in secretion and assembly of these fimbriae have a high level of homology with the corresponding proteins of the *f1* gene cluster. In the following sections structure and function of the *f1* genes and corresponding products will be presented in more details.

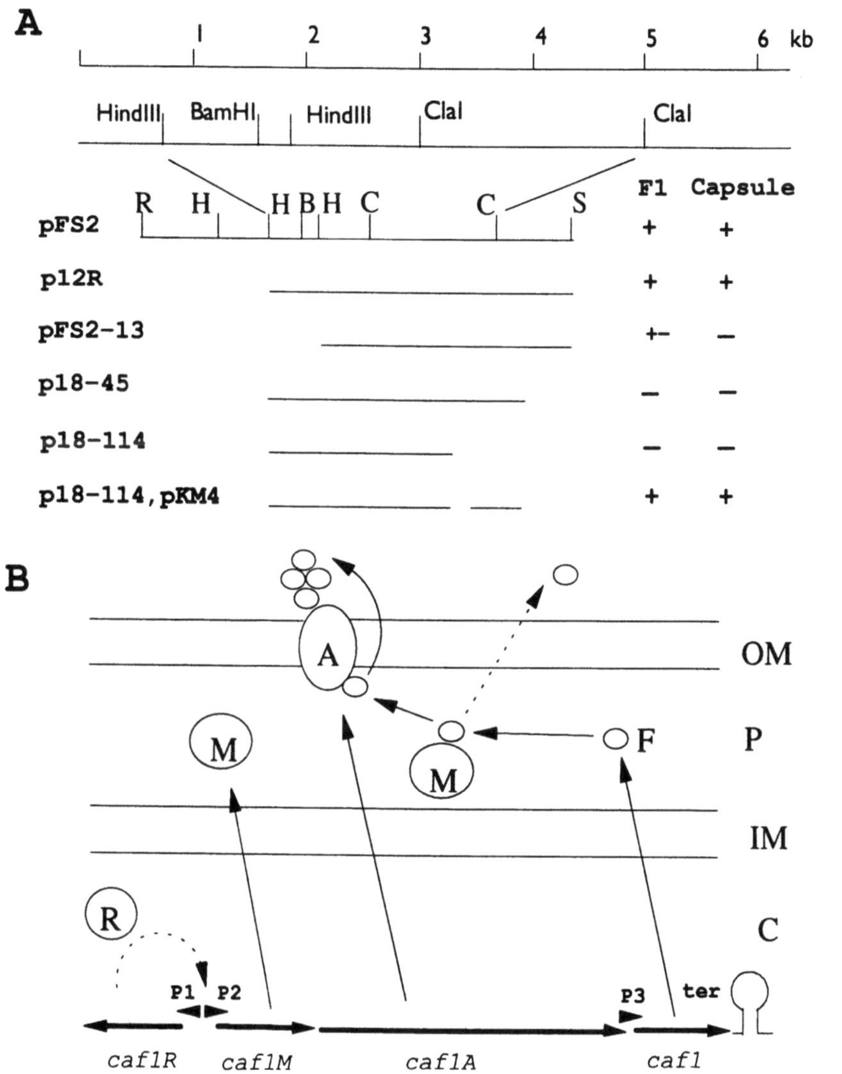

Fig.1. Physical and genetic map of the gene cluster involved in the biogenesis of *Y.pestis* capsule.

A. Structure of different deletion derivatives of pFS2 plasmid and the phenotypes of *E.coli* strains harbouring corresponding plasmids. R-*Eco*RI, S-*Sal*GI, B-*Bam*HI, C-*Cla*I, H-*Hind*III.

B. Genetic organization and proposed regulation of expression of the gene cluster and schematic representation of the involvement of the corresponding gene products in secretion of the subunits and in formation of capsule. Products of genes are: Caf1M (M), Caf1A (A), Caf1 (F) and Caf1R (R). OM-outer membrane, P-periplasmic space, IM-inner membrane, C-cytoplasm.

3. Characterization of *f1* genes and corresponding gene products.

3.1. *Caf1* gene.

The *caf1* gene (for capsule antigen F1) encodes the capsule subunit of. It has been localized by *Exo*III-Mung bean nuclease deletion analysis. Cells carrying pFS2 derivatives lacking more than 1 kb from *Sal*GI site (Fig.1) are not capable of synthesizing the F1 antigen. Nucleotide sequencing of this region revealed an ORF encoding a protein (Mr=15.5Kda) which posesses an N terminal leader sequence 21 amino acid residues long (Galyov et al., 1990). Cloning of this region in pUC19 plasmid under lac-promoter resulted in production of the F1 antigen but in low yields. One of the possible explanations of the low yield of the F1 antigen is the absence of a specific protein (CaF1M, see below) which probably prevents degradation of the subunits by periplasmic proteases. The multiple alignment of amino acid sequences of Caf1 protein and of some fimbria subunits revealed that the C-terminal parts of these proteins are higly concerved and could be involved in the recognition by periplasmic chaperones (Bertin et al., 1993).

3.2. *Caf1M* gene.

Deletion, insertion and frame-shift mutagenesis of a region about 2.5-3.5 kb upstream from the caf1 gene dramatically decreased the level of F1 antigen production (Galyov et al., 1991). The corresponding DNA region has been cloned and sequenced. The putative product of this gene Caf1M (for capsule antigen F1 mediator) has been shown to have significant homology in amino acid sequence with PapD family of chaperones. These chaperones recognize periplasmic intermediates of secreted proteins and prevent them from degradation and polimerization in the periplasmic space. Some of the hypervariable regions in these proteins are possibly involved in the recognition and binding of the secreted proteins. This process of recognition seems to be very similar to that during antigen and antibody interaction. Structural similarity between PapD and Ig domains (Holmgren et al., 1989) supports this idea.

3.3. *Caf1A* gene.

A 2,5 kb DNA region between the *caf1* and the *caf1M* genes has been sequenced and found to contain a large ORF encoding a protein of 833 amino acid residues long (Karlyshev et al., 1992a). It has been found from subsequent experiments that the corresponding gene product

```
        Source          GenBank Accession N
Caf1A   Y.pestis                          239 GETYSDSSIFDSIPIK
FaeD    E.coli          X03675            236 GKTSTGDSLLGSTGTY
FimD    E.coli          X51655            264 GDGYTQGDIFDGINFR
FimD    S.typhimurium   M37853            228 GKSQTRSALFSDFGFY
FimD    B.pertussis     X64876            248 GDVFSSGEYFAPYSMR
MrkD    K.pneumonia     M55912            250 GQIFTNGEFFDTIGLR
PapC    E.coli          Y00529            255 GENNINSDIFRSWSYT
                                              *

Caf1A   GIKIASDESMVPYYQWNFAPVVRGIARTQARVEVLRDGYTVSNELVPSGPFELAN
FaeD    GVSLSRNNSMKP-GNLGYTPVFSGIANGPSRVTLTQNGRLLHSEMVPAGPFSITD
FimDE   GAQLASDDNMLPDSQRGFAPVIHGIARGTAQVTIKQNGYDIYNSTVPPGPFTIND
FimDS   GAALRSNSNMLPWEARGYAPLITGVANSTSRVTISQNGYAVYSKVVPPGPYQLDD
FimDB   GMLVGSDTAMLPYSERLYRPTIRGVARTRANVKVYQAGVLVFQDAVPPGPFAIDD
MrkD    GVNLATDDNMFPDGMRSYAPEIRDVAQSNALATVRQGSNIIYQTTVPPGPFTLQD
PapC    GASLESDDRMLPPRLRGYAPQITGIAETNARVVVSQQGRVLYDSMVPAGPFSIQD
        *  .  .  * *     . *   ..*   .   . . .  .    **.**. . .
```

Fig.2. A fragment of multiple aligment of amino acid sequences of some outer membrane proteins, involved in the fimbria formation. of *E.coli*, *S.typhimurium*, *B.pertussis* and *K.pneumonia* and in the assembly of *Y.pestis* capsule using CLUSTALV program of GCG package.

is involved in capsule assembly and has been named *caf1A* (for capsule antigen F1 assembly). *E.coli* cells containing the *caf1* gene on pUC19 plasmid have been transformed by pACYC177 derivatives containing either the *caf1M* gene alone or in combination with the native or a truncated *caf1A* gene. Although no significant changes have been found in the F1 antigen synthesis and secretion among these variants, the capsule has been detected only in the cells containing an intact *caf1A* gene. The amino acid sequence of Caf1A protein derived from the DNA sequence bears stretches of homology with amino acid sequences of outer membrane proteins involved in the assembly of some fimbria (Fig.2). Some regions with high level of homology may be involved in assembly into the outer membrane, while some variable regions could be involved in specific recognition of the secreted protein. This hypothesis is supported by recent evidence for the direct recognition of some fimbria subunits in chaperone-subunit complexes by outer membrane protein PapC (Dodson et al. 1993) which seems to play the same role in fimbria assembly as Caf1A protein plays in the assembly of *Y.pestis* capsule.

3.4. *Caf1R* gene.

We have found that the DNA region upstream from the *caf1M* gene is necessary for the regulation of capsule formation (Karlyshev et al., 1992b). Sequencing of this region has revealed an ORF 903 bp long named *caf1R* (for capsule antigen F1 regulation). It has been found from unidirectional DNA deletion experiments that elimination of the gene leads to dramatic decrease in F1 antigene yield, whereas a fragment of the gene encoding only 81 amino terminal residues is sufficient for effective antigen synthesis and capsule formation. This region has been found to have extensive homology in amino acid sequence with the XylS/AraC family of transcription regulators (Gallegos et al., 1993). It contains a putative DNA binding domain as well as a consensus sequence outside this domain characteristic of

```
                      SWISS-PROT
      Source           access.N
Caf1R Y.pestis                    1 MLKQMTVNS IIQYIEENLE SKFINIDCLV
SoxS  E.coli          P22539      1 MSHQKIIQD LIAWIDEHID QPL.NIDVVA
AdaA  B.subtilis      P19219     89 MPDSEWVDL ITEYIDKNFT EKL.TLESLA
AraC  E.coli          P03021    167 PPMDNRVRE ACQYISDHLA DSNFDIASVA
MmsR  P.aerug.        P28809    190 GSLD..LDG LHAYMREHL. HARLELERLA
VirF  Y.enterocol.    P13225    155 .LGNRPEER LQKFMEENYL QGW.KLSKFA
XylS  P.putida        P07859    201 SKGNPSFER VVQFIEENLK RNI.SLERLA
                                                           L---A

Caf1R  LYSGFSRRYL QISFKEYVGM PIGTYIRVRR ASRAAALLRL TR...LTIIE
SoxS   KKSGYSKWYL QRMFRTVTHQ TLGDYIRQRR LLLAAVELRT TE...RPIFD
AdaA   DICHGSPYHM HRTFKKIKGI TLVEYIQQVR VHAAKKYLIQ TN...KAIGD
AraCE  QHVCLSPSRL SHLFRQQLGI SVLSWREDQR ISQAKLLLST TR...MPIAT
MmsR   AFCNLSKFHF VSRYKAITGR TPIQHFLHLK IEYACQLLDS SD...QSVAR
VirF   REFGMGLTTF KELFGTVYGI SPRAWISERR ILYAHQLLLN GK...MSIVD
XylS   ELAMMSPRSL YNLFEKHAGT TPKNYIRNRK LESIRACLND PSANVRSITE
       ------S--- ---F----G- T----I---R ---A--LL-- -------I-D

Caf1R  ISAKLFYDSQ QTFTREFKKI FGYTPRQYRM IPFWSFKGLL GRREINCEYL
SoxS   IAMDLGYVSQ QTFSRVFRRQ FDRTPSDYRH RL........ ..........
AdaA   IAICVGIANA PYFITLFKKK TGQTPARFRQ MSKMEETYNG NK........
AraCE  VGRNVGFDDQ LYFSRVFKKC TGASPSEFRA GCEEKVNDVA VKLS......
MmsR   VGQAVGYDDS YYFSRLFSKV MGLSPSAYRQ RVRQGEGGA. ..........
VirF   IAMEAGFSSQ SYFTQSYRRR FGCTPSQAR. LTKIATTG.. ..........
XylS   IALDYGFLHL GRFAENYRSA FGELPSDTLR QCKKEVA... ..........
       IA---GF-S- -YF---F--- -G-TPS--R
```

Fig.3. Multiple alignment of amino acid sequences of Caf1R gene regulator and some transcription regulators of XylS/AraC family using PILEUP program of the GCG package. The putative DNA binding region indicated by line and the consensus sequence beneath are from data on XylS/AraC regulators (Gallegos et al., 1993).

all regulators of this type (Fig.3). However, in contrast to Caf1R, the DNA-binding regions of most of these proteins are located in C-terminal parts. The role of the C-terminal part of Caf1R remains to be elucidated. It could serve as a sensor in the process of thermoregulation of capsule formation.

4. Intergenic regions and gene regulation.

The *f1* gene cluster consists of three transcription units: one unit contains the *caf1* gene, second - the *caf1M* and the *caf1A* genes and the third one is oriented in the opposite direction and contains *caf1R* gene (Fig.1B). The only potential rho-independent transcription terminator has been identified at the end of the *caf1* gene. Deletion of this region resulted in a sharp decrease of F1 antigen synthesis perhaps due to the formation of unstable mRNA (data not shown). A relatively large intergenic region exists between the *caf1A* and *caf1R* genes and is likely to be a key regulatory element in the process of capsule formation. This region contains two kinds of repeats: inverted repeats close to the *caf1M* gene and five short direct repeats close to the *caf1R* gene (Karlyshev et al.,1992b). It is of interest, that short repeats are at the same distance (33-36) bp from each other corresponding to 1 coil of dsDNA. These two kinds of repeats may well be recognition sites of two different regulation proteins. We also propose that inverted repeats are recognised by the Caf1R protein, whereas tandem direct repeats may be recognized by some other proteins regulating expression of Caf1R protein itself. It has been recently found that inverted repeats of regions responsible for regulation of formation of some kinds of *E.coli* pili contain GATC sequences which are differentially methylated by deoxyadenosine methylase (dam) (Nou et al., 1993). Since the *caf1M-caf1R* intergenic region does not contain any GATC sequences, the mechanism of regulation in this case should be quite different from the one mentioned above.

5. Summary.

DNA sequencing and gene expression studies in *E.coli* has revealed organization of the *Y.pestis* gene cluster involved in the capsule formation. Two proteins, Caf1M and Caf1A, have been found to be involved at different stages of subunit translocation through the outer membrane. Capsule formation has been shown to be regulated by the positive transcription regulator Caf1R. The detailed mechanism of interaction between Caf1M and Caf1A proteins with Caf1 subunit is not known. We believe that Caf1M-Caf1-Caf1A system could serve as a good model for the understanding the phenomenon of protein-protein recognition as well as the study of protein secretion mechanisms in gram-negative bacteria.

Acknowledgement: We are grateful to Dr. S.MacIntyre for critical reading of the manuscript and helpful discussion.

REFERENCES:

Amies S (1951) The envelope substance of *Y.pestis*. Brit. J. Exp. Pathol. 32: 259-262.

Bertin Y, Girardeau J-P, Der Vartanian M and Martin C (1993) The ClpE protein involved in biogenesis of capsule-like antigen is a member of a periplasmic chaperone family in gram-negative bacteria. FEMS Microbiol. Lett., 108: 59-68.

Dodson KW, Jacob-Dubuisson F, Striker RT and Hultgren SJ (1993) Outer membrane PapC molecular usher discriminately recognizes periplasmic chaperone-pilus subunit complexes. Proc. Nat. Acad. Sci. USA 90: 3670-3674.

Gallegos M-T, Michan C and Ramos JL (1993) The XylS/AraC family of regulators. Nucl. Acids Res., 21: 807-810.

Galyov EE, Smirnov OY, Karlyshev AV, Volkovoy KI, Denesyuk AI, Nazimov IV, Rubtsov KS, Abramov VM, Dalvadyanz SM and Zav'yalov VP (1990) Nucleotide sequence of the *Y.pestis* gene encoding F1 antigen and the primery structure of the protein. FEBS Lett., 277: 230-232.

Galyov EE, Karlyshev AV, Chernovskaya TV, Dolgikh DA, Smirnov OY, Volkovoy KI, Abramov VM and Zav'yalov VP (1991) Expression of the envelope antigen F1 of *Y.pestis* is mediated by the product of *caf1M* having homology with the chaperone protein PapD of *E.coli*. FEBS Lett., 286:79-82.

Glosnicka R & Gruszkiewitcz (1980) Chemical composition and bioactivity of *Y.pestis* envelope substance. Infect. Immun., 10: 506-612.

Holmgren A & Brauden C (1989) Crystal structure of chaperon protein PapD reveals an immunoglobulin fold. Nature, 342: 248-251.

Kitazato S (1894) The baciilus of bubonic plague. Lancet, 2: 428-430.

Karlyshev AV, Galyov EE, Smirnov OY, Guzaev AP, Abramov VM and Zav'yalov VP (1992a) A new gene of the *f1* operon of *Y.pestis* involved in the capsule biogenesis. FEBS Lett. 297: 77-80.

Karlyshev AV, Galyov EE, Abramov VM and Zav'yalov VP (1992b) *Caf1R* gene and its role in the regulation of capsule formation of *Y.pestis*. FEBS Lett., 305: 37-40.

Nou X, Skinner B, Braaten B, Blyn L, Hirsh D and Low D (1993) Regulation of pielonephritis-associated pili phase variation in *E.coli*: binding of the PapI and the Lrp regulatiry proteins is controlled by DNA methilation. Mol. Microbiol. 7: 545-553.

THE TOXIN OF THE MARINE ALGA *PRYMNESIUM PATELLIFERUM* INCREASES CYTOSOLIC CA^{2+} IN SYNAPTOSOMES AND VOLTAGE SENSITIVE CA^{2+}-CURRENTS IN CULTURED PITUITARY CELLS

A-S Meldahl, S Eriksen[1], V A T Thorsen, O Sand[1] and F Fonnum
Norwegian Defence Research Establishment
Division for Environmental Toxicology
P.O. Box 25, N-2007 Kjeller, Norway

Introduction

During the last four years the closely related ichthyotoxic marine flagellates *Prymnesium patelliferum* and *Prymnesium parvum* have appeared in blooms in the brackish waters in the fjords of south-western Norway. The algal blooms have caused the death of large amounts of salmon and rainbow trout in fish farms (varying between 150 - 750 metric tons per year). In highly toxic blooms the fish die within 5-30 minutes after spasmodic movements. The precise mechanism of action of the algal toxin is not clear although the gills seems to be the primary organ affected (Shilo, 1971). Studies have been carried out on various biological test systems, and in the litterature the effects of the *Prymnesium*-toxin is divided into four categories: ichthyotoxic effects, hemolytic effects, cytotoxic effects and neurotoxic effects (Yariv and Hestrin, 1961; Dafni and Giberman, 1971; Parnas and Abbott, 1965); which are all results of interaction with biological membranes. Addition of high doses of toxin for prolonged periods leads to membrane lysis. Using liposomes Imai and Inoue (1974) found that the amount of toxin required for lysis increased with decreasing content of cholesterol in the membrane. Furthermore, the net charge of the liposome membrane was not important for the interaction with the algal toxin. The chemical structure of the algal toxin is not yet fully described, and it is not clear whether the observed effects are due to one or several toxin components. A substance of *P.parvum* screened for its hemolytic activity has been isolated and identified as a mixture of two polar digalactomonoglycerides: 1'-O-octadecatetraenoyl-3'-O-(6-O-β-D-galactopyranosyl-β-D-galactopyranosyl)-glycerol and 1'-O-octadecapentaenoyl-3'-O-(6-O-β-D-galactopyranosyl-β-D-galactopyranosyl)-glycerol (Kozakai et al, 1982).

[1]Department of Biology
University of Oslo
P.O. Box 1051, N-0316 Oslo, Norway

NATO ASI Series, Vol. H 82
Biological Membranes:
Structure, Biogenesis and Dynamics
Edited by Jos A. F. Op den Kamp

Recently we reported the inhibitory effect of an extract of *Prymnesium* on the high affinity uptake of the neurotransmitters glutamate and γ-amino butyric acid (GABA) in synaptosomes and the low affinity uptake of glutamate, GABA and dopamine in synaptic vesicles (Meldahl and Fonnum, 1993). The synaptosomal reuptake of glutamate and GABA is strongly dependent on external sodium, and the uptake is coupled to a Na^+-gradient which is maintaied by the Na^+/K^+-ATPase and influenced by the membrane potential (for review see Kanner, 1983). Similarly, the mechanisms for the uptake of amino acids in synaptic vesicles is dependent on an electrochemical H^+-gradient maintained by the Mg^{2+}-dependent H^+-ATPase (Fykse et al, 1989). Interference with either of the factors involved in the uptake driving force would, therefore, be expected to cause a decrease in the accumulation of the amino acid neurotransmitters. In order to approach the specific mechanism of action of this algal toxin we have examined the effect on the membrane potential and cytosolic calcium concentration in synaptosomes. Our results show that the *Prymnesium*-toxin, at a concentration that decreases the accumulation of glutamate, induces synaptosomal depolarization followed by an increase in cytosolic Ca^{2+}. In addition we show that the algal toxin selectively interferes with the voltage dependent Ca^{2+}-channels in clonal rat anterior pituitary cells, thereby increasing the influx of Ca^{2+}.

Methods

Algal cultures and toxin extraction

Axenic cultures of *Prymnesium patelliferum* were grown and extracted as described previously (Meldahl and Fonnum, 1993). The lipophilic algal extract (toxin) was dissolved in methanol. Methanol was used as control in all experiments and had no significant effect on any of the test systems. The toxin concentration is indicated as the number of cells extracted.

High affinity uptake of glutamate, membrane potential and cytosolic calcium in synaptosomes

Synaptosomes were isolated from a sucrose homogenate of whole rat brain as described in principle by Gray and Wittaker (1962). For the measurements of the high affinity uptake of glutamate a crude synaptosomal fraction (P_2) was used. For the studies of membrane potential and intracellular calcium concentration this fraction was further purified by fractionation on a sucrose gradient (P_2B).

The high affinity uptake of glutamate was measured as described by Fonnum et al (1980). The algal toxin was incubated with the synaptosomes (final conc. 4 μg protein/ml) in Tris-Krebs buffer for 15 minutes at 25 °C. Radiolabelled glutamate was added and the uptake was terminated after 3 minutes by filtration onto a glass-fiber filter and the filters were counted by liquid scintillation.

The synaptosomal membrane potential was determined with the cationic membrane potential-sensitive dye $DilC_1(5)$. The synaptosomes, resuspended in Hepes buffer with 10 mM glucose, were added to a solution of $DilC_1(5)$ (final conc. 100 μg protein/ml). Fluorescence measurements were performed on a computerized Perkin-Elmer LS 50 luminescense spectrometer, excitation wavelength 620 nm and emission wavelength 665 nm.

Changes in the synaptosomal cytoplasmic calcium was measured in fura-2 loaded synaptosomes resuspended in calcium free Hepes buffer to final concentration 50 μg protein/ml, as described by Yates *et al* (1992), employing a computerized Perkin-Elmer LS-5B luminescence spectrometer, with excitation changing between 340 and 380 nm and recording at 510 nm.

Low affinity uptake of glutamate in synaptic vesicles

Synaptic vesicles were prepared as previously described (Fykse and Fonnum, 1988). The P_2 fraction, obtained as described above, was osmotically shocked by resuspension in 10 mM Tris-maleate and 0.1 mM EGTA (pH 7.4) and the vesicle fraction was isolated by differential centrifugation. Vesicular uptake was assayed as described by Fykse and Fonnum (1988). Synaptic vesicles (0.3 mg protein/ml) were incubated with the toxin extract for 15 minutes at 30 °C in 0.32 M sucrose, 10 mM Tris-maleate (pH 7.4) and 4 mM $MgCl_2$. Radiolabelled glutamate and ATP were added and the uptake was terminated after 3 minutes by filtration onto a Millipore HAWP filter, and the filters were counted by liquid scintillation.

Recordings of membrane potential and calcium current in clonal anterior pituitary cells

The measurements were carried out as described by Sand *et al* (1989) using the patch-clamp method (Hammil *et al*, 1981) on prolactin-producing rat anterior pituitary GH_4C_1 cells. To record the membrane potential and the current responses of the membrane, we used whole-cell recordings in either the current-clamp or the voltage-clamp configuration. For recordings of Ca^{2+}-currents, the potassium salts in the internal solution were replaced with the corresponding caesium salts to eliminate current through the K^+-channels, and 1 μM TTX was added to the external solution to block current through the Na^+-channels. To facilitate the Ca^{2+}-currents the Ca^{2+} concentration of the external saline was raised to 10 mM, with a corresponding reduction of NaCl. During the voltage clamp recordings, the holding potential was -60 mV and the potential was clamped to different levels between -100 mV and +50 mV for 300 ms.

Results

The crude algal toxin strongly inhibited, in a dose dependent manner, the high affinity uptake of the neurotransmitter glutamate into synaptosomes and the low affinity uptake of glutamate into synaptic vesicles (figure 1). The half-inhibitory concentration (IC_{50}) was 4 000 cells/ml and 80 000 cells/ml respectively, where the number indicates the number of algal cells extracted. This difference in IC_{50} values could partly be due to the amount of tissue protein necessary to measure the uptakes, 4 µg/ml and 100 µg/ml, respectively, and partly to the different composition of membranes.

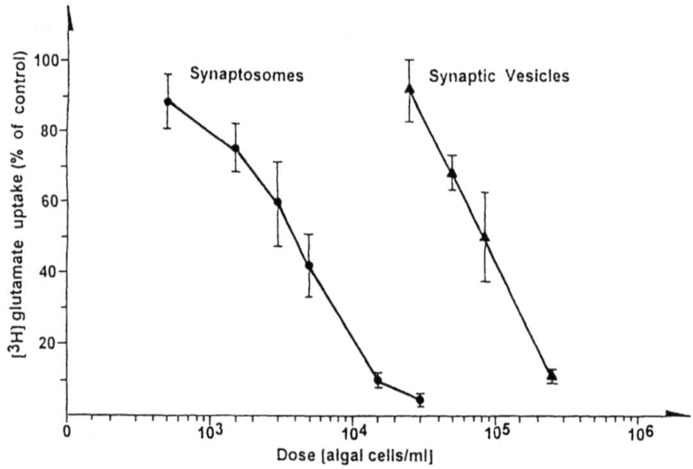

Figure 1. The inhibitory effect of toxin extract on the high affinity uptake of [3H]glutamate in synaptosomes (•) and the low affinity uptake of [^{3}H]glutamate in synaptic vesicles (▲) of rat brain. The dose is the number of algal cells extracted. Bars indicate s.d., n=3 or 4.

The inhibition of the uptake systems could be a due to interference with factors involved in the uptake mechanism. Since both transport systems are dependent on a potential gradient across the membrane, negative for the synaptosmes and positive for the synaptic vesicles, a depolarization would be critical for the transmitter uptake. Figure 2 shows typical curves for the effect of toxin on the synaptosomal membrane potential. Methanol, the toxin solvent, had no effect whereas the algal toxin of 5 000 and 10 000 cells/ml caused a 32 % and 56% reduction in the membrane potential, respectively, relative to the depolarization by high potassium concentration (40 mM). A consequence of the depolarization would be opening of voltage dependent ion channels such as Na^+- and Ca^{2+}-channels. In fura-2 loaded synaptosomes algal extract of 5 000 and 10 000 cells/ml gave a 28 % and 68% increase in flourescence ratio respectively, relative to the increase indued by depolarisation with high potassium concentration, which indicates an increase in cytosolic [Ca^{2+}] (figure 3). Methanol, the toxin solvent, had no effect on the fluorescence ratio measurements .

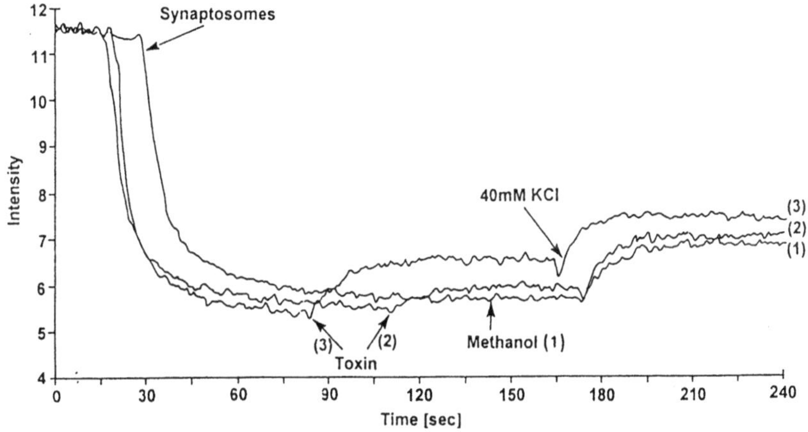

Figure 2. Changes in membrane potential of synaptosomes determined by the fluorescent probe $DilC_1(5)$. Each experiment was conducted in triplicate, with representative traces shown. (1) 10 µl methanol added, (2) toxin dissoved in 10 µl methanol added to a final toxin concentration of 5 000 cell/ml and (3) toxin dissoved in 10 µl methanol added to a final toxin concentration of 10 000 cell/ml.

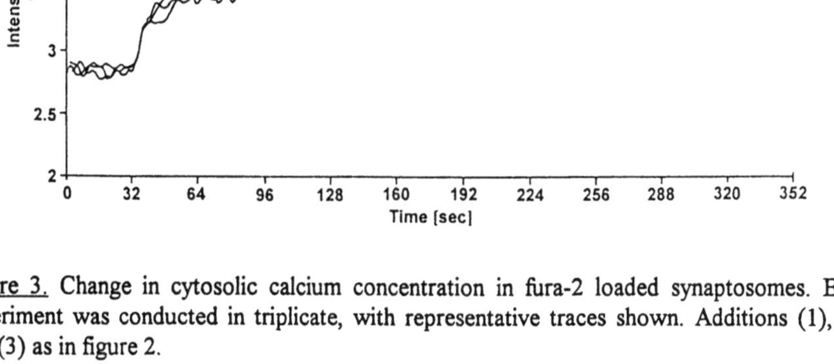

Figure 3. Change in cytosolic calcium concentration in fura-2 loaded synaptosomes. Each experiment was conducted in triplicate, with representative traces shown. Additions (1), (2) and (3) as in figure 2.

In electrophysiological studies using the patch-clamp method we recorded the toxin effects on the resting membrane potential, the input resistance and the trans-membrane Ca^{2+}-currents. could study separately. We used clonal pituitary cells that possess both T- and L-type Ca^{2+}-channels. Addition of toxin at a final concentration of 50 000 - 70 000 cells/ml in the culture medium did not acutely disturb the resting membrane potential or resistance. However, after 30 to 60 minutes a dramatic reduction in these membrane parameters was observed, probably due to cell lysis. To test the effect on the voltage dependent calcium channels the cells were stepwise voltage clamped at varying potentials from the holding potential of -60 mV. Figure 4 shows the total current (through both T- and L-channels) in two different cells depolarized to 0 mV. After 5-15 minutes incubation with algal toxin and until lysis both the T-current, measured as the initial peak of the Ca^{2+}-current, and the L-current, measured as the value at the end of tyhe clamping pulse, increased 2 - 3 times compared to controls (incubated with methanol). The current-voltage relations for the separate channel currents are also presented in figure 4. The voltage sensitivity of the Ca^{2+}-currents was not affected by the toxin, indicating that the enhanced currents are due to increased influx through the normal endogenous Ca^{2+}-channels.

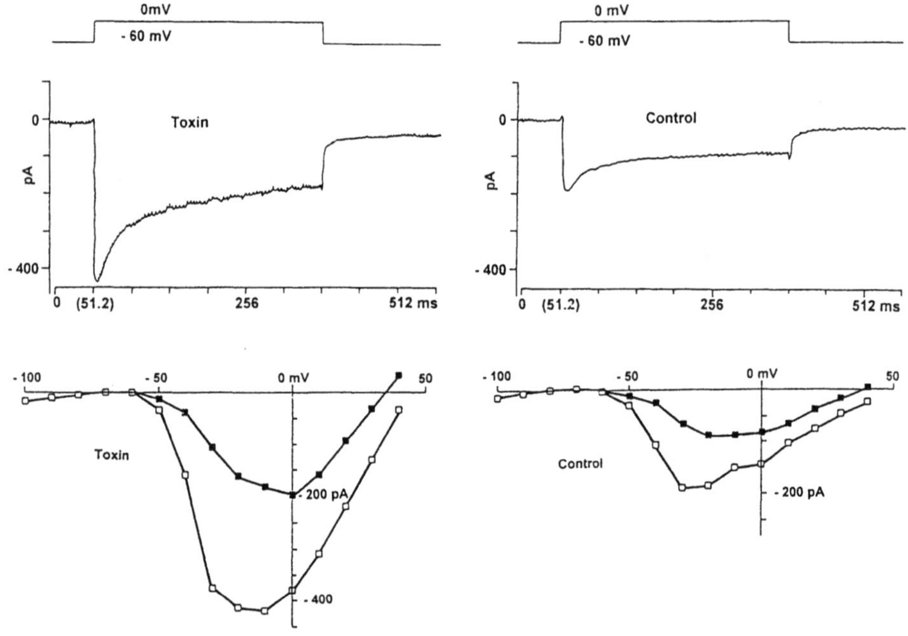

Figure 4. Recordings of voltage dependent Ca^{2+}-current in whole-cell voltage clamped clonal anterior pituitary cells. Holding potential was -60 mV. The upper recordings show the ion currents during depolarization of the cell membrane to 0 mV. Lower diagrams show the current-voltage relation for the currents through the T-type (open symbols) and L-type (filled symbols) Ca^{2+}-channels. Toxin concentration was 70 000 cells/ml. The recordings are from two representative cells. The exposed cell was recorded 20 min after toxin addition.

Discussion

Synaptosomes are pinched-off nerve cell terminals that behave similarly to intact nerve endings when suspended in physiological salt solution. They respire and synthesize ATP, extrude sodium and retain potassium, contain functional action potential sodium channels, release neurotransmitters by calcium dependent mechanisms and exhibit sodium-dependent high affinity uptake of neurotransmitters. This preparation has therefore been useful as a model for studies of cell membrane functions. The establishment of and maintenance of a transmembrane potential is crucial for normal uptake of amino acids and intraterminal ion homeostasis. The present results show that an extract of the marine alga *Prymnesium patelliferum* in a dose-dependent manner inhibited the high affinity uptake of glutamate and caused a reduction in the membrane potential in rat brain synaptosomes. The 50% effective toxin dose for membrane depolarization was slightly higher (about 10 000 cells/ml) than the IC_{50} dose for the uptake of glutamate (about 4 000 cells/ml). This may be due to the high amount of synaptosomes required for the measurements of membrane potential compared to the high affinity uptake. Thus, it is likely that the decrease in transmembrane electrical potential is of sufficient magnitude to explain the decline in the maximal accumulation of glutamate. Since the Na^+/K^+-ATPase was not affected directly at the toxin concentration we used (results not shown), it is possible that the mechanism for the effect of the toxin is the increased permeability to ions. Measurements of intrasynaptosomal Ca^{2+} showed that the cytoplasmic level of this ion increased in the presence of the algal toxin at the relevant concentrations. Moreover, it should be added that a membrane depolarization was observed also in calcium-free medium (results not shown) although to a slightly lesser extent than in the normal medium. It is possible that the increase in permeability to Ca^{2+} is partly responsible for membrane depolarization. A disruption of the synaptosomal membrane is unlikely since there was no leakage of the cyosolic marker lactate dehydrogenase (results not shown). However, this does not exclude the possibility of formation of smaller ion pores.

The electrophysiological experiments on the living pituitary cells does not fully support the effects that were observed in the synaptosomes. First, at the high dose of 50 000 - 70 000 cells/ml there was no effect on the resting potential during the period prior to cell lysis. On the other hand, when the membrane of the pituitary cell was depolarised during voltage clamp, the toxin induced an increase in both T-and L- calcium currents. The fact that the voltage dependency was the same both before and after addition of toxin, shows that only the normal endogenous Ca^{2+}-channels were involved. It is unlikely that the toxin formed a channel with exactly the same voltage dependency as the endogenous Ca-channels. It is possible that the toxin revealed inactive voltage dependent Ca^{2+}-channels already present in the pituitary cell membrane.

In conclusion, our preliminary studies show that the inhibitory effect of the extract of *Prymnesium patelliferum* on glutamate uptake results from a depolarization of the synaptosomal membrane caused partly by its increased permeability to calcium ions. The inhibition of the low affinity uptake in synaptic vesicles may also be due to the depolarization of the vesicular membrane. The lack of effect on the membrane potential in the cultured pituitary cells could be due to differences in membrane lipid composition.

Acknowledgement

The authors thank Professor E Paasche, Department of Biology, University of Oslo, for the donation of the clone of *Prymnesium patelliferum*. This work was supported by a research grant from The Norwegian Research Council, Department of Fisheries Reseach, under the program of Harmful Algae.

References

Dafni Z and Giberman E (1972) Nature of the initial damage to Erlich ascites cells caused by *Prymnesium parvum* toxin. Biochem. Biophys. Acta 255: 380 - 385

Fonnum F, Lund Karlsen R, Malthe- Sörensen D, Sterri S and Walaas I (1980) High affinity transport systems and their role in transmitter action. In The cell surface and neuronal function (Cotman C W, Poste G and Nicolson G L, eds) pp. 445 - 504, Elsevier/North-Holland Biomedical Press, Amsterdam

Fykse E M, Christensen H and Fonnum F (1989) Comparison of the properties of γ-aminobutyric acid and L-glutamate uptake into synaptic vesicles isolated from rat brain. J. Neurochem. 52: 946 - 951

Fykse E M and Fonnum F (1988) Uptake of γ-aminobutyric acid by a synaptic vesicle fraction isolated from rat brain. J Neurochem. 50: 1237 - 1242

Gray E G and Whittaker V P (1962) The isolation of nerve endings from brain: An electron microscope study on cell fragments derived by homogenization and centrifugation. J Anat. 96: 79 - 88

Hammil O P, Marty A, Neher E, Sakmann B and Sigworth F J (1981) Improved patch-clamp techniques for high-resolution current recording from cells and cell-free membrane patches. Pflügers Arch. 391: 85 -100

Imai M and Inoue K (1974) The mechanism of the action of *Prymnesium* toxin on membranes. Biochim. Biophys. Acta 352: 344 - 348

Kanner B I (1983) Bioenergetics of neurotransmitter transport. Biochim. Biophys. Acta 726: 293 - 316

Kozakai H, Oshima Y and Yasumoto T (1982) Isolation and structural elucidation of hemolysin from the phytoflagellate *Prymnesium parvum*. Agric. Biol. Chem. 46: 233 - 236

Meldahl A-S and Fonnum F (1993) The effect of toxins of *Prymnesium patelliferum* on neurotransmitter transport mechanisms. The development of a sensitive test method. J. Environ. Toxicol. Health. 38: 57 - 67

Parnas I and Abbott B C (1965) Physiological activity of the ichthyotoxin from *Prymnesium parvum*. Toxicon 3: 133 - 145

Sand O, Chen B, Li Q, Karlsen H E, Bjøro T and Haug E (1989) Vasoactive intestinal peptide (VIP) may reduce the removal rate of cytosolic Ca^{2+} after transient elevations in clonal rat lactothrophs. Acta Physiol Scand. 137: 113 - 123

Shilo M (1971) Toxins of Chrysophyceae. In Microbial Toxins (Kadis S, Ciegler A and Ajl S J, eds) pp. 67 - 103, Academic Press, New York

Yariv J and Hestrin S (1961). Toxicity of the extracellular phase of *Prymnesium parvum* cultures. J. gen. Microbiol. 24: 165 - 175

Yates S L, Fluhler E N and Lippiello P M (1992) Advances in the use of the fluorescent probe fura-2 for the estimation of intrasynaptosomal calcium. J Neurosc Res. 32: 255 - 260

SEMLIKI FOREST VIRUS INDUCED CELL-CELL FUSION AND PORE FORMATION

Markus Lanzrein, Magda Spycher-Burger, and Christoph Kempf
Institute of Biochemistry
University of Bern
Freiestrasse 3,
3012 Bern, Switzerland

Introduction

Membrane fusion is an important and ubiquitous process in cell biology: In a living cell membrane-bound compartments are continuously either separated or united through fusion reactions. The fusion of cell membranes represents a crucial step in myogenesis, osteogenesis and fertilization (White and Blobel, 1989). Many enveloped animal viruses enter host cells by endocytosis and subsequent membrane fusion of the virion envelope and the endosomal membrane. In most cases, this fusion is triggered by the low intracompartmental pH (for reviews see White et al., 1983; Stegmann et al., 1989; White 1992). One of the best studied viruses with respect to entry is Semliki Forest Virus (SFV) (reviewed in Kielian and Helenius, 1986). The SFV envelope contains 80 copies of a fusogenic protein spike (von Bonsdorff and Harrison, 1978). The spike protein is homotrimeric, each monomer consisting of three subunits E1, E2 and E3 (M_r 51, 52 and 11 kDa, respectively). In mildly acidic pH (< 6.2), the SFV spike proteins are irreversibly changed to their fusion-active conformation (Kielian and Helenius, 1985). E1 is considered to be the fusion protein, since virus particles devoid of E2 and E3 are still infectious (Omar and Koblet, 1988) and E2/E3 expressed in the absence of E1 is fusion negative (Kondor-Koch et al., 1983). The notion is further supported by the fact that the E1 protein contains near the N-terminus a hydrophobic sequence highly conserved among alphaviruses (Garoff et al., 1980). This sequence could represent the "fusion peptide" that may insert into the target membrane and thereby initiate fusion. Site directed mutagenesis in the fusion peptide results in altered pH thresholds and efficiency of fusion (Levy-Mintz and Kielian, 1991) suggesting a crucial role for this peptide in membrane fusion. Recently, it was reported that E1 subunits, when exposed to mildly acidic pH, dissociate from the E1/E2/E3 complexes and reassemble into homotrimeric or higher oligomeric structures which are thought

NATO ASI Series, Vol. H 82
Biological Membranes:
Structure, Biogenesis and Dynamics
Edited by Jos A. F. Op den Kamp

to be involved in the fusion reaction (Wahlberg and Garoff, 1992; Wahlberg et al., 1992; Bron et al., 1993).

The mechanism of membrane fusion is still far from being fully understood. Amongst the many unresolved issues, one significant question is whether membrane joining is accompanied by changes in membrane permeability or whether it is tight. Currently, there is rather a general agreement that virus-induced fusion is a non-leaky process (Stegmann et al., 1989). Tightness of fusion has been presumed for Influenza as well as for SFV (Spruce et al., 1991; White and Helenius, 1980; Young et al., 1983). However, it is well established that viruses can alter the properties of the host cell plasma membrane early in infection, i. e. upon fusion with the plasma membrane. Such phenomena have been described for Sendai virus (Pasternak and Micklem 1973), Influenza virus at low pH (Patel and Pasternak, 1983; 1985), Vesicular Stomatitis virus and SFV (Carrasco, 1981). The changes consisted primarily in modifications in membrane permeability to ions and solutes. The leakage may be a consequence of the incorporation of an inherently leaky membrane patch (Wyke et al., 1980). This would mean that either the virion membrane becomes leaky upon receptor- or low pH activation of the fusogenic proteins or is permanently leaky. Alternatively, the permeability change could result from spillover at the site of membrane fusion. Some characteristics of these viral effects on membrane permeability are in common with those of certain bacterial toxins, activated complement and other cytotoxic agents (reviewed in Pasternak, 1985; 1992). Typical features are namely the sequential induction of permeability to ions, metabolites and proteins, the sensitivity to divalent cations and changes in ionic strength and the ability of cells to recover spontaneously at low ratios of agent (Bashford et al., 1986).

The purpose of the present study was to investigate early stages of cell membrane fusion and to characterize concomitant changes in membrane permeability which were found to appear under certain conditions.

Early events in virus induced cell membrane Fusion

Aedes albopictus cells (a mosquito cell line), when infected with SFV, express large amounts of fusogenic SFV spike proteins in their plasma membrane at 16 hours post infection (hpi). Exposure of such cells to low pH (< 6.2) leads to cell-cell fusion. This provides an excellent model to study virus-induced membrane fusion. We have used this system to study cell-cell fusion using a novel approach. It involved whole cell patch clamp recording in conjunction with cell pairs. Two separate voltage clamp amplifiers were used to hold each cell at a defined membrane potential. Junctional current flow (I_j) resulting from the potential difference between cells was then continuously recorded (Weingart, 1986; Bukauskas and

Weingart, 1993). This method allowed to monitor the early events of cell membrane fusion at high resolution. 1 mM $ZnCl_2$ was added to the solutions to block low pH induced changes in g_m (see below). As shown in Fig. 1, exposure of a cell pair to low pH (5.8) was not immediately followed by the development of intercellular conductance g_j, instead, there was a lag time of 24 s in the present recording. This delay time was smaller than 60 s (median = 32 s) in most experiments (18 of 21) and ranged in its extremes from 3 to 138 s.

After the lag time, the beginning of cell membrane fusion was reflected in a stepwise increase in g_j. The conversion between two stable conductance levels (step increases) was very abrupt, i. e. it was completed within less than 1 ms. Such fast transition is analogous to the opening of an ion channel. It is unlikely that a single pore dilates in such abrupt steps. Therefore, each step increase probably represents the opening of a separate small fusion pore. The term fusion pore is used here to describe a small conductive pathway between two cells that is intermediately formed under fusion conditions. No allusion to any structured type of channel is intended so far. The fusion pore of SFV has a defined size, since analysis of all conductance steps (n=66) revealed an unitary conductance of ≈300 pS, corresponding to a pore diameter of approximately 2 nm. The definite size together with the ion channel like, fast opening suggests the fusion pore being of proteinacious nature. The period of fusion pore formation was followed by a period of more gradual increase in g_j, which may represent the dilation of the pores due to continuous lipid flow into the pore area and a consecutive pore

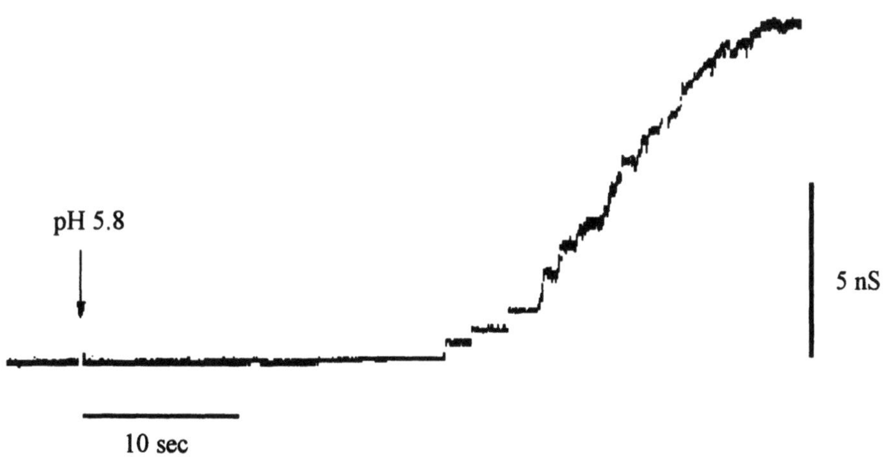

Fig 1. *Time course of the junctional conductance g_j in a cell pair of SFV-infected Aedes albopictus cells:* The arrow marks the start of influx of low pH solution.

widening. The absence of I_j flickering after pore formation implies that SFV induced fusion is not reversible. This finding is in contrast to the results obtained with Influenza virus, where reversible on-off oscillations in g_j have been described (Spruce et al., 1989, 1991). Gap junctions are the only other interbilayer pores currently known. Aedes cells have gap junctions with a single channel conductance of 365 pS (Bukauskas and Weingart, 1993). However, they exhibit voltage gating, subconductance states and are sensitive to heptanol (Bukauskas and Weingart, 1993). Such properties were not found in SFV fusion pores and served to discriminate both structures.

pH-dependent changes in membrane permeability

According to the suggestions of diverse authors (for reviews see Pasternak, 1985; 1992; Carrasco et al., 1989; 1993), viruses can change the membrane permeability early or late in their infectious cycle. We have tested this possibility in SFV-infected Aedes cells (at 16-20 hpi) using whole cell patch clamp recording (Lanzrein et al., 1993). The results of these studies are summarized in Table I. SFV-infected cells, which express large amounts of SFV-spike proteins, exhibited no difference to uninfected cells with respect to membrane conductance g_m at pH 7.4; only the resting potential was depolarized. However, when cells were exposed to pH 5.6 during the experiment, there was a marked increase in g_m, which developed within 5-10 sec. The current at low pH was not voltage dependent and the resting potential dropped down to zero. Hence, the increase in g_m is caused by unspecific pores or lesions formed upon exposure to pH 5.6. Interestingly, the effects of low pH on g_m and V_m were reversibly blocked by millimolar concentrations of Zn^{2+} and Ca^{2+}.

Table I: *Electrical properties of Aedes cells in neutral and mildly acidic pH*

	membrane conductance g_m	membrane potential V_m
uninfected pH 7.4	0.48 ± 0.17 nS	-66 ± 4 mV
uninfected pH 5.6	0.49 ± 0.21 nS	-33 ± 10 mV
SFV-infected pH 7.4	0.48 ± 0.09 nS	-41 ± 5 mV
SFV-infected pH 5.6	14.2 ± 10.8 nS	-2 ± 7 mV

The change in membrane permeability induced at low pH was further characterized by using biochemical methods (Lanzrein et al., 1992). For this purpose, efflux or uptake of various marker substances was measured. Confirming the results obtained by patch clamp recording, the efflux of molecules with a molecular weight below 900 Da was increased at low pH in SFV-infected Aedes cells but not in uninfected cells, and the effect was reversibly blocked by Ca^{2+} and Zn^{2+}. The molecular weight cut off point of 900 Da allowed to roughly estimate the size of the putative pores formed at low pH to be 1-1.5 nm. As discussed elsewhere (Lanzrein et al., 1992; 1993), several lines of evidence indicated that these pores were formed by viral membrane proteins: i) The change in membrane permeability occurred only in infected cells ii) it was completely blocked by compounds that block SFV induced fusion and virus penetration, e. g. Trypanblue. iii) The permeability change was first observed at 4 hpi, coinciding with the first appearance of viral structural proteins.

We were interested to know which components of the SFV envelope proteins were involved in the above described pore formation. Thus, the effects of monoclonal antibodies

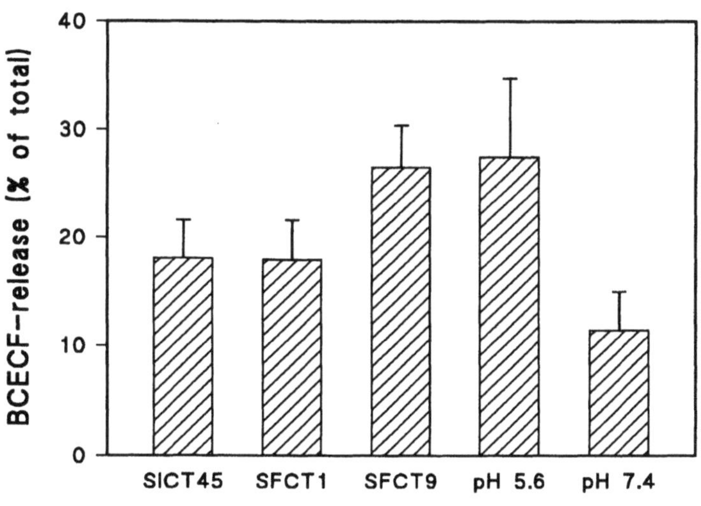

Fig 2. *Effects of various monoclonal antibodies on release of BCECF.* Monolayers of SFV-infected Aedes cells were preincubated with the appropriate monoclonal antibody for 30 min, washed and loaded with BCECF. Release of BCECF to the medium was measured after 10 min incubation in PBS pH 5.6. Control samples were incubated at pH 5.6 or 7.4 without pretreatment (two rightmost bars). SICT45: monoclonal anti-Sindbis-E1, crossreacting with SFV (Böttcher, 1991); SFCT1: monoclonal anti-SFV-E1; SFCT9: monoclonal anti-SFV-E2. Unpaired two-tailed t test revealed that the mean values of SICT45 and SFCT1 but not SFCT9 were significantly different ($p < 0.0001$) from control (pH 5.6 w/o pretreatment).

raised against E1 and E2 (Böttcher, 1991) on the low pH induced permeability change were tested. Briefly, SFV-infected cells were preincubated with antibodies for 30 min, loaded with BCECF and subsequently the efflux was measured. Fig. 2 shows that SICT45 and SFCT1, directed against E1, significantly hampered BCECF-efflux compared to control (pH 5.6 without pretreatment). These two antibodies were previously found to hamper cell-cell fusion as judged by light microscopy. Neither SFCT9 the anti-E2-antibody nor a nonsense monoclonal mouse IgG of the same subclass as SICT45 and SFCT9 which was used as a further control (not shown), had any effect. These results suggest that E1, which is also considered to be the fusion protein (Omar and Koblet, 1988; Levy-Mintz and Kielian, 1991),.is involved in pore formation. One could assume that the permeability change reflects a spillover of contents from the two fusing compartments to the exterior during the process of membrane joining. However, the change in membrane permeability was measured also by patch clamp recording in single cells where fusion was excluded because the cells were separated from each other. Moreover, the permeability change was not abolished at temperatures below 10 ^{o}C, conditions that block membrane fusion. Therefore, the permeability change is not a consequence of membrane fusion but occurs independently upon induction of the fusogenic conformation of the viral spike. Our laboratory has provided evidence that the envelope membrane of purified virions is becoming permeable to protons at low pH and that this effect is mediated by the spike protein ectodomains (Schlegel et al., 1991). In case of the purified virion, only detection of proton fluxes was methodologically feasible, but the permeability change of the virion envelope presumably features increased permeabilities to ions and solutes as well. It is reasonable to assume that both the permeability change in the virion membrane and in the cell membrane are the same. Hence, the possibility of E1 activating other cellular proteins or channels is not likely.

Summarizing, exposure of SFV-infected Aedes cells to low pH (5.8) leads to formation of leakage pores within 10 sec. Thereafter, with a delay of about 30 sec, formation of fusion pores begins. Conceivably, leakage pores appear approximately within the time required for the irreversible conformational change of the SFV spike which is completed within 10 sec (Kielian and Helenius, 1985). The delay of fusion may reflect the time needed for the approach of two membrane patches designated to fuse. The formation of leakage pores, as well as cell-cell fusion, were both significantly hampered by a monoclonal antibody against E1, suggesting that E1 is involved in both fusion and leakage pore formation. Moreover, leakage- and fusion pores were found to have roughly the same diameter (1.5 vs 2 nm). Thus, it is tempting to speculate that both structures are partly identical, possibly involving oligomers of E1 forming either a leakage pore or, in case of contact to a target membrane, a fusion pore.

As delineated elsewhere (Lanzrein et al., 1993), the viral leakage pores formed at low pH might have a function in viral uncoating. Apparently, an influx of protons into the virion during its stay in the acidic endocytotic compartment triggers the disassembly of the rigid structure of the nucleocapsids, which in turn liberates the genome and thus initiates replication. In vitro studies have shown that isolated nucleocapsids are destabilized by acid (Soederlund et al., 1972; Mauracher et al., 1991; Schlegel et al., unpublished observation). A similar model was also postulated for Influenza A virus (reviewed in Helenius, 1992) after evidence was obtained that Influenza M2 protein functions as a pH sensitive cation channel (Pinto et al., 1992).

Acknowledgements

We thank Drs. J. Böttcher and I. Greiser-Wilke for generous gift of the alphavirus monoclonal antibodies. This work was supported in part by the Swiss National Science Foundation (Grant No. 31-25732.88 to C.K.).

References

Bashford CL, Alder GL, Menestrina G, Micklem KJ, Murphy JJ, and Pasternak CA (1986) Membrane Damage by Hemolytic Viruses, Toxins, Complement, and Other Cytotoxic Agents. J Biol Chem 261 (20): 9300-9308

Böttcher J (1991) Die Kreuzneutralisation von Alphaviren: Identifizierung und Charakterisierung konservierter Epitope auf den Hüllglykoproteinen. Ph. D. Thesis, Tierärztliche Hochschule Hannover, FRG

Bron R, Wahlberg JM, Garoff H, and Wilschut J (1993) Membrane fusion of Semliki Forest virus in a model system: correlation between fusion kinetics and structural changes in the envelope glycoprotein. EMBO J 12 (2): 693-701

Bukauskas F, and Weingart R (1993) Multiple conductance states of newly formed single gap junction channels between insect cells. Pflüg Arch (in press)

Carrasco L (1981) Modification of Membrane Permeability Induced by Animal Viruses Early in Infection. Virology 113: 623- 629

Carrasco L, Otero MJ, and Castrillo JL (1989) Modification of Membrane Permeability by Animal Viruses. Parmac Ther 40 (2): 171-212

Carrasco L, Pérez L, Irurzun A, Lama J, Martinez-Abarca F, Rodriguez P, Guinea R, Castrillo JL, Sanz MA, and Ayala MJ (1993) In: Regulation of Gene Expression in Animal Viruses. Carrasco L, Sonenberg N, and Wimmer E (eds) Plenum, New York

Garoff H, Frischauf AM, Simons K, Lehrach H, and Delius H (1980) Nucleotide sequence of cDNA coding for Semliki Forest virus membrane Glycoproteins. Nature 288: 236-241

Helenius A (1992) Unpacking the Incoming Influenza Virus. Cell 69: 577-578

Kielian M, and Helenius A (1985) pH-Induced Alterations in the Fusogenic Spike Protein of Semliki Forest Virus. J Cell Biol 101: 2284-2291

Kielian M, and Helenius A (1986) Entry of Alphaviruses. In: "The Togaviridae and Flaviviridae" (SS Schlesinger and MJ Schlesinger, eds) 91-119

Kondor-Koch C, Burke B, and Garoff H (1983) Expression of Semliki forest virus proteins from cloned complementary DNA I. The fusion activity of the spike glycoprotein. J Cell Biol 97: 644-651

Lanzrein M, Käsermann N, and Kempf C (1992) Changes in Membrane Permeability During Semliki Forest Virus Induced Cell Fusion. Biosci Rep 12 (3): 221-236

Lanzrein M, Weingart R, and Kempf C (1993) PH-Dependent Pore Formation in Semliki Forest Virus-Infected Aedes Albopictus Cells. Virology 193: 296-302

Levy-Mintz P, and Kielian M (1991) Mutagenesis of the Putative Fusion Domain of the Semliki Forest Virus Spike Protein. J Virol 65 (8): 4292-4300

Mauracher CA, Gillam S, Shukin R, and Tingle AJ (1991) pH-Dependent Solubility Shift of Rubella Virus Capsid Protein. Virology 181: 773-777

Omar A, and Koblet H (1988) Semliki Forest virus particles containing only the E1 envelope glycoprotein are infectious and can induce cell-cell fusion. Virology 166: 17-23

Pasternak CA (1985) Virus, Toxin, Complement: Common Actions and Their Prevention by Ca2+ or Zn2+. BioEssays 6 (1): 14-19

Pasternak CA, Alder GM, Bashford CL, Korchev YE, Pederzolli C, and Rostovtseva TK (1992) Membrane damage: common mechanisms of induction and prevention. FEMS Microbiol Immunol 105: 83-92

Pasternak CA, and Micklem KJ (1973) Permeability Changes during Cell Fusion. J Membrane Biol 14: 293-303

Patel K, and Pasternak CA (1983) Ca2+-sensitive permeability changes caused by influenza virus. Biosci Rep 3: 749-755

Patel K, and Pasternak CA (1985) Permeability Changes Elicited by Influenza and Sendai Viruses: Separation of Fusion and leakage by pH-jump Experiments. J Gen Virol 66: 767-775

Pinto LH, Holsinger LJ, and Lamb RA (1992) Influenza Virus M2 Protein has Ion Channel Activity. Cell 69: 517-528

Schlegel A, Omar A, Jentsch P, Morell A, and Kempf C (1991) Semliki Forest Virus Envelope Proteins Function as Proton Channels. Biosci Rep 11 (5): 243-255

Soederlund H, Kaeaeriaeinen L, von Bonsdorff CH, and Weckstroem P (1972) Properties of Semliki Forest virus nucleocapsid. Virology 47: 753-760

Spruce AE, Iwata A, and Almers W (1991) The first milliseconds of the pore formed by a fusogenic viral envelope protein during membrane fusion. Proc Natl Acad Sci 88, 3623-3627

Spruce AE, Iwata A, White JM, and Almers W (1989) Patch clamp studies of single cell-fusion events mediated by a viral fusion protein. Nature 342: 555-558

Stegmann T, Doms RW, and Helenius A (1989) Protein- Mediated Membrane Fusion. Annu Rev Biophys Biophys Chem 18: 187-211

von Bonsdorff CH, and Harrison SC (1978) Hexagonal glycoprotein arrays from Sindbis virus membranes. J Virol 28: 578-583

Wahlberg, JM, and Garoff H (1992) Membrane Fusion Process of Semliki Forest Virus I: Low pH-induced Rearrangement in Spike Protein Quaternary Structure Precedes Virus Penetration into Cells. J Cell Biol 116 (2): 339-348

Wahlberg JM, Bron R, Wilschut J, and Garoff H (1992) Membrane Fusion of Semliki Forest Virus Involves Homotrimers of the Fusion Protein. J Virol 66 (12): 7309-7318

Weingart R (1986) Electrical properties of the nexal membrane studied in rat ventricular cell pairs. J Physiol 370: 267-284

White JM (1992) Membrane Fusion.Science 258: 917-924

White JM, and Blobel CP (1989) Cell-to-cell fusion. Current Opinion in Cell Biology 1: 934-939

White JM, and Helenius A (1980) pH-dependent fusion between the Semliki Forest virus membrane and liposomes. Proc Natl Acad Sci USA 77 (6) 3273-3277

White J, Kielian M, and Helenius A (1983) Membrane fusion proteins of enveloped animal viruses. Quarterly Reviews of Biophysics 16 (2): 151-195

Wyke AM, Impraim CC, Knutton S, and Pasternak CA (1980) Components involved in virally mediated membrane fusion and permeability changes. Biochem J 190: 625-638

Young JD-E, Young GPH, Cohn ZA, and Lenard J (1983) Interaction of Enveloped Viruses with Planar Bilayer Membranes: Observations on Sendai, Influenza, Vesicular Stomatitis, and Semliki Forest Viruses. Virology 128: 186-194

SUBJECT INDEX

NATO ASI Series H

Vol. 1: **Biology and Molecular Biology of Plant-Pathogen Interactions.** Edited by J.A. Bailey. 415 pages. 1986.

Vol. 2: **Glial-Neuronal Communication in Development and Regeneration.** Edited by H.H. Althaus and W. Seifert. 865 pages. 1987.

Vol. 3: **Nicotinic Acetylcholine Receptor: Structure and Function.** Edited by A. Maelicke. 489 pages. 1986.

Vol. 4: **Recognition in Microbe-Plant Symbiotic and Pathogenic Interactions.** Edited by B. Lugtenberg. 449 pages. 1986.

Vol. 5: **Mesenchymal-Epithelial Interactions in Neural Development.** Edited by J. R. Wolff, J. Sievers, and M. Berry. 428 pages. 1987.

Vol. 6: **Molecular Mechanisms of Desensitization to Signal Molecules.** Edited by T M. Konijn, P J. M. Van Haastert, H. Van der Starre, H. Van der Wel, and M.D. Houslay. 336 pages. 1987.

Vol. 7: **Gangliosides and Modulation of Neuronal Functions.** Edited by H. Rahmann. 647 pages. 1987.

Vol. 8: **Molecular and Cellular Aspects of Erythropoietin and Erythropoiesis.** Edited by I.N. Rich. 460 pages. 1987.

Vol. 9: **Modification of Cell to Cell Signals During Normal and Pathological Aging.** Edited by S. Govoni and F. Battaini. 297 pages. 1987.

Vol. 10: **Plant Hormone Receptors.** Edited by D. Klämbt. 319 pages. 1987.

Vol. 11: **Host-Parasite Cellular and Molecular Interactions in Protozoal Infections.** Edited by K.-P. Chang and D. Snary. 425 pages. 1987.

Vol. 12: **The Cell Surface in Signal Transduction.** Edited by E. Wagner, H. Greppin, and B. Millet. 243 pages. 1987.

Vol. 13: **Toxicology of Pesticides: Experimental, Clinical and Regulatory Perspectives.** Edited by L.G. Costa, C.L. Galli, and S.D. Murphy. 320 pages. 1987.

Vol. 14: **Genetics of Translation. New Approaches.** Edited by M.F. Tuite, M. Picard, and M. Bolotin-Fukuhara. 524 pages. 1988.

Vol. 15: **Photosensitisation. Molecular, Cellular and Medical Aspects.** Edited by G. Moreno, R. H. Pottier, and T. G. Truscott. 521 pages. 1988.

Vol. 16: **Membrane Biogenesis.** Edited by J.A.F Op den Kamp. 477 pages. 1988.

Vol. 17: **Cell to Cell Signals in Plant, Animal and Microbial Symbiosis.** Edited by S. Scannerini, D. Smith, P. Bonfante-Fasolo, and V. Gianinazzi-Pearson. 414 pages. 1988.

Vol. 18: **Plant Cell Biotechnology.** Edited by M.S.S. Pais, F. Mavituna, and J. M. Novais. 500 pages. 1988.

Vol. 19: **Modulation of Synaptic Transmission and Plasticity in Nervous Systems.** Edited by G. Hertting and H.-C. Spatz. 457 pages. 1988.

Vol. 20: **Amino Acid Availability and Brain Function in Health and Disease.** Edited by G. Huether. 487 pages. 1988.

NATO ASI Series H

Vol. 21: **Cellular and Molecular Basis of Synaptic Transmission.**
Edited by H. Zimmermann. 547 pages. 1988.

Vol. 22: **Neural Development and Regeneration. Cellular and Molecular Aspects.**
Edited by A. Gorio, J. R. Perez-Polo, J. de Vellis, and B. Haber. 711 pages. 1988.

Vol. 23: **The Semiotics of Cellular Communication in the Immune System.**
Edited by E.E. Sercarz, F. Celada, N.A. Mitchison, and T. Tada. 326 pages. 1988.

Vol. 24: **Bacteria, Complement and the Phagocytic Cell.**
Edited by F. C. Cabello und C. Pruzzo. 372 pages. 1988.

Vol. 25: **Nicotinic Acetylcholine Receptors in the Nervous System.**
Edited by F. Clementi, C. Gotti, and E. Sher. 424 pages. 1988.

Vol. 26: **Cell to Cell Signals in Mammalian Development.**
Edited by S.W. de Laat, J.G. Bluemink, and C.L. Mummery. 322 pages. 1989.

Vol. 27: **Phytotoxins and Plant Pathogenesis.**
Edited by A. Graniti, R. D. Durbin, and A. Ballio. 508 pages. 1989.

Vol. 28: **Vascular Wilt Diseases of Plants. Basic Studies and Control.**
Edited by E. C. Tjamos and C. H. Beckman. 590 pages. 1989.

Vol. 29: **Receptors, Membrane Transport and Signal Transduction.**
Edited by A. E. Evangelopoulos, J. P. Changeux, L. Packer, T. G. Sotiroudis, and K.W.A. Wirtz. 387 pages. 1989.

Vol. 30: **Effects of Mineral Dusts on Cells.**
Edited by B.T. Mossman and R.O. Begin. 470 pages. 1989.

Vol. 31: **Neurobiology of the Inner Retina.**
Edited by R. Weiler and N.N. Osborne. 529 pages. 1989.

Vol. 32: **Molecular Biology of Neuroreceptors and Ion Channels.**
Edited by A. Maelicke. 675 pages. 1989.

Vol. 33: **Regulatory Mechanisms of Neuron to Vessel Communication in Brain.**
Edited by F. Battaini, S. Govoni, M.S. Magnoni, and M. Trabucchi. 416 pages. 1989.

Vol. 34: **Vectors asTools for the Study of Normal and Abnormal Growth and Differentiation.**
Edited by H. Lother, R. Dernick, and W. Ostertag. 477 pages. 1989.

Vol. 35: **Cell Separation in Plants: Physiology, Biochemistry and Molecular Biology.** Edited by D. J. Osborne and M. B. Jackson. 449 pages. 1989.

Vol. 36: **Signal Molecules in Plants and Plant-Microbe Interactions.**
Edited by B.J.J. Lugtenberg. 425 pages. 1989.

Vol. 37: **Tin-Based Antitumour Drugs.** Edited by M. Gielen. 226 pages. 1990.

Vol. 38: **The Molecular Biology of Autoimmune Disease.**
Edited by A.G. Demaine, J-P. Banga, and A.M. McGregor. 404 pages. 1990.

NATO ASI Series H

NATO ASI Series H

Vol. 57: **Molecular Techniques in Taxonomy.**
Edited by G . M . Hewitt, A. W. B. Johnston, and J. P. W. Young .
420 pages. 1991.

Vol. 58: **Neurocytochemical Methods.**
Edited by A. Calas and D. Eugene. 352 pages. 1991.

Vol. 59: **Molecular Evolution of the Major Histocompatibility Complex.**
Edited by J. Klein and D. Klein. 522 pages. 1991.

Vol. 60: **Intracellular Regulation of Ion Channels.**
Edited by M. Morad and Z. Agus. 261 pages. 1992.

Vol. 61: **Prader-Willi Syndrome and Other Chromosome 15q Deletion Disorders.**
Edited by S. B. Cassidy. 277 pages. 1992.

Vol. 62: **Endocytosis. From Cell Biology to Health, Disease and Therapie.**
Edited by P. J. Courtoy. 547 pages. 1992.

Vol. 63: **Dynamics of Membrane Assembly.**
Edited by J. A. F. Op den Kamp. 402 pages. 1992.

Vol. 64: **Mechanics of Swelling. From Clays to Living Cells and Tissues.**
Edited by T. K. Karalis. 802 pages. 1992.

Vol. 65: **Bacteriocins, Microcins and Lantibiotics.**
Edited by R. James, C. Lazdunski, and F. Pattus. 530 pages. 1992.

Vol. 66: **Theoretical and Experimental Insights into Immunology.**
Edited by A. S. Perelson and G. Weisbuch. 497 pages. 1992.

Vol. 67: **Flow Cytometry. New Developments.**
Edited by A. Jacquemin-Sablon. 1993.

Vol. 68: **Biomarkers. Research and Application in the Assessment of Environmental Health.** Edited by D. B. Peakall and L. R. Shugart.
138 pages. 1993.

Vol. 69: **Molecular Biology and its Application to Medical Mycology.**
Edited by B. Maresca, G. S. Kobayashi, and H. Yamaguchi. 271 pages.
1993.

Vol. 70: **Phospholipids and Signal Transmission.**
Edited by R. Massarelli, L. A. Horrocks, J. N. Kanfer, and K. Löffelholz.
448 pages. 1993.

Vol. 71: **Protein Synthesis and Targeting in Yeast.**
Edited by A. J. P. Brown, M. F. Tuite, and J. E. G. McCarthy. 425 pages.
1993.

Vol. 72: **Chromosome Segregation and Aneuploidy.**
Edited by B. K. Vig. 425 pages. 1993.

Vol. 73: **Human Apolipoprotein Mutants III. In Diagnosis and Treatment.**
Edited by C. R. Sirtori, G. Franceschini, B. H. Brewer Jr. 302 pages. 1993.

NATO ASI Series H

Springer-Verlag and the Environment

We at Springer-Verlag firmly believe that an international science publisher has a special obligation to the environment, and our corporate policies consistently reflect this conviction.

We also expect our business partners – paper mills, printers, packaging manufacturers, etc. – to commit themselves to using environmentally friendly materials and production processes.

The paper in this book is made from low- or no-chlorine pulp and is acid free, in conformance with international standards for paper permanency.

MIX
Papier aus verantwortungsvollen Quellen
Paper from responsible sources
FSC® C105338

If you have any concerns about our products,
you can contact us on
ProductSafety@springernature.com

In case Publisher is established outside the EU,
the EU authorized representative is:
Springer Nature Customer Service Center GmbH
Europaplatz 3, 69115 Heidelberg, Germany

Printed by Libri Plureos GmbH
in Hamburg, Germany